D0782596

WELDING PROCESSES

AND

POWER SOURCES

Third Edition

by

Edward R. Pierre
Welding Specialist

Ed Pierre was formerly associated with:

Lockheed Missiles and Space Corporation
Aluminum Company of America
Airco Pacific Co.
United States Navy
Miller Electric Manufacturing Company
Liquid Air Corporation

Burgess Publishing Company
Minneapolis, Minnesota

Development Editor: Nancy Crochiere
Assistant Development Editor: Charlene Brown
Production Coordinator: Barbara Horwitz
Production: Judy Vicars, Pat Barnes, Morris Lundin
Keyliner: Melinda Radtke
Compositor: Gloria Otremba

Printed in the United States of America

Library of Congress Cataloging in Publication Data

Pierre, Edward R.
Welding processes and power sources.

Includes index.
1. Electric welding. 2. Electric welding—Power supply. 3. Welding. I. Title.
TK4660.P526 1984 671.5'2 84-14239
ISBN 0-8087-3369-9

Burgess Publishing Company
7108 Ohms Lane
Minneapolis, MN 55435

J I H G F E D C B A

CONTENTS

PREFACE

This Third Edition textbook is a total revision in format and content of the book *Welding Processes and Power Sources, Second Edition.* I thank the hundreds of welding instructors who have used the *WP&PS* text and workbook over the past 15 years. Their comments and suggestions have provided the impetus to proceed with this new textbook and workbook.

The fundamentals that are necessary for the student to learn do not change; they are only ''added to'' as new technology is developed. To help the welding instructor and the student, this book is arranged to present the material in the most logical sequence beginning with some very fundamental concepts and continuing to more difficult subjects.

The data charts at the back of the text are designed to be useful to the student during the course of instruction and afterwards in the welding shop as a reference base.

The separate workbook and exercise book are designed to take the student from basic oxy-fuel gas welding and brazing through air carbon-arc procedures. Each lesson is based on information learned in the course. Applications require the student to have mastered a basic welding method before progressing to more difficult welding processes, welding positions, and techniques.

— Ed Pierre, 1984

''Education is the progressive discovery of our own ignorance.''

— Will Durant, Historian

ACKNOWLEDGMENTS

Grateful acknowledgement is made to the many people who have contributed to the preparation and production of this book. I am indebted to the many Community College and Vocational-Technical Institute Welding Instructors who contributed ideas and advised on subject matter and chapter layout. They are exceptional people as well as outstanding men in the welding profession.

The following organizations provided assistance with technical data and illustrations which make the explanations in this text much easier to follow:

Air Products & Chemicals, Inc.; AIRCO; American Optical Company; American Society for Metals; American Welding Society; Arcair, Inc.; Alloy Rods Corporation; California Department of Community Colleges; Combustion Engineering Corporation; Hobart Brothers Company; Inco, Inc. (Dr. Norman Kenyon); Lincoln Electric Co.; Linde Company; Liquid Air Corporation; Miller Electric Mfg. Co.; North Carolina Department of Community Colleges; PowCon Corporation; The Simpson Company; Smith Tescom Corporation; S. E. Nebraska Technical Community College, Milford Campus; Thermal Dynamics Corporation; Tweco; University of Wisconsin-Madison (Dr. Richard Moll); Victor Equipment Company.

To all the individuals and companies who have encouraged me through the years of data gathering, writing, and editing this book I say a very grateful "Thank You!"

— Ed Pierre

CHAPTER 1

Communications in Common Terms

To learn is the art of communicating and understanding words and pictures. It is often difficult for us to say or show what we mean to another person. Very often we use terms that are very clear in their meaning to us but which our listeners don't seem to understand. In such cases, there is a communications problem.

In our everyday work-life, terms are used to describe various jobs, tools, materials, and many other things. The unfortunate thing is that people often use or hear words with which they are not familiar. Rather than appear uninformed or unknowing, they just nod their heads and say nothing. Their intention is usually to find out what it means at a later time but "later" never comes.

A glossary of terms is usually found in the back of technical text books such as this. Most students seldom take the time, or make the effort, to use the glossary of terms. This is unfortunate.

It is necessary for the student of welding to have some knowledge of the terms used in this profession. The sooner they are learned, the easier it will be to understand the course materials.

With this thought in mind, and because of the prime necessity of knowing what you are talking about when discussing welding, this chapter is devoted to practical definitions of common terms used in welding and metal shops. The work is alphabetized for easy reference.

The terms used are based on their general use in electricity, metallurgy, testing methods and welding in general. There are many other terms that could, and possibly should, be in this list. Add these to your vocabulary and the others will come in time.

1. Acetylene

Acetylene is a colorless fuel gas which is created when appropriate amounts of water and calcium carbide are mixed together. Acetylene gas has a rather sharp and pungent odor. It is classed as a hydro-carbon fuel. When mixed with oxygen in the proper proportions, and ignited, acetylene is capable of producing the hottest welding flame of any of the commercial oxygen-fuel gas mixtures. Acetylene is an unstable gas which should never be used at working pressures exceeding 15 psi.

2. Air Carbon-Arc Process (AAC)

AAC is an arc cutting and gouging process which uses a special electrode holder. An arc is created between a carbon electrode and the base metal. The molten metal is literally blown away by a jet of high pressure air (80-100 psi).

This process is often used for weld joint preparation. It is also used for welding inspection and the removal of unwanted metal. AAC will work with most weldable metals although special techniques are used for some metals which oxidize rapidly, especially when heated.

3. Alternating Current (AC)

Alternating current is an electrical current that has both positive and negative half-cycles alternately. Current flows in a specific direction for one complete half-cycle, stops at the instant of time that it passes through the "zero" line, then reverses direction of flow for the next half-cycle. The term "alternating current" is derived from the alternation of direction of current flow each half-cycle.

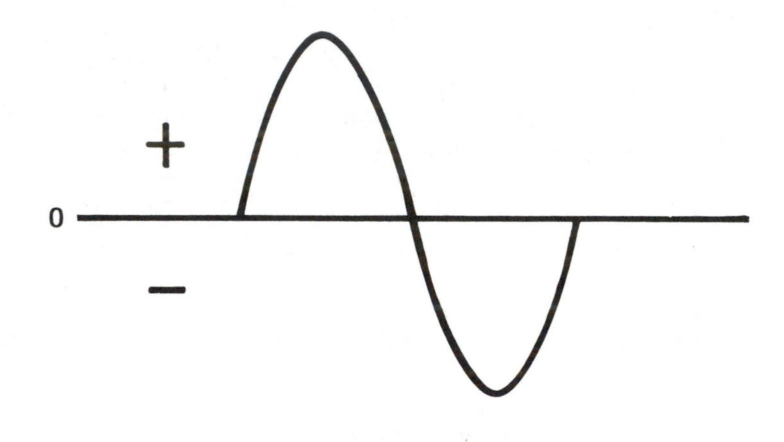

Figure 1. One Complete AC Sine Wave Form.

4. Ampere (I)

The ampere is another name for electrical current. It is the **unit of electrical rate measurement.** This means that a certain amount of electrical current is moving past a given point in an electrical conductor every second. It is amperage, or current, that moves in an electrical conductor.

5. Ampere-Turn

An ampere-turn is a current of one ampere flowing in a coil of one turn. Its primary use is in coil design for electrical apparatus.

6. Angstrom

An angstrom is a unit of length based on the metric system. It has a mathematical value of 1×10^{-8} cm.

7. Anode (+)

The positive pole in a welding arc is called the anode. It is symbolized by the (+) sign. For example, the electrode is the anode when using direct current reverse polarity.

8. Arc (for welding)

A welding arc can be sustained when there is sufficient voltage and amperage to overcome the natural resistance to current flow between an electrode tip and the work surface. With the shielded metal arc welding (SMAW) process, the arc is normally initiated by touching the electrode tip to the work and then withdrawing the electrode about 1/8"-1/4".

9. Arc Blow

The deflection or distortion of the welding arc column from its normal path because of magnetic forces in or around the arc is called **magnetic arc blow.** Magnetic arc blow is most noticeable with DC welding power on ferrous metals.

Another form of arc blow is **atmospheric arc blow.** It is normally caused by vagrant breezes or winds blowing in the welding area. Atmospheric arc blow occurs with both AC and DC welding power.

10. Arc Time

Arc time is considered to mean the actual time the welding arc is maintained when making an arc weld. It is also referred to as "operator duty cycle" or "welding time".

11. Arc Voltage

Arc voltage is the actual voltage force, or electrical pressure, measured across the welding arc between the electrode tip and the base metal surface. The electrical resistance in the arc is caused by the air space between the electrode tip and the base metal.

Arc length and arc voltage are normally correlated. The greater the arc length, the higher the arc voltage. In some gas shielded welding processes, arc voltage may vary for a specific arc length due to the difference in densities of the various shielding gases. Arc voltage can only be measured at the welding arc.

12. As-Welded

Deposited weld metal is in the "as-welded" condition before any other work is done to the part. If a weld is heat-treated, or has any type of mechanical or chemical treatment done to it, the weld is no longer in the as-welded condition.

13. Atomic Hydrogen Welding Process

Atomic hydrogen welding is an arc welding process in which the arc is maintained between two tungsten electrodes in an atmosphere of hydrogen. The circuit is completed between the electrodes and no electrical grounding of the workpiece is required.

The hydrogen atoms separate into sub-atomic particles in the heat of the welding arc. In so doing they absorb heat energy. When the sub-atomic particles strike the relatively cold work piece they recombine and release the stored heat energy which is transmitted to the workpiece. This process is now used primarily for welding repair of hot dies. The atomic hydrogen process has been superseded in industry by the gas tungsten arc, and other, welding processes.

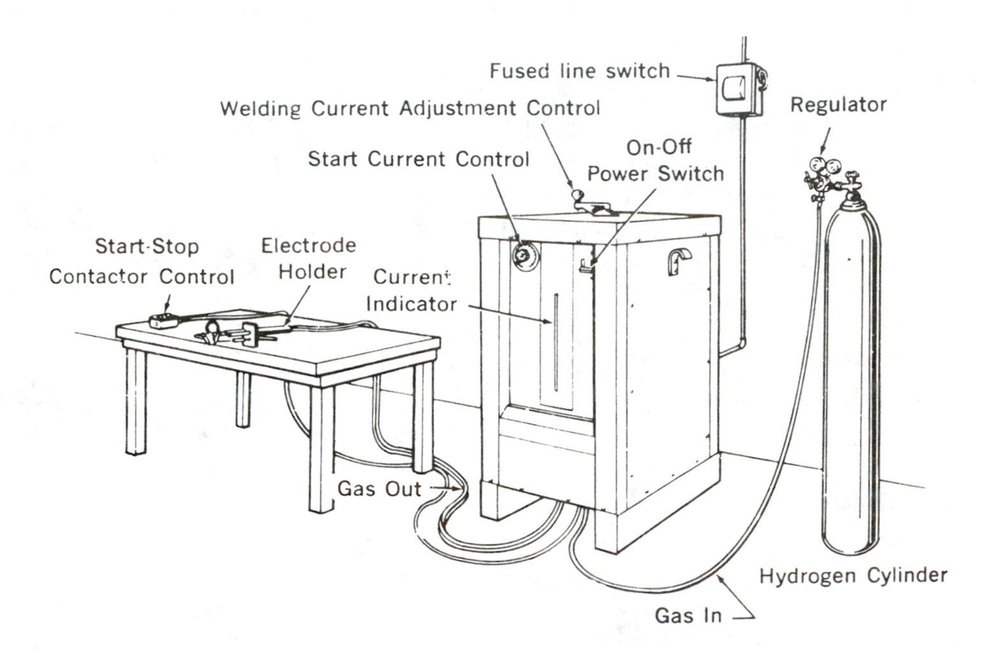

Figure 2. Atomic Hydrogen Welding Equipment.

14. Backfire

A "backfire" occurs when the tip of an oxy-fuel gas torch overheats. This will usually occur when the tip is held too close to the molten weld puddle. The backfire is apparent as a momentary disappearance of the welding flame within the torch tip followed immediately by either the reappearance of the flame or the total extinction of the flame.

The backfire gets its name from the "popping" sound made when it occurs. In almost all cases the backfire is caused by poor welding or cutting practices.

15. Back Gouging

When a Code calls for a full penetration weld in an open butt joint, it is necessary to weld both sides of the joint. Back gouging is a means of making a groove design on the backside of a welded joint root. At the same time, the groove is made deep enough to assure that the deposit from the back side will provide a full penetration weld. Back gouging is normally done with the AAC process.

16. Backhand Welding Technique

The welding technique where the torch flame, electrode, or gas shielded process electrode holder is pointed away from the direction of travel and toward the welding puddle is termed the "backhand technique" of welding. This technique has also been referred to as the "lag angle" or "drag angle". Backhand welding is often used where deeper weld metal penetration is required. This technique of welding requires slower travel speeds than the forehand welding technique. More heat energy per linear inch of weld is put into the base metal using this technique.

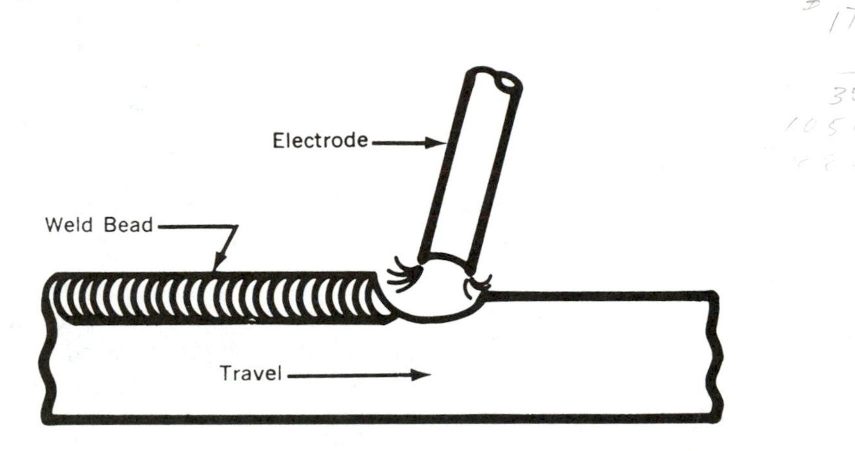

Figure 3. Backhand Welding Technique.

17. Backing, or Backup

Backing is the use of some kind of material (either metallic or non-metallic) behind or under the joint to promote better weld quality in the root of the weld. Backup bars may be designed and used to regulate the weld deposit cross sectional shapes. Some backing bars are made to provide a purge gas at the back of the weld. This is intended to prevent oxidation of the weld root underbead.

18. Backstep Welding Sequence

Use of the "backstep welding sequence" is usually done to control heat input and minimize distortion. It is typically accomplished as follows: a segment of weld bead is welded for

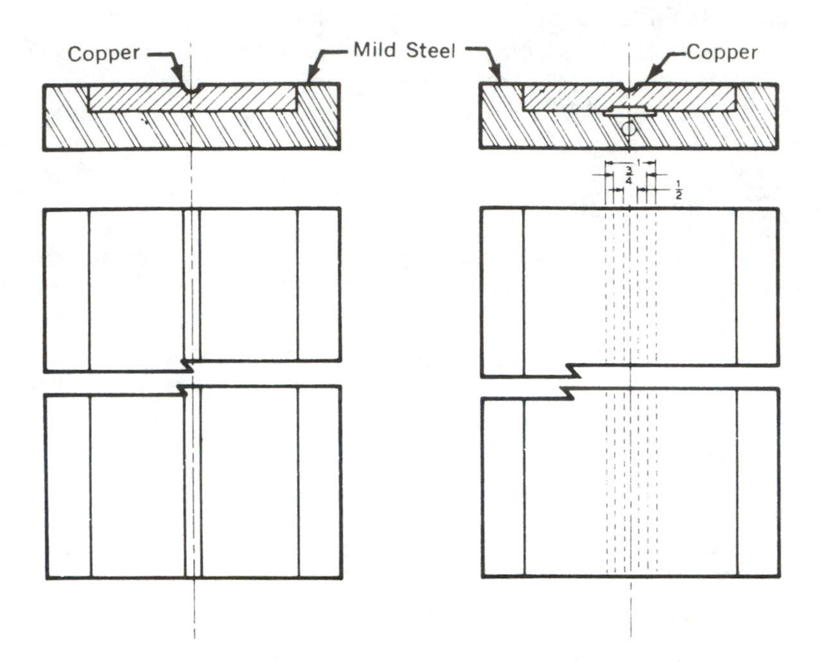

Figure 4. Some Typical Backing Bar Shapes.

some distance on a longitudinal plane. After cleaning the weld, the starting point is ground to a tapered flare from about 3/4'' into the weld back to the root of the original joint.

The next increment of weld is then made starting at a point a specified distance from the start of the original weld bead. As the welder approaches the connection point with the original weld, there is a groove for him to weld into and whip back for a crater fill. The intent of grinding the flare groove is to prevent unsightly bulges where the two welds connect.

19. Bare Electrode

A bare electrode is a solid consumable electrode wire which may be either a cut length ''stick'' electrode or a continuous wire electrode as used for the gas metal arc welding process. The electrode material has no coating other than that occurring incidental to the wire drawing operation. The electrode must be electrically energized to perform its function.

20. Base Metal

a. In welding applications, ''base metal'' refers to the metal to be welded. It is sometimes referred to as the parent metal.

b. In metallurgy, the ''base metal'' in an alloy is the metal having the highest percentage of content in the alloy. For example, brass is a copper-based alloy with copper the dominant, or base, metal.

21. Boxing

The continuation of a fillet weld around the corner of a structural member as an extension of a principal weld is called "boxing". It is normally done to minimize notch effects and for strength. The term "end return" is sometimes used to describe the boxing technique.

22. Braze

A braze is a method of joining solid base metals with a dissimilar molten filler metal. The base metal must be heated to a usable working temperature. It is joined by a filler metal having a melting temperature above 800° F. but below the melting temperature of the base metal. The filler metal flows by capillary attraction between the close fitting surfaces of the heated base metal. Common brazing filler metals include silver alloys and several types of brass and bronze alloys.

23. Buttering

Depositing weld metal on the surface of a joint to improve weldability of the base metal is called "buttering". This welding technique is often used with cast iron base metal to reduce the dilution of the weld metal deposit with the base metal. Buttering is also used to provide a buffer, or transition point, between dissimilar base metals.

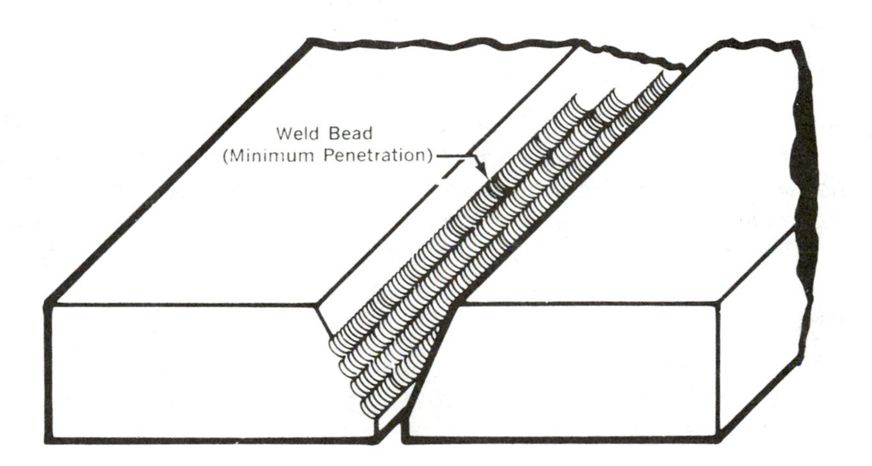

Figure 5. Partial Buttering of a Joint.

24. Capacitor

A capacitor is an electrical device which, when connected to an AC circuit, will cause current to lead voltage by 90 electrical degrees in time. It is a device that can store some electrical energy momentarily while power is on the circuit. The peak of the current sine wave form will reach maximum amplitude, or strength, 90 electrical degrees before the voltage sine wave trace. This is the result of the storage, and discharge, of electrical energy by the capacitor.

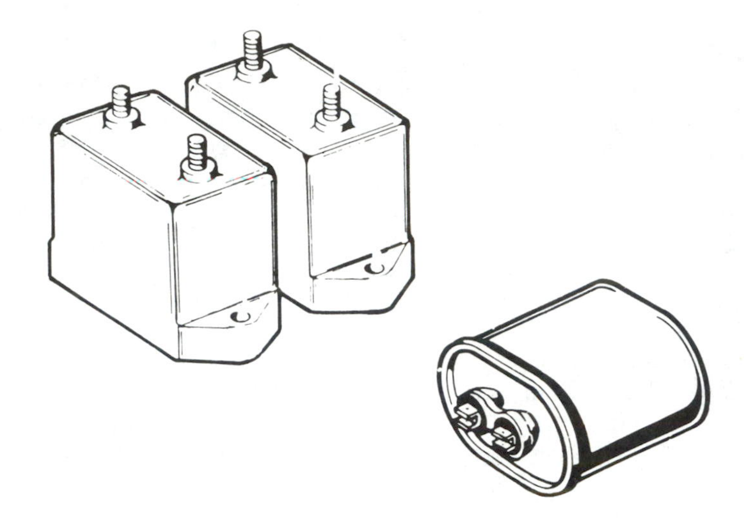

Figure 6. Some Capacitor Shapes.

25. Capacitance

The properties in a system of dielectrics and conductors that results in the storage of an electrical charge is called capacitance. It is the result of the capacitors ability to store electrical energy. **Capacitance is always an electrostatic effect and is voltage induced.**

The unit of measure of capacitance is the farad. The farad has too great a value for most applications. The micro-farad (one millionth of a farad) is more commonly used in industry.

26. Capillary Attraction

The combination of adhesion and cohesion forces which will cause molten metal to flow between closely spaced solid metal surfaces, even against the force of gravity, is called capillary attraction. The term is normally used in brazing, especially silver alloy brazing.

27. Cathode (–)

The negative pole in a welding arc is called the cathode. In DC welding the electrode is the cathode when using straight polarity (DCSP). For DC reverse polarity (DCRP) the base metal workpiece is the cathode.

28. CFH (Cubic Feet per Hour)

The various shielding gases and gas mixtures used for the gas shielded welding and cutting processes are normally measured in cubic feet per hour (CFH). This is done with gas measuring devices called **flowmeters.**

29. Charpy Test

The Charpy Test is a pendulum-type impact test in which the specimen is usually notched to control the point of fracture. The specimen is supported at both ends as a simple beam. It is broken by the force of the falling pendulum. The energy absorbed, as indicated by the subsequent rise of the pendulum past the point of specimen fracture, is a measure of the impact strength, or notch toughness, of the material.

30. Chipping

a. Chipping is the mechanical removal of metal from a weld joint using an impact tool and a chisel. Chipping may also be used in the edge preparation of some types of weld joint designs.

b. Chipping is also the term used to describe the removal of flux, or slag, residues from welds made with flux coated electrodes. For this purpose, air operated multineedle impact tools are often used.

31. Circuit, Electrical

Any system of electrical conductors that is designed to carry current is termed an electrical circuit. A circuit performs its function when the proper voltage and amperage is impressed on the electrical system.

32. Coil, Electrical

An electrical coil is normally made from insulated copper or aluminum conductor wire. Copper is the preferred conductor for most design uses. Aluminum has about 62% of the current carrying capacity of copper and would require about 40% greater cross sectional area than copper to carry the same amount of current.

A coil is wound on a "coil form" which determines the inside dimensions of the coil. The design of a coil determines how many electrical "turns" the coil will have in order to perform its proper function. An electrical turn is one complete wrap of the coil wire around the outer edge, or periphery, of the coil.

33. Conductor

Any electrical path is considered to be a conductor. Good conductors are normally thought of as those materials that offer the least amount of electrical resistance to current flow. For example, most metals are considered good electrical conductors. A few, such as silver, copper and aluminum, are excellent conductors.

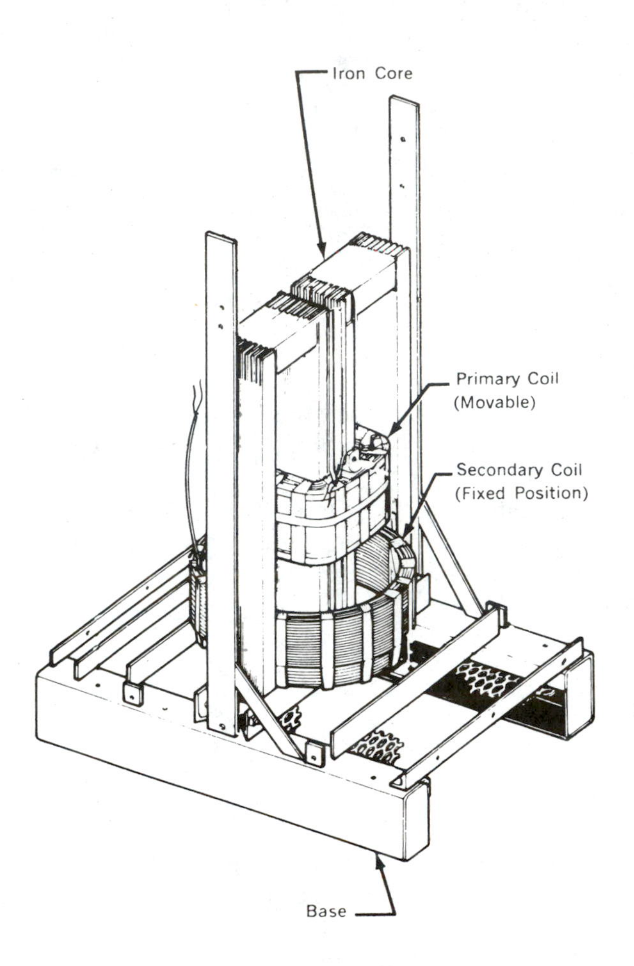

Figure 7. Primary and Secondary Coils.

34. Constant Current Power Source (CC)

Constant current power sources for welding have a substantially negative volt-ampere output characteristic. The limited maximum short circuit current of CC power sources is normally about 150% of the amperage rating of the unit. Constant current type power sources are specifically designed for use with the Shielded Metal Arc, and Gas Tungsten Arc, welding processes. This type of power source may be used with other arc welding and cutting processes also.

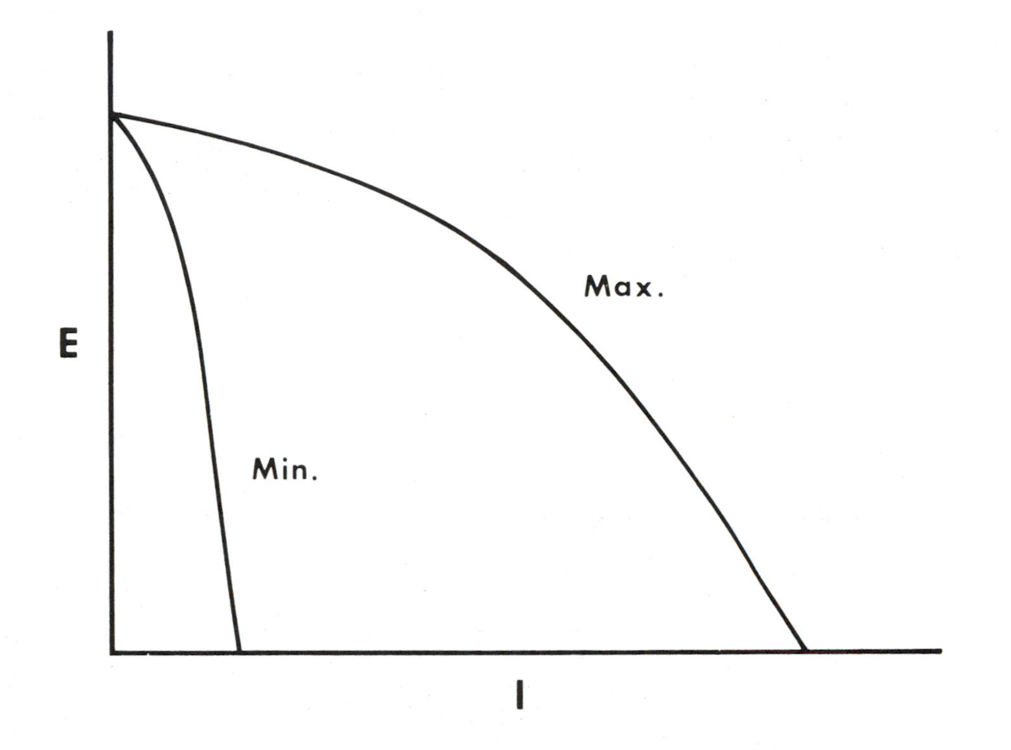

Figure 8. Typical Constant Current Volt-Ampere Curve.

35. Constant Potential/Constant Voltage Power Source (CP/CV)

In welding terminology, the terms "potential" and "voltage" are synonymous in their meaning. Constant voltage power sources provide a relatively flat volt-ampere output characteristic with a slight voltage drop per hundred amperes of welding output.

CP/CV type power sources are designed for use with the Gas Metal Arc Welding, Flux Cored Arc Welding, and AAC processes. They may also be used with certain other welding processes. CP/CV type power sources should never be used for the SMAW welding process.

36. Constricted Arc

A "constricted arc" is the term used to describe the plasma arc column used with plasma arc cutting (PAC) and plasma arc welding (PAW) processes. The constricted arc is shaped by the constricting orifice, or small hole, in the gas nozzle of the torch.

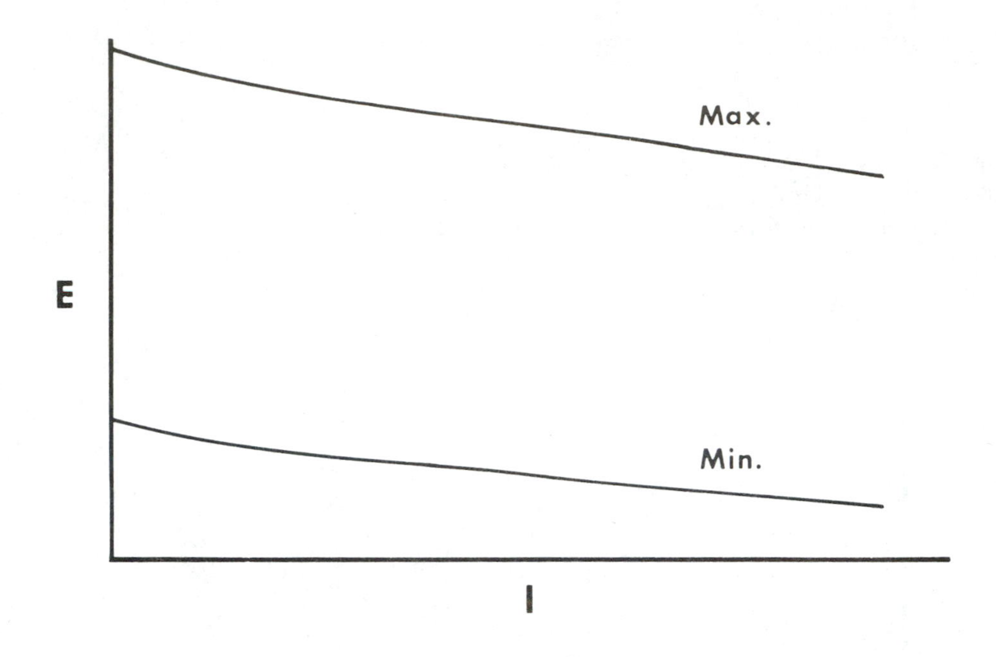

Figure 9. Typical Constant Voltage Volt-Ampere Curve.

37. Consumable Insert

Consumable inserts are designed to be used as preplaced filler metal at the root of weld joints in pipe. The root pass is made by melting and fusing the consumable insert and the edges of the joint. Consumable inserts are normally used with the Gas Tungsten Arc Welding process. The AWS filler metal specification A5.30 "Specification for Consumable Inserts" shows the various forms and requirements for consumable inserts.

38. Contact Tube

The term "contact tube" refers to the copper tip in the gun or torch used with the GMAW and FCAW processes. The continuous electrode picks up the welding current from the contact tube. Contact tubes are threaded into the head of the gun or torch. It is important to use the correct ID contact tube for the electrode diameter used.

39. Contactor

A contactor is simply a type of switch. It may be located in either the primary or secondary circuit of the power source.

A **primary contactor** is located in the relatively high voltage, low amperage primary circuit of the power source. When the contactor points are in the "open" position, there is no primary power to the main transformer of the power source. When the contactor coil is energized, it closes the contactor points and the electrical circuit is energized. Power now goes to the main transformer, the welding power source is energized, and open circuit voltage is apparent at the output terminals of the power source.

A **secondary contactor** may be located internally in the power source or externally in the welding circuit of the power source. In some solid state electronic controlled power sources, silicon controlled rectifiers (SCR's) act as secondary contactors to control welding power output.

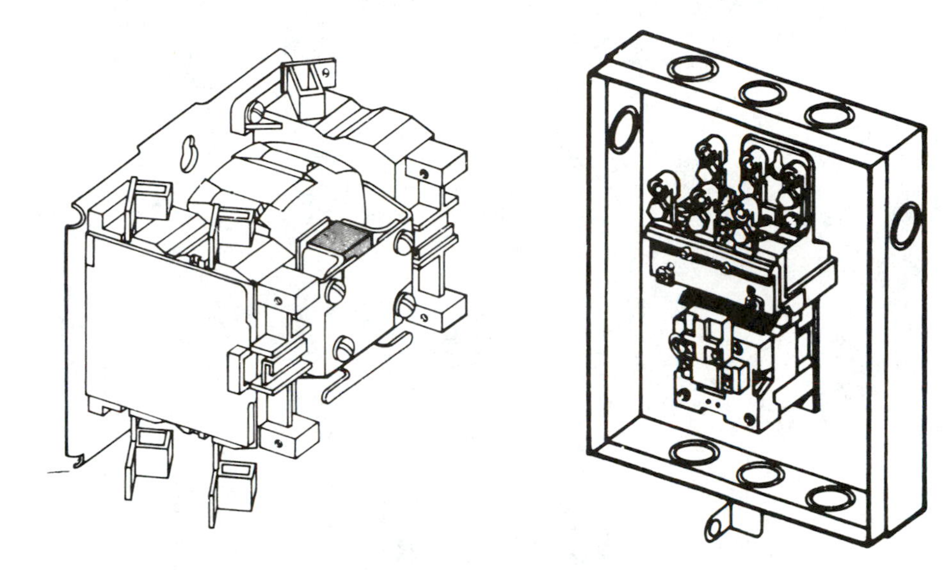

Figure 10. Some Types of Contactors.

40. Core, Iron

Iron cores are made of laminated and film-insulated electrical steel. The typical lamination thickness is 0.018"-0.020". The steel laminations are insulated on both sides so each will be an electrically isolated entity. This is to prevent the formation of eddy currents in the iron cores. Electrical steel is used to minimize hysteresis in the iron core.

41. Cored Solder

A manufactured wire solder that has a flux as a core. The solder may be either rosin core or acid core.

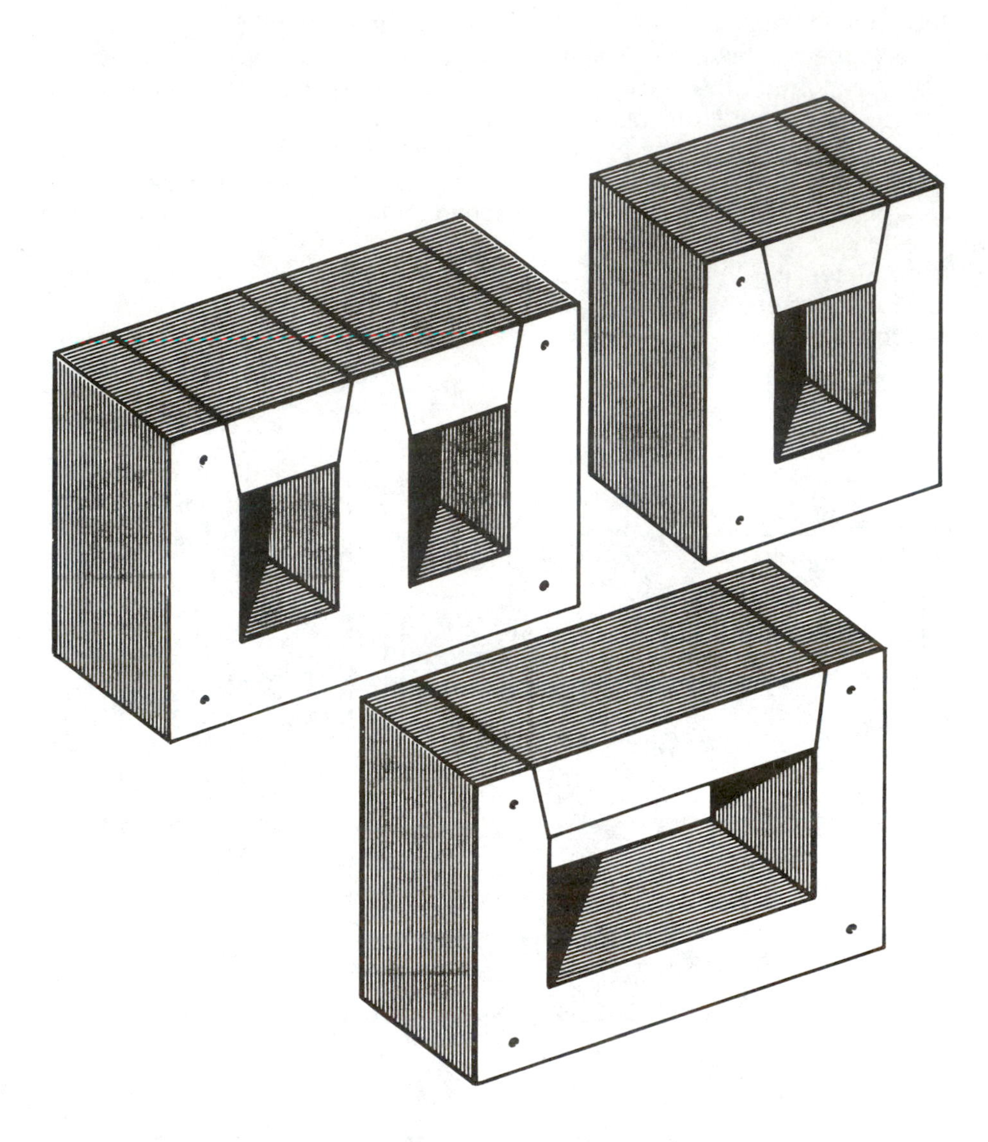

Figure 11. Some Iron Core Configurations.

42. Counter Electro-Motive Force

The term ''electro-motive force'' is another name for voltage. Counter electro-motive force (induced voltage) opposes the impressed AC voltage in an electrical circuit. Lenz's Law states that ''an induced voltage will always be opposite in force direction to the effect that created it''. This electrical concept is used in several power source output control designs such as the saturable reactor, magnetic amplifier control, and the reactor slope control.

43. Crater, Weld

A crater is a depression at the end of an arc weld. It is caused by poor welding techniques. Correct welding techniques require that any possible weld-end crater be filled with weld metal to eliminate crater cracking.

44. Current Density

Current density is the amount of current per unit of cross sectional area of the electrode. In the United States, it is the amount of current per square inch of electrode cross sectional area. To calculate current density for any electrode diameter, divide the welding current value by the electrode cross sectional area in square inches (inches^2). For example:

0.030'' dia. electrode = 0.00071 inches^2
welding current = 100 amperes

$0.00071''^2 \overline{)100}$ = 141,000 amperes per inch^2
approx. current density.

A data chart showing current density calculation figures is included in the appendix of this book.

45. Current Flow

The question of which direction current flows is academic until the subject of electronic circuits is discussed. It has been proven that electrons must flow from **negative (–) to positive (+)** in electron tubes. This is the basis for the electron theory.

46. Cycles per Second (Hertz)

An electrical cycle is one double alternation of AC power. The AC sine wave trace equals 360 electrical degrees. The sine wave form is created around the "zero" horizontal line which is a function of time. Above the zero line is considered to be the positive half-cycle. Below the zero line is considered to be the negative half-cycle. Electrical power generated in the United States is 60 cycle per second power where one complete cycle equals 1/60th of a second.

The term "hertz" means, and is equal to, the same as the term "cycle(s) per second".

As illustrated, the sine wave form starts at "0" electrical degrees and ascends to its maximum amplitude (strength) at 90 electrical degrees. **Note that a definite measure of time has elapsed in getting from 0 to 90 electrical degrees.** At the 90 degree mark one quarter cycle has been completed.

The sine wave curve then descends from maximum strength to the "zero" line (180 electrical degrees). At the precise instant of time that the sine wave curve line touches the zero line, the first half-cycle (positive) is completed (1/120 of a second in time has elapsed). At the same instant of time, there is no electrical power. Current flow is in one direction only during the entire first half-cycle.

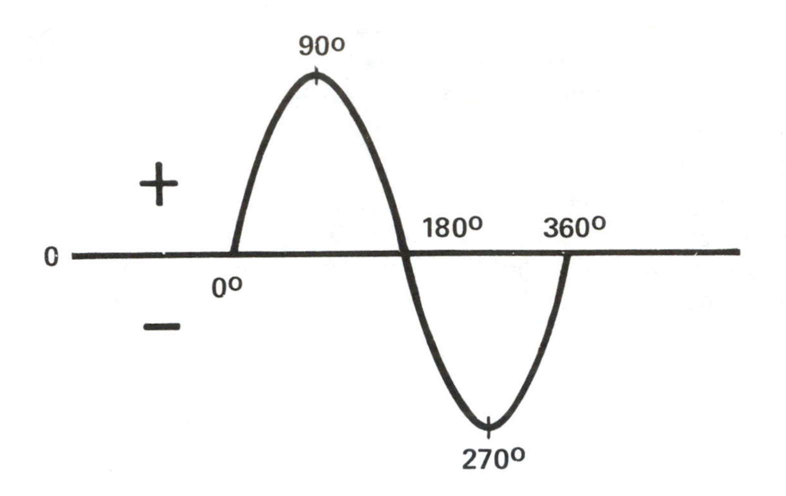

Figure 12. One AC Cycle with Electrical Degrees.

At the beginning of the second half-cycle there is a reversal, or alternation, of current flow direction. The sine wave trace starts at the 180 degree mark but this time it extends below the zero line (the negative half-cycle). The sine wave trace builds up to its maximum strength at 270 electrical degrees. At that point, the sine wave decreases in strength until it reaches the zero line at 360 electrical degrees. At this instant of time one full cycle has been completed and 1/60 of a second in time has elapsed.

47. Cylinder, Gas

All industrial gases used in the various welding processes must be contained in some type of storage device. A large percentage of the storage devices are compressed gas cylinders. The cylinders are designed for the specific types of gases they are to contain.

Most gas cylinders are capable of storing gases at very high pressures. For example, a standard oxygen cylinder contains approximately 2,200 psi when full of oxygen. Controlled release of the gas is achieved through a regulator attached to the cylinder valve.

Compressed gas cylinders are often improperly called "bottles".

48. Defect, Weld

Weld defects are essentially a form of discontinuity that decreases the strength and integrity of a weld. Some discontinuities may be allowable under a specific Code. Defects are discontinuities that exceed Code requirements and must be repaired. Defects may include porosity, nonmetallic inclusions, cracks, lack of fusion, undercut, and many others.

49. Deoxidizer

A deoxidizer is an element or compound added to an electrode flux or core wire chemistry to scavenge and remove oxygen and its derivatives from the weld. The same element may also function as an alloy in the deposited weld metal. A basic deoxidizer found in steels is the element silicon.

50. Deposition Efficiency

The term deposition efficiency means the ratio of weld metal deposited, by weight, to the total purchased weight of the electrodes. For example, a typical 5/32" diameter SMAW electrode will have the following approximate loss by weight when welding:

Rod stub loss, based on 2" rod stub ends	=	17%
Flux and spatter losses, by weight	=	27%
total loss, by weight	=	44% approx.

The best possible deposition efficiency would be about 56%—and this is based on 2" electrode stub ends, something not often found in welding shops. Professional estimators normally use 30%-35% deposition efficiency for SMAW electrodes.

51. Dielectric

A dielectric material will not conduct direct current (DC). Dielectric materials are used in the manufacture of capacitors, for example.

52. Direct Current (DC)

Direct current is electric current that flows in one direction only. DC has either a positive or negative polarity.

53. Direct Current, Electrode Negative (DCEN)

The term "DC Electrode Negative" is the same as saying "DC Straight Polarity". DCEN is more meaningful to the welder since it describes the actual electrical connection of the welding cables.

54. Direct Current, Electrode Positive (DCEP)

The term "DC Electrode Positive" is the same as saying "DC Reverse Polarity". DCEP is more descriptive of the actual electrical cable connections for welding.

55. Ductility

Ductility is the metallurgical property of a metal that permits it to deform under stress without rupturing. Metals with good ductility may be stretched, formed or drawn without tearing or cracking. Gold, silver, copper and some iron alloys are examples of metals that exhibit good ductility. A ductile metal is not necessarily a soft metal.

56. Duty Cycle

Duty cycle is based on a ten minute period of time for most welding power sources. The National Electrical Manufacturers Association (NEMA) Standard EW-1, latest edition, sets the U.S. Standards for electric welding power sources of all types. All legitimate manufacturers of welding power sources in the United States conform to the NEMA Standards.

Duty cycle is not accumulative in a power source. For example, a welding power source that has an output rating of 300 amperes, 32 load volts, 60 percent duty cycle, is designed to

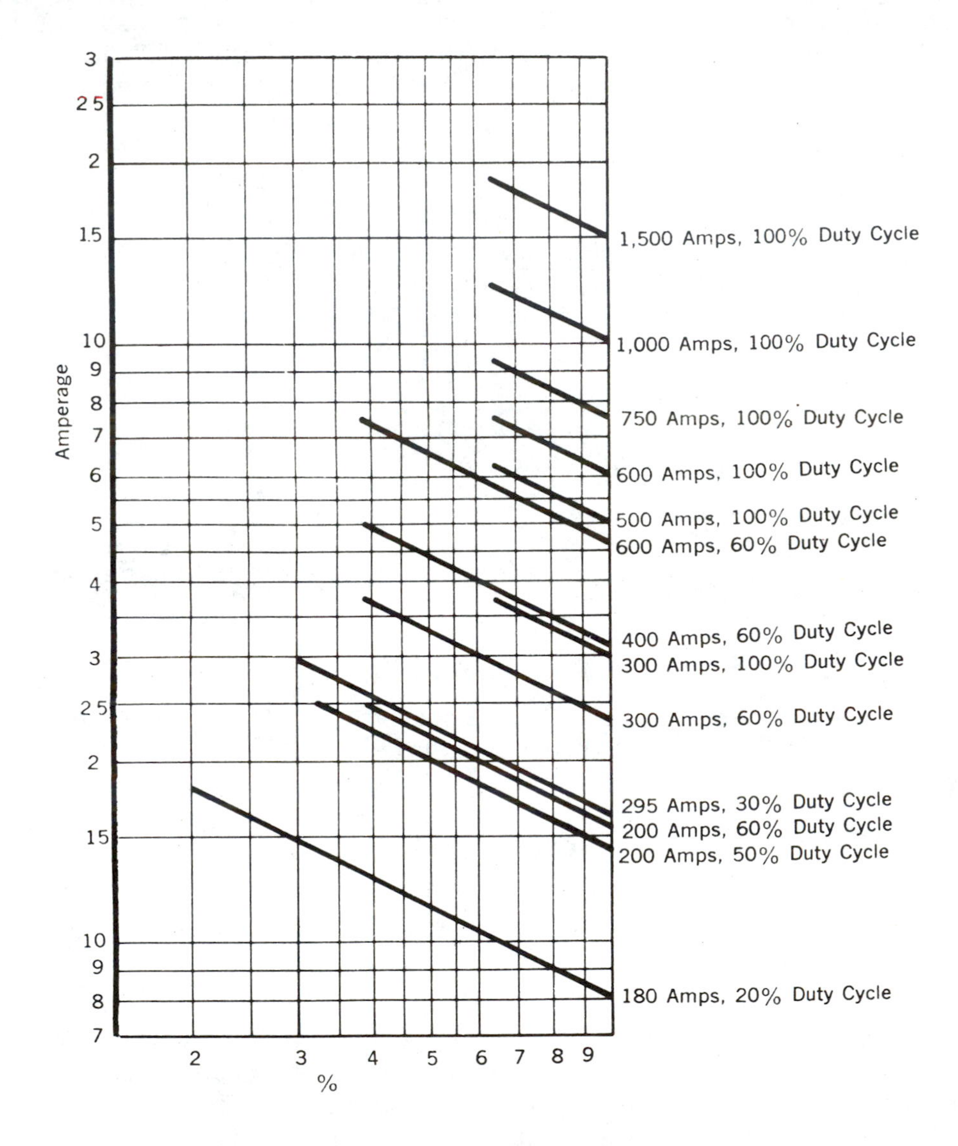

Figure 13. Typical Duty Cycle Chart.

supply the rated amperage, at the rated load voltage, for six minutes out of every ten minutes of time. The other four minutes out of every ten, the power source must idle and cool. The six minutes of welding time and the four minutes of idle and cooling time equal the ten minute duty cycle. The cooling time is necessary to prevent damage to the insulation materials on the various component parts of the power source.

57. Dynamic Electricity

The term dynamic electricity means electricity at work. An excellent example of electrical dynamics is the welding arc. The current and voltage in the arc are constantly fluctuating. There is a constant change in the power ratio of volts and amperes.

58. Effective Value (AC)

Alternating current is constantly changing its direction of current flow so that the total **net current flow** is zero. In practice, measured values of AC volts and AC amperes are taken as **effective values** unless otherwise specified. Often the effective value is referred to as the Root Mean Square, or "RMS", value. It is shown as:

$$\text{RMS value} = 0.707 \times (\text{maximum, or peak, value}).$$

59. Elasticity

In metals, elasticity is the ability of a metal to return to its original shape and dimensions after being deformed in some manner and then having the load removed.

60. Electrical Charges (Voltage)

Electrical charges are termed "positive" (+) and "negative" (−). A basic electrical law is, "**Dissimilar charges attract and similar charges repel**". For example, two negative electrical charges would repel each other. A positive and a negative charge would attract each other.

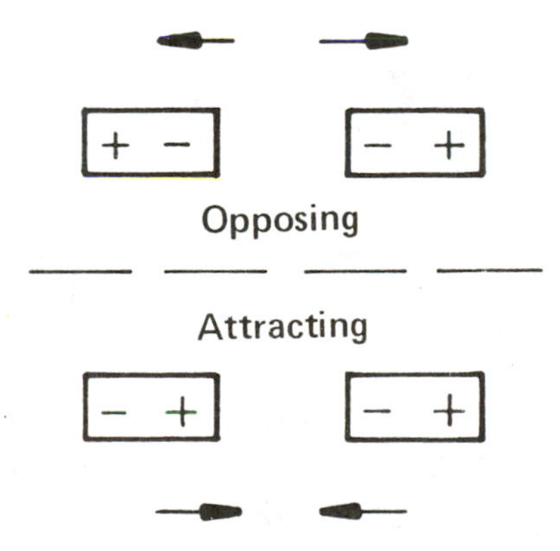

Figure 14. Electrical Charges Attract and Repel.

61. Electrical Conductivity

Materials that promote the flow of electrical current are considered to have good electrical conductivity. Some metals, such as aluminum and copper, are used as electrical conductors because they have the ability to release great quantities of free electrons when voltage is impressed on them. It is the movement of free electrons in a conductor that constitutes the flow of electrical current in a circuit.

62. Electrical Degree

An electrical degree is applied as a unit of measure of time for alternating current. One AC cycle equals 360 electrical degrees.

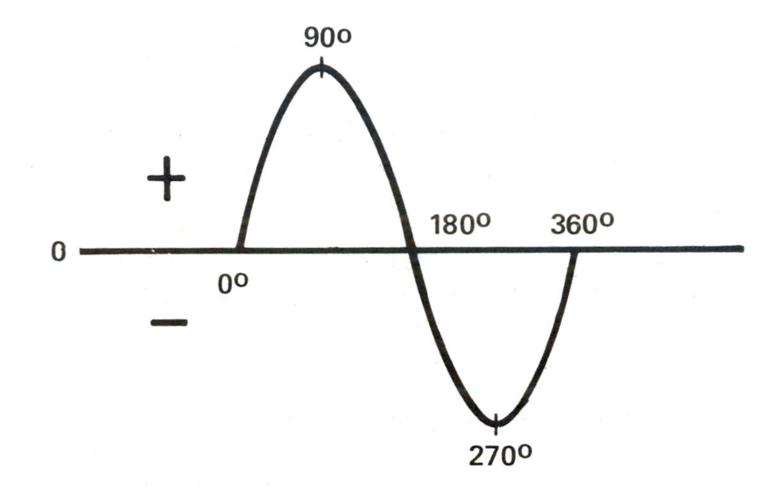

Figure 15. Electrical Degrees in One AC Cycle.

63. Electrode

The term "electrode" is used in arc welding to describe the conducting material between the electrode holder and the welding arc. An electrode carries the welding current and functions as one electrical pole in the welding arc. The base metal is the other electrical pole in the arc.

The two basic types of electrodes are consumable and non-consumable. A consumable electrode is designed to be an integral part of the weld metal deposit. A non-consumable electrode is designed to carry current only and *NOT* be a part of the weld deposit. The two most common non-consumable electrode materials are tungsten and carbon.

64. Electrode Extension

The term "electrode extension" refers to the unmelted electrode length extending beyond the copper **contact tube.** It is applicable to the gas metal arc welding (GMAW) and the flux cored arc welding (FCAW) processes. The term "stickout" is sometimes used to describe electrode extension.

65. Electrode Holder

The electrode holder is a device for conducting welding current to the electrode while the electrode is in contact with the holder. It may be a simple mechanical device for holding covered SMAW electrodes or it may be more complex such as those used for the gas shielded welding processes.

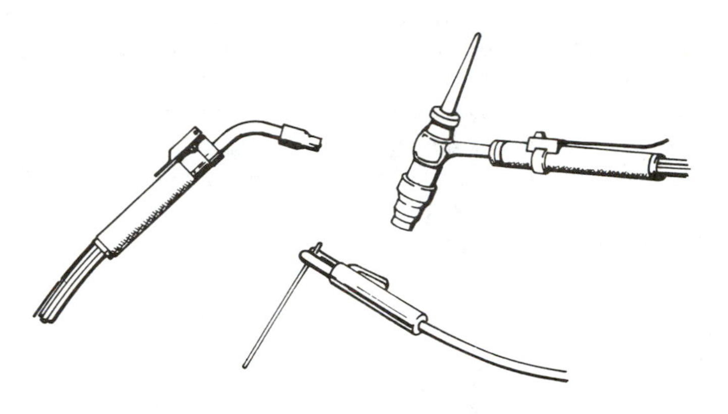

Figure 16. Electrode Holders for Arc Welding.

66. Electron

The electron is a sub-atomic particle which is considered to be the fundamental unit of negative electricity. It is the smallest particle of matter known to man that carries a negative electrical charge. Electrons are capable of moving from one place to another within atomic structures. It is the electrons that move when electrical current flows in an electrical conductor.

67. Electron Bond

Electron bond is the term used to describe the relative ability of electrons to move within the atomic structure of a material. Materials that have a "loose" electron bond allow electrons to leave the atom easier, and with less potential force being applied, than do materials which have a "tight" electron bond.

All of the metals have a relatively loose electron bond while wood and rubber, for example, have tight electron bonds. As you might surmise, the metals are relatively good electrical conductors while wood and rubber are classed as poor electrical conductors.

68. Embrittlement

Embrittlement describes the reduction or loss of normal ductility in a metal due to physical or chemical changes. In welding, hydrogen embrittlement can occur if improper welding procedures are used. It is for this reason that low hydrogen (LH) type electrodes were developed. LH electrodes have a closely controlled flux chemical analysis and moisture content to prevent the formation of hydrogen gas while welding. Low hydrogen electrodes must be maintained in an electrode oven after the shipping container is opened. The nominal oven temperature should be about 250°-300° F. for safety.

69. Flashback

A pre-ignition of the oxy-fuel gas mixture in the mixing chamber of a welding or cutting torch is called a flashback. Its occurrence can be extremely dangerous since the flame could follow the fuel gas back to the cylinder. The resulting explosion could cause severe injury and damage. If a flashback occurs in the torch, the gas cylinders should be shut off immediately, fuel gas first, to prevent further combustion.

A flashback is usually caused by an overheated or clogged torch tip. If a flashback occurs, the torch tip should be immediately examined for damage and proper mechanical operation before it is used again. In particular, check the tip orifice and the tip seat and mixer connection. For tips that have an integral mixer, check the seat of the mixer and torch connection. The torch needle valves should also be checked for burned or damaged packing before proceeding.

70. Flowmeter

A gas flowmeter is a device for measuring the flow of shielding gases while welding. The measurement may be made in cubic feet per hour (CFH) or liters per hour. The flowmeter may be attached to a manifold piping system (line flowmeter) or it may be part of a metering regulator that attaches to a compressed gas cylinder.

71. Flux

a. In electrical terms, flux is another name for magnetic lines of force. Magnetic lines of force are located in a magnetic field.

b. For some forms of welding, brazing, and soldering, various types of chemical fluxes are used. The fluxes are intended to break down and remove metal oxides from the surface of the weld area. Fluxes also reduce surface tension and permit better "wetting" of the weld metal.

72. Flux Cored Arc Welding (FCAW)

The flux cored arc welding process employs a continuous tubular electrode which has granular flux within the tube. It is a fast weld metal deposition process. Flux cored electrodes in diameters of 5/64" and larger are normally used in the flat groove, or horizontal fillet, positions. The flux cored electrodes may be either self-shielded or gas shielded with an externally supplied gas. Better mechanical and physical weld properties can normally be expected from the gas shielded flux cored electrodes.

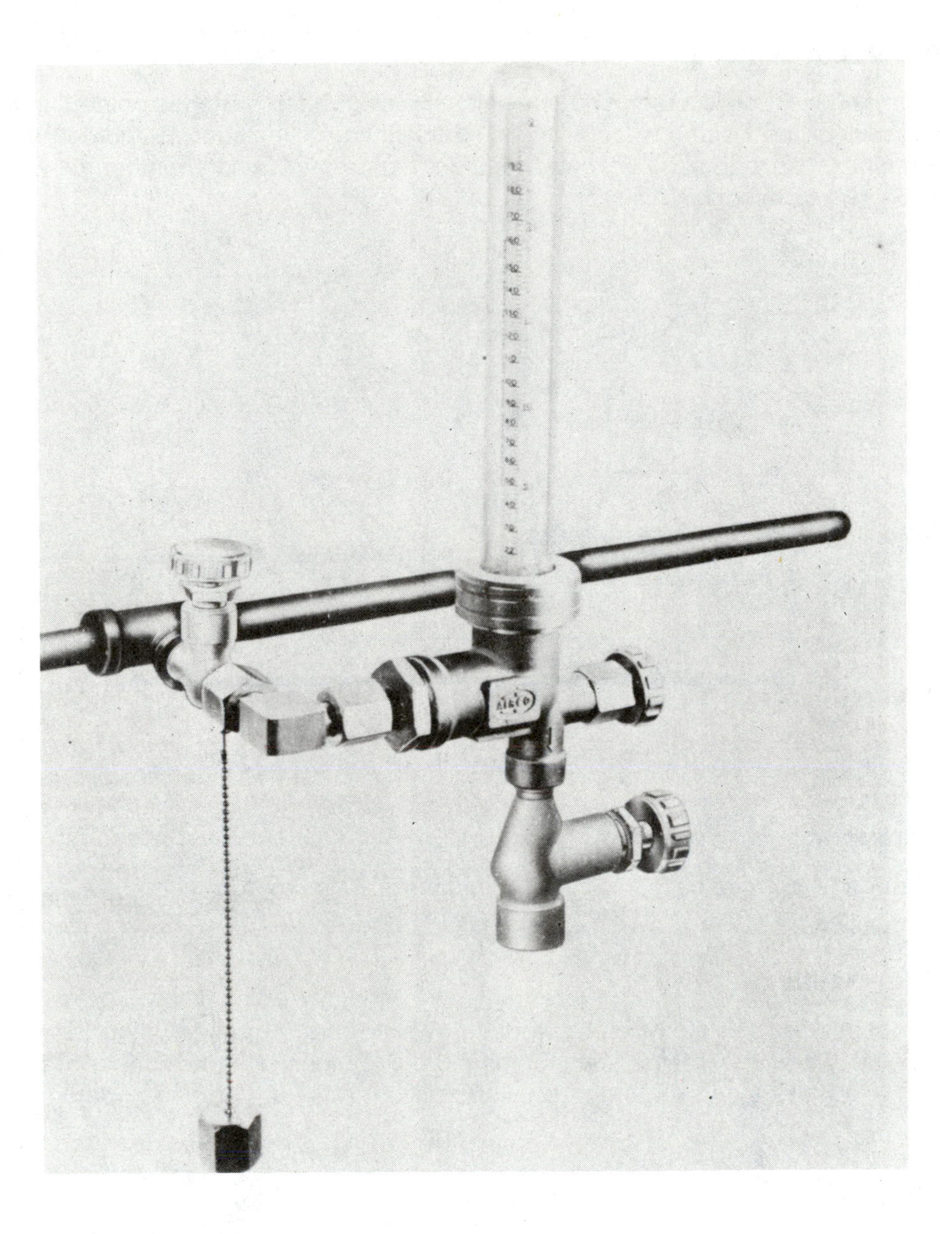

Figure 17. Typical Line Flowmeter.

Several manufacturers have produced small diameter carbon steel gas shielded flux cored electrodes that may be used for out-of-position welding. These electrodes are available in 1/16'', 0.045'', and 0.035'' diameters.

73. Forehand Welding

Forehand welding is a technique in which the electrode or torch is pointed in the direction of travel of the weld. It is faster than the backhand technique of welding but the weld penetration is less than with backhand welding. The terms "push angle" and "lead angle" are also used to describe this technique of welding.

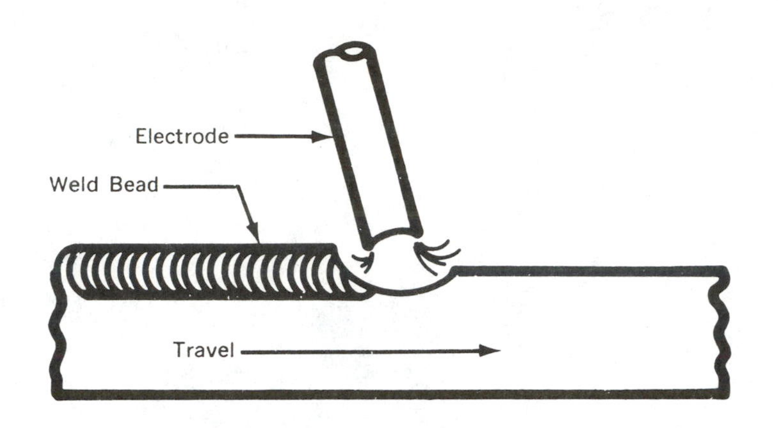

Figure 18. The Forehand Welding Technique.

74. Frequency

Frequency is the rate at which alternating current makes complete cycles of reversals. Frequency is expressed as "Hertz" or "Cycle(s) per second".

75. Fuel Gases

The gases used with oxygen to produce welding, brazing, and cutting flame temperatures are known as fuel gases. Some of the more commonly used fuel gases are acetylene, propylene, methylacetylene propadiene stabilized (MAPS), propane, natural gas and hydrogen.

76. Fusion Welding

Fusion welding processes are those where welding is done without pressure and where the base metal is caused to melt and flow together in a weld joint. Filler metal may, or may not, be used. As the deposited weld metal solidifies, it produces a solid fusion weld.

77. Gage

a. For sheet metals, gage denotes the thickness of a metal sheet.

b. For industrial gases, a gage is a device for measuring pressures of gases. Gages are part of gas pressure regulators. There are usually two gages on a gas regulator. One gage measures the pressure of the remaining gas in a cylinder and is called the "cylinder pressure gage". The second gage measures the amount of pressure being delivered to the torch. It is called the "working pressure gage". The spelling may be either "gage" or "gauge" since both are correct.

78. Gas Metal Arc Welding (GMAW)

The gas metal arc welding process (GMAW) was patented in 1950. The process produces an arc between the continuous consumable solid electrode wire tip and the surface of the base metal. The weld metal is protected by an externally supplied shielding gas which may, or may not, be inert to the products of the weld zone. The process name may be explained in the following manner:

Gas	=	externally supplied shielding gas
Metal Arc	=	consumable electrode
Welding	=	arc welding

Several types of metal transfer are possible with the GMAW process.

79. Gas Pocket

A gas pocket is a spherical shaped cavity often found in weld metal deposits. It is a weld defect in the form of porosity. A hole at the surface of a weld deposit usually indicates piping porosity.

80. Gas Tungsten Arc Welding (GTAW)

The gas tungsten arc welding process (GTAW) was patented in 1942. The welding arc is maintained between a non-consumable tungsten electrode and the base metal. The arc is shielded with an inert gas such as argon or helium. The process name may be explained in this manner:

Gas	=	externally supplied shielding gas
Tungsten Arc	=	non-consumable tungsten electrode
Welding	=	arc welding

Gas tungsten arc welding provides the highest quality welds of any of the manual welding processes.

81. Groove Angle

The term "groove angle" describes the total included angle of a prepared weld joint groove between two parts to be welded. It is often referred to as the "included angle" of the joint.

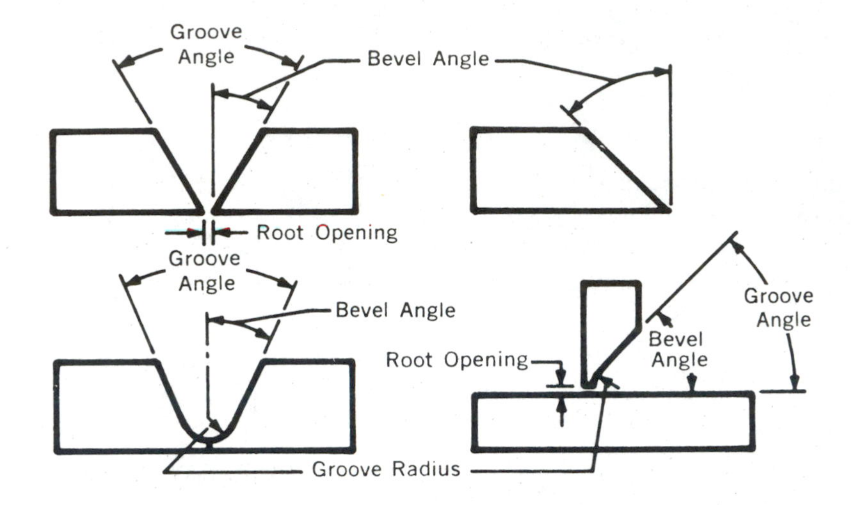

Figure 19. Typical Groove Angles of a Joint.

82. Gun

In the GMAW and FCAW processes, there are two distinct shapes of devices for directing the continuous electrode to the weld area. One device is shaped like a pistol with a "grip" form of handle. It is normally called a "gun" because of it's pistol shape. The other device is shaped like a gas welding torch with tip. It is commonly called a "torch" because of it's shape. Electrode wire and shielding gas, if used, is directed at the weld by the welding gun.

83. Hard Facing

The process of applying a hard metal surface to a softer metal base is called "hard facing". The hard material is normally applied to reduce erosion and wear of the part due to abrasion and impact.

84. Heat-Affected Zone (HAZ)

The heat-affected zone is that area of the base metal not melted during the welding operation but whose physical characteristics and properties were altered by the heat induced from the welded joint. The heat-affected zone begins at the interface of the weld deposit and the base metal. It ends in the base metal where no physical changes in metal structures or properties has occurred.

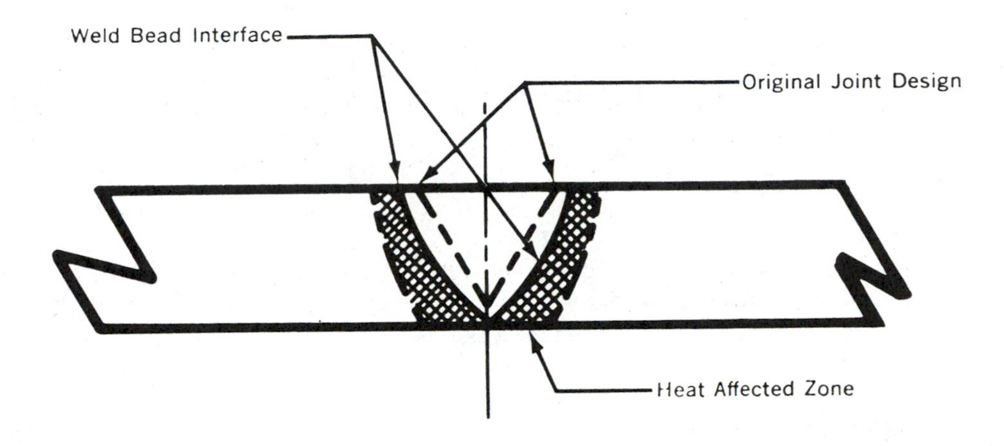

Figure 20. Weld With Heat Affected Zone (HAZ).

85. Hertz

The term "hertz" has replaced the term "cycles per second" in electrical terminology. Hertz refers to "cycles per second" frequency of AC. Hertz is abbreviated Hz.

86. High Frequency (HF)

High frequency covers the entire frequency spectrum above approximately 50,000 cycles per second. The numerically high frequency rate causes the current to flow on the surface of electrical conductors. This is known as the "skin effect" of high frequency power transfer. High frequency is used primarily as a means of bringing safe high voltage to the gas tungsten arc electrode tip. Improved non-touch arc ignition and better arc stabilization with AC welding power are other benefits.

87. Hold Time

"Hold time" is a term employed with the resistance welding process. It describes the time when pressure is maintained on the workpiece after the resistance welding current has ceased to flow. The purpose of hold time is to maintain a constant pressure on the weld joint until solidification of the weld nugget has been completed.

88. Horsepower

The measure of *rate of work* may be made in horsepower. Electrically, one horsepower equals 746 watts. Mechanically, one horsepower is equal to lifting a 33,000 pound weight to a height of one foot in one minute of time.

89. Hysteresis

Hysteresis is the resistance, or reluctance, of magnetic particles to polar orientation when they are subjected to a magnetic field. As the direction of current flow changes each half-cycle with alternating current, so must the molecules in the iron core materials in a transformer change polarity each half-cycle. Since work is being done energy is being expended. The energy is expended as heat in the iron core material, causing it to lose electrical efficiency.

90. Impedance

Electrical impedance is a combination of resistance and reactance which opposes the flow of current in an alternating current circuit. The resistance value is fixed according to the type and diameter of conductor material used and is considered to be "real" and measurable. The reactance value is entirely dependent on other factors in the circuit. It may have a maximum value or it may have no value at all. Reactance values are considered as "apparent" values.

91. Inclusions

Inclusions are usually non-metallic particles that appear in a weld deposit. They may be slag residue that is trapped in the weld because of fast solidification characteristics inherent in the weld metal. Most inclusions must be removed since they are a form of weld defect that will weaken the weld.

92. Induced Voltage

Induced voltage is created by an electromagnetic action through devices such as welding power source transformers and reactors. An induced voltage is always opposite in force direction to the effect that created it, according to Lenz's Law.

93. Inductance

Inductance is the electrical phenomenon that causes voltage and amperage to be apparent in the secondary circuit of a welding transformer power source. It is the electrical influence exerted by current flow in a conductor, through a magnetic field, on adjacent conductors. There is no physical contact between the two conductors. Inductance is always a magnetic effect and is current induced.

For example, current is applied to the primary coil of a welding transformer. The primary coil is located around one leg of the iron transformer core. A magnetic field is created when current is caused to flow through the primary coil. Energy is used to create the magnetic field. The energy in the magnetic field is induced into the secondary coil of the transformer without physical contact between the coils.

94. Inert Gas

An inert gas will not combine with any known element. There are six inert gases presently known. They are: argon, helium, xenon, radon, neon, and krypton. Only argon and helium are used as shielding gases for welding.

95. Interpass Temperature

Control of interpass temperature is critical to many multipass welding applications. Interpass temperature of the deposited weld metal may be shown as either a maximum or minimum value on the welding procedure.

In many cases, a maximum interpass temperature is specified to minimize distortion of the part. It is also a method of controlling heat input to the weldment. When welding materials such as heavy sections of copper, a minimum interpass temperature may be specified to permit the weld deposit to wet out and fuse with the base metal.

96. Ion

In gas shielded welding applications, the gas atoms are equal in number of electrons (−) and protons (+). By causing an electron to leave the gas atom, there is created an electrical imbalance in the gas atom. For example, a helium atom has two electrons and two protons. The protons are in the nucleus of the atom and do not move. The electrons are in motion around the nucleus and may move freely.

Removing an electron leaves one electron (−) and two protons (+). The gas atom is electrically charged positive (+). An electrically charged gas atom is called an ion.

97. Ionization Potential

Ionization potential is the energy necessary to remove an electron from the gas atom thereby making it an ion. The potential (voltage) energy required will depend on the material to be ionized. For example, the ionization potential for argon gas is 15.7 electron volts (eV). For helium, the ionization potential is 24.5 eV.

98. Izod Test

The Izod test is a pendulum-type impact test in which the specimen is notched, held at one end, and broken by the falling pendulum. The energy absorbed, as measured by the subsequent rise of the pendulum, is a measure of impact strength, or notch toughness, of the material.

99. Joint

A joint, as used in welding terminology, is the junction of two or more pieces of metal that have been, or are to be, welded together. Joint design is critical in many weldments. A joint design shows the actual geometry of the joint with angles and dimensions of the joint. Some common groove joint designs are illustrated.

100. Joint Efficiency

Joint efficiency is the strength of a welded joint expressed as a percentage of the guaranteed minimum strength of the unwelded base metal. Joint efficiency is calculated with the top and underbeads flush with the surface of the base metal.

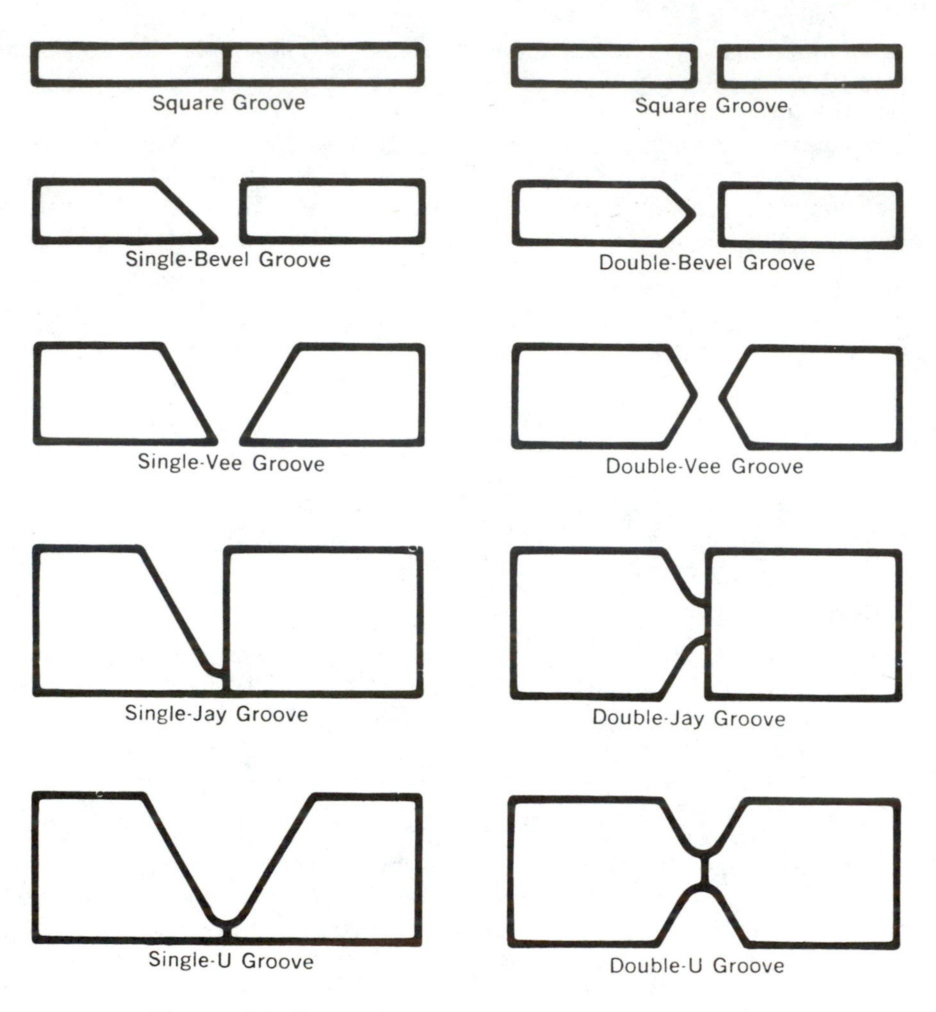

Figure 21. Some Common Groove Joint Designs.

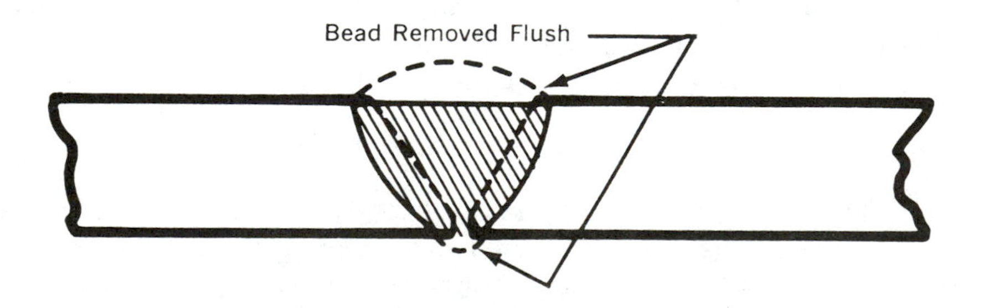

Figure 22. Specimen Prepared for Joint Efficiency Test.

101. Joint Penetration

Joint penetration is the distance fused weld metal extends into the base metal of the welded joint. Penetration begins at the surface of the weld joint. For example, maximum joint penetration with the shielded metal arc welding process, using an E6010 electrode, is approximately 1/8''.

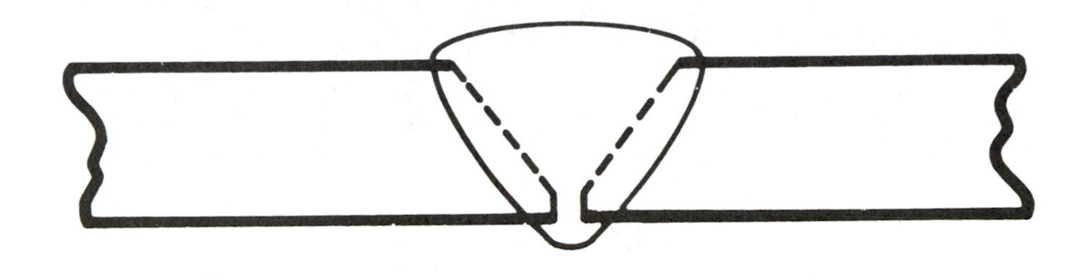

Figure 23. Joint Penetration Diagram.

102. Joule

For welding, the ''joule'' is a designation of heat energy input to the welding area. The joule is the unit of power in the metric system of measurement. The equation for determining joules of energy input is:

$$H \text{ (joules per inch)} = \frac{E \text{ (volts)} \times I \text{ (amperes)} \times 60}{S \text{ (speed in inches per minute)}}$$

Specifically, joule energy input is the heat energy imparted to the weld puddle by the welding arc.

103. Kerf

The width of a cut made in the base metal during a flame or arc cutting process. It is the space formerly occupied by the actual metal removed.

104. Leg of a Fillet Weld

The exact distance from the root of the fillet weld to the toe of the fillet weld is called the leg of the fillet weld.

105. Load Voltage

Load voltage is measured at the output terminals of a welding power source while welding. It is the total voltage load, including arc voltage and the voltage drop through the welding cables, that the power source senses. It is measured with a voltmeter.

106. Magnetic Coupling

Magnetic coupling is the term used to describe the relationship between two coils so located that a magnetic field created in one of them will interact on the other one. Another term that describes this effect is inductive coupling. In welding transformer power circuits an iron core is normally the intermediate link between the two coils.

Some welding power sources have "loose" magnetic coupling in the primary-secondary coil arrangement of the main transformer. These are the constant current type power sources normally used for shielded metal arc welding and gas tungsten arc welding. They have a substantial negative output power curve.

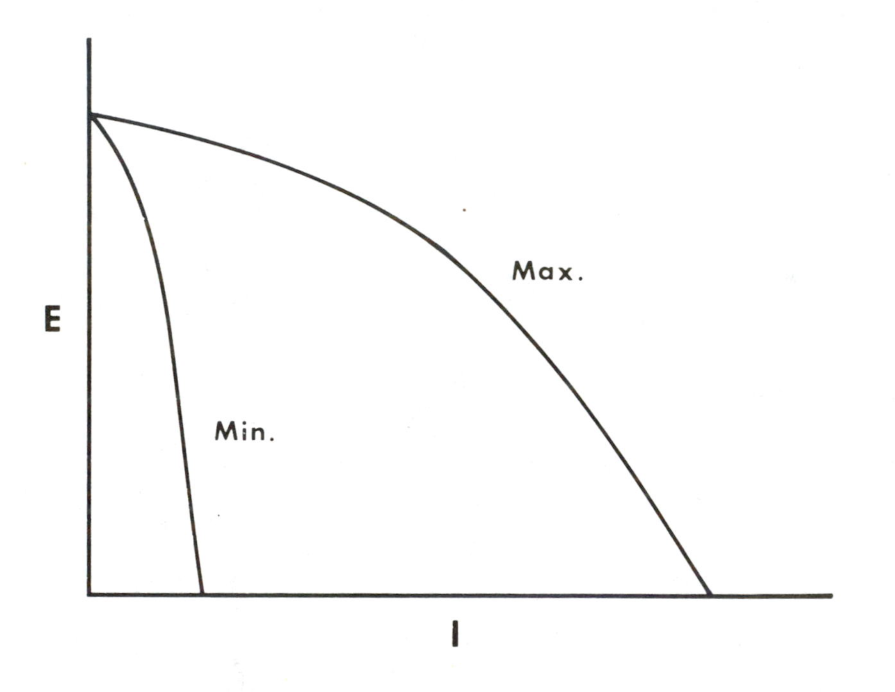

Figure 24. Typical Constant Current Volt-Ampere Curve.

Other arc welding power sources have a "tight" magnetic coupling between the primary and secondary coils. Called constant voltage, or constant potential, type power sources, they have a relatively flat volt-ampere output curve. This type of power source is used mostly for the gas metal arc, and flux cored arc, welding processes.

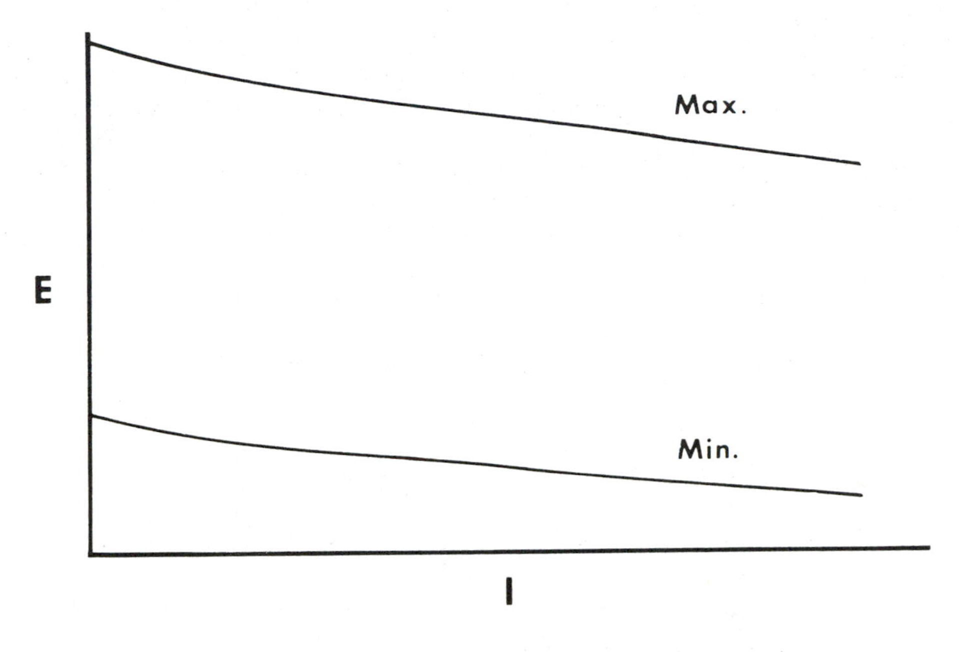

Figure 25. Typical Constant Voltage Volt-Ampere Curve.

107. Magnetic Field

A magnetic field will be created when current is caused to flow through a coil wrapped around an iron core. The strength of the magnetic field will depend on three factors:

a. The mass and type of iron in the core.
b. The number of effective electrical turns in the coil.
c. The amount of current flowing in the coil.

If the mass of iron and the number of effective electrical turns remain constant in a welding transformer, the only variable will be the amount of current in the circuit. The magnetic field strength will, therefore, follow the alternating current sine wave trace, increasing in strength when current increases and decreasing in strength when current decreases.

The lines of force in a magnetic field are always at right angles (90°) to the direction of current flow.

108. Malleability

Malleability is the characteristic of metals and other materials that permits plastic deformation of the material without rupturing. It indicates the pliability of a metal in forming operations. A good example of malleability in metals is the deep-drawing of household utensils such as cooking pans.

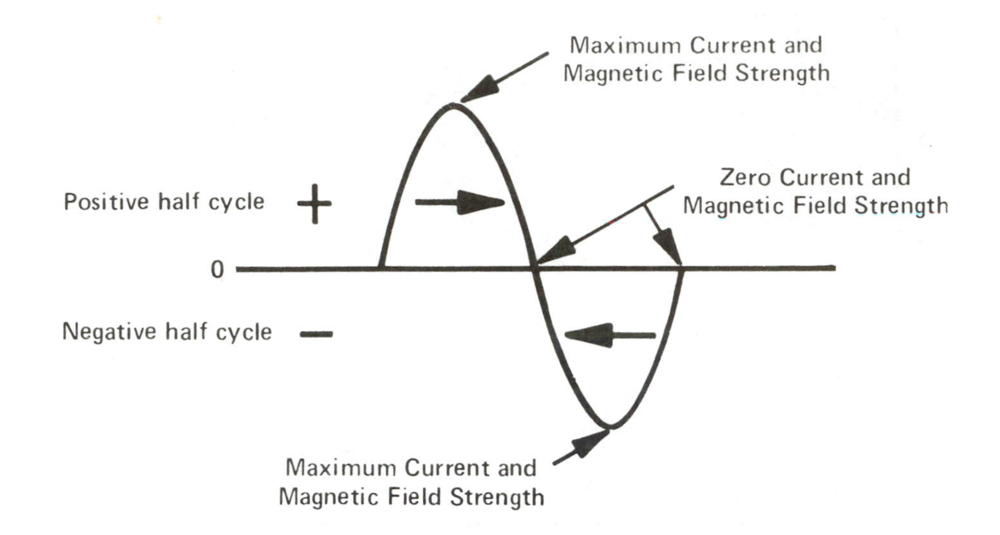

Figure 26. Sine Wave Form and Magnetic Field Strength.

109. Mild Steel

Mild steel is a name given to carbon steel having approximately 0.15%-0.30% carbon content. Mild steel is considered to be non-heat treatable for hardness. It may be normalized and stress relieved after welding with no problem.

110. Mixer

The "mixer" is the part of an oxy-fuel gas torch where the gases come together and are mixed. The mixer is also referred to as the mixing chamber of the torch.

111. Negative Charge (−)

"Negative charge" is the term used to indicate the type of electrical charge carried by an electron. It is illustrated by the negative, or minus, (−) sign.

112. NEMA

NEMA is the abbreviation for the National Electrical Manufacturers Association. It is a self-regulating group of electrical manufacturers whose purpose is to promote safe, common standards of manufacture for electrical apparatus. It is entirely supported by member companies. NEMA is the regulating agency for standards for welding power sources, wire feeder/control units and other electrical apparatus.

113. Nucleus

In physics, the nucleus is the heavy central core of an atom. Most of the atomic mass and weight, and all of the positive electrical charge, is contained in the nucleus of an atom. This fact is most important when considering the gas shielded welding processes.

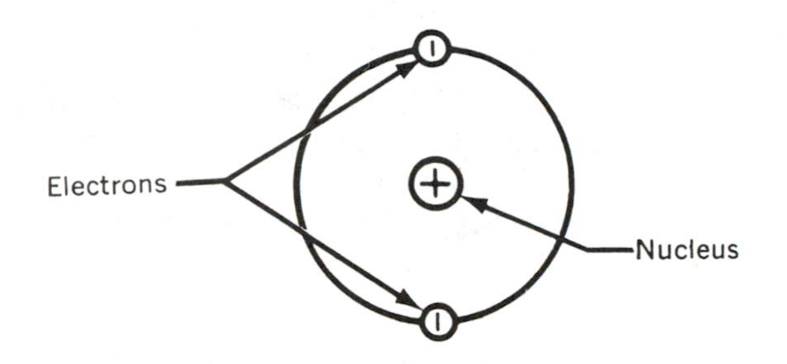

Figure 27. Helium Atom, Diagrammatic View.

A diagrammatic view of a helium atom is shown in the illustration. Note that the nucleus is not shown in proportional size to the electrons. Its mass is many times greater than the mass of the orbiting electrons.

114. Nugget

A nugget is the fused metal in a resistance spot, seam or projection weld. It is sometimes mistakenly used to describe a fusion weld deposit. A nugget is considered to be a solid lump of fused weld metal rather than a continuous weld seam.

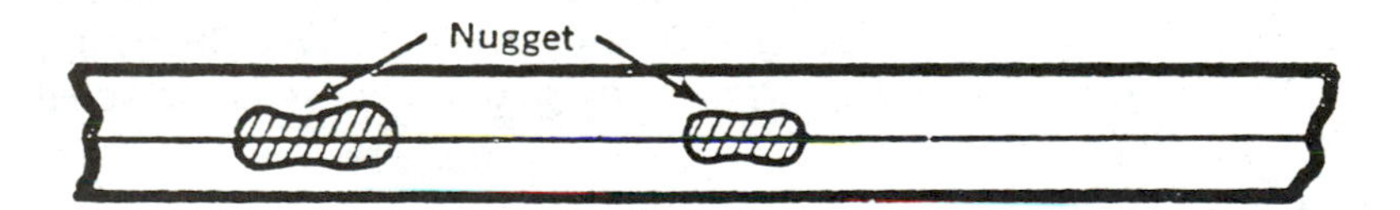

Figure 28. Resistance Spot Weld Nugget.

115. Ohm

The ohm is the practical unit of electrical resistance. There is one ohm resistance when a pressure of one volt causes a current of one ampere to flow in a circuit. Ohm's Law is presented as follows:

$$E = IR \text{ or, } I = \frac{E}{R} \text{ or, } R = \frac{E}{I}$$

where
E = Volts
I = Amperes
R = Resistance in ohms.

116. Open Circuit Voltage

Open circuit voltage is always measured across the output terminals of a welding power source when the unit is energized but under no welding load. The electrical measurement is made with a voltmeter.

117. Oxygen

Oxygen is an odorless, colorless invisible gas which comprises approximately 21% of the earth's atmosphere. Oxygen is necessary to, and apparent in, all living things including human life. It has the ability to support combustion in all burnable materials. In concentrated quantities, oxygen can support combustion in theoretically non-flammable materials.

118. Permeability

The word permeability comes from the word "permeate" meaning to go through, saturate or pass through by diffusion. In electric welding power sources, **magnetic permeability** is a measure of the ease with which a material, such as electrical steel, can accept magnetic lines of force.

119. Plasma

Although this term applies to all arc welding, it is especially applicable to the plasma arc welding (PAW) and plasma arc cutting (PAC) processes. The term "plasma" refers to an industrial gas that has been heated sufficiently to cause it to ionize. The ionized gas is electrically charged, of course, and is capable of conducting current.

120. Polarity

Electrical polarity primarily concerns DC welding power. Direct current flows in one direction only and has polarity. Polarity means that current has stabilized directional flow.

The terms DC Straight Polarity (DCSP) and DC Reverse Polarity (DCRP) have been used for many years. A more descriptive set of terms is now in general use to replace these terms. The new terms are:

DC *Electrode Negative* (DCEN) instead of DCSP.
DC *Electrode Positive* (DCEP) instead of DCRP.

These terms will make it easier for welders and others in the welding profession to set up power source connections for either polarity.

121. Porosity

Porosity is a defect often found in welds. Porosity is usually caused by trapped gases in the deposited weld metal. It will always have a spherical shape.

The type of porosity will often indicate its basic cause. In-line porosity normally indicates a lack of sufficient heat input to the weld area. Random porosity may mean that welding speed of travel was too fast. In this case, the weld metal solidifies before the gases can evolve out of the weld.

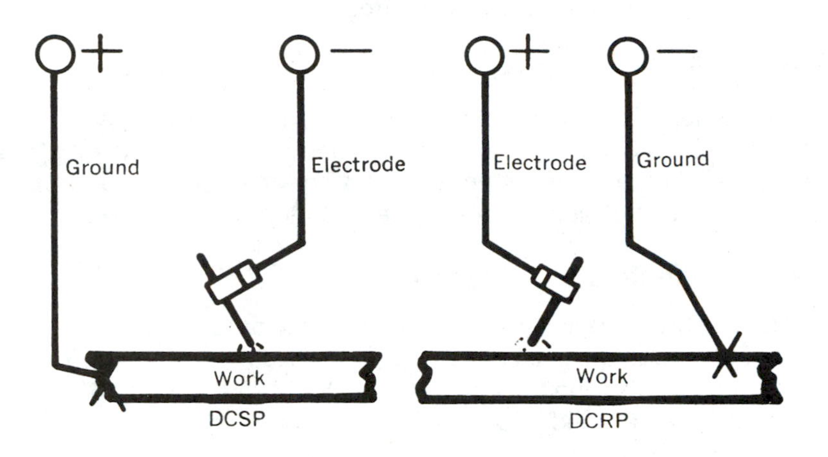

Figure 29. Polarity Connections for Direct Current.

122. Positive Charge (+)

"Positive charge" is the term used to indicate the type of electrical charge carried by a proton. The symbol for a positive electrical charge is the plus (+) sign.

123. Power Factor

Power factor is the measure of time phase difference between the voltage and amperage in an alternating current circuit. It is expressed as a percentage of power used (primary KW) to total power drawn (KVA) from the primary power system.

The ratio of used power in KW to the total power in KVA is called power factor. Dividing the primary KW by the total power KVA will give the power factor percentage of a welding power source. Power factor may be improved by adding capacitors to the primary power circuit. The addition of capacitors to an inductive circuit improves power factor by **demanding less primary amperage.** The addition of capacitors is called **power factor correction.**

124. Preheat Temperature

This term is often called out on welding procedure specifications as part of the procedure. "Preheat temperature" is a required temperature which the base metal must attain prior to any welding operation being done. In many applications, an interpass temperature is also called out on the welding procedure. This is usually done to retard the cooling rate of the metal for some metallurgical reason.

125. Procedure Qualification

A Welding Procedure Specification (WPS) is developed to establish the parameters for a required weld. The Procedure Qualification Record (PQR) is a record made of the "procedure qualification". The procedure qualification assures that welds made with the procedure will meet a required Code or standard.

126. Proton

The proton is the fundamental unit of positive electricity. It has approximately 1800 times the mass and weight of an electron. The electrical charge of a proton is equal, but opposite, the electrical charge of an electron. If an atom has an equal number of protons and electrons, it is electrically neutral and is called a neutral atom.

127. Reactance, Inductive

The characteristic of a current conducting coil wrapped around an iron core which causes the current to lag behind voltage in time phase peak is called "reactance". The current wave trace reaches its maximum strength, or amplitude, later in time than the voltage wave trace. The time lag is between "0" and "90" electrical degrees.

128. Reactor

A reactor is an electrical device consisting of at least one current conducting coil wrapped around an iron core. Normally located in the secondary AC portion of a welding power source circuitry, a reactor's function is to change the effective reactance of an AC circuit. Reactance may be either inductive or capacitive.

129. Rectifier

A rectifier is an electrical device that permits the flow of electrical current in one direction only. Its basic function is to change alternating current (AC) to direct current (DC). The two major types of rectifiers in welding power sources are selenium and silicon.

130. Refractory Material

A refractory material is considered to be any material that has a melting point in excess of 3,600° F. For example, tungsten is a refractory element since it has a melting temperature of approximately 6,170° F. Some known refractory elements are listed in the illustration.

Material	M.P.,	° F	Material	M.P.,	° F.
Boron	3690	(approx.)	Osmium	4900	(approx.)
Carbon	6740	"	Rhenium	5755	"
Columbium	4474	"	Ruthenium	4530	"
Hafnium	4032	"	Tantalum	5425	"
Iridium	4449	"	Technetium	3870	"
Molybdenum	4730	"	Tungsten	6170	"

Figure 30. Some Known Refractory Elements

131. Regulator

A gas regulator is a mechanical device used to control the output pressure of compressed gases for welding operations. The two basic types of regulators are single stage and two stage. The function of any regulator is to furnish a constant flow of gas at a prescribed rate and pressure regardless of the actual pressure at the source of supply.

132. Relay

A relay is a type of switch operated by electro-mechanical force. It is usually used as a control mechanism for electrical circuits. The "normal" position for the relay contact points will be shown in the circuit diagram for the circuit in which it exists. "Normal" is considered to be the de-energized condition.

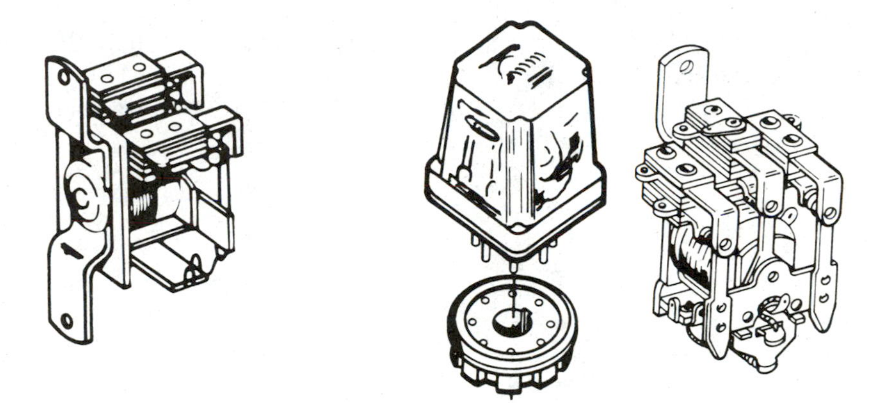

Figure 31. Sealed and Unsealed Relays.

133. Reluctance

The term reluctance refers to that characteristic of a magnetic path, or material, which resists the flow of magnetic lines of force through its body. Metals and other materials with high reluctance values have poor magnetic permeability.

134. Residual Magnetic Field

A residual magnetic field is the magnetic field remaining in a ferrous metal part after the source of the magnetic field has been removed.

135. Resistance

The properties in an electrical conductor that oppose the passage of current are called electrical resistance. The ohm is the practical unit of measure of electrical resistance. The energy dissipated in overcoming electrical resistance in a conductor is apparent as heat in the conductor.

136. Resistance Welding

Resistance welding is a method of electric welding which uses both resistance heating of the weld area and pressure. The lap joint of the workpiece is an integral part of the welding electrical circuit. The resistance of the base metal at the interface of the joint, where current flows, creates a localized heating effect. This factor, combined with the applied localized pressure, causes a fusion bond to be made. Sometimes called "spot" welding, resistance welding may be used with most of the commonly welded metals.

137. Reverse Polarity (DCRP)

Reverse polarity is a term used with DC welding. It is synonymous with the term "Direct Current Electrode Positive" (DCEP).

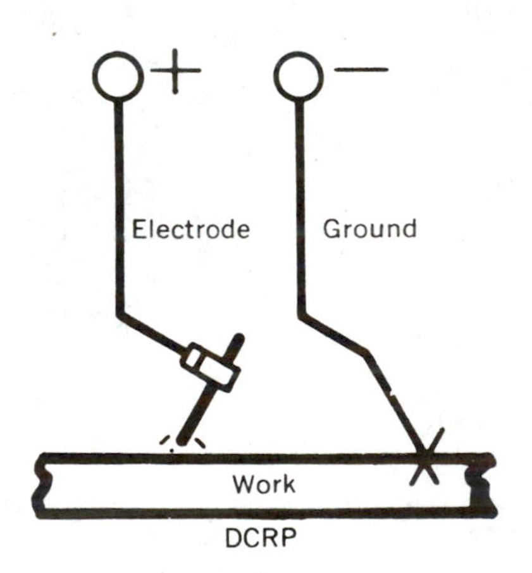

Figure 32. Connections for DC Reverse Polarity—DCEP.

138. Root Pass

The root pass is the first, and most important, pass made in a multipass weld joint. The root pass must provide full penetration and fusion of the root area of the weld joint. The root pass must be of excellent quality because the balance of the weld passes are essentially fill-in of the joint.

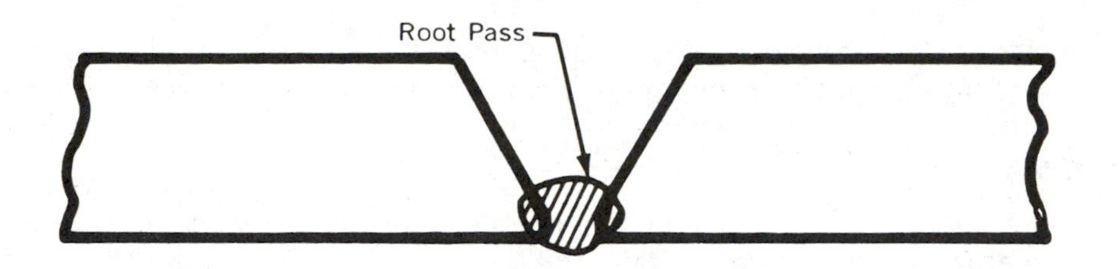

Figure 33. Root Pass in Joint.

139. Seal Weld

A seal weld is any weld that is made in a vessel to prevent leakage. It may, or may not, be a strength weld.

140. Secondary Circuit

The secondary circuit refers to the welding leads and other welding apparatus between the output terminals of the power source. It is the total electric welding circuit external to the power source.

141. Secondary Coil

The secondary coil of a welding transformer is located on the main transformer core. The energy in the magnetic field, created by the current flow in the primary coil, is induced into the secondary coil. There is no direct connection between the primary and secondary coils of the transformer. The induced voltage may be measured at the output terminals of the power source as open circuit voltage.

142. Semiautomatic Arc Welding

GMAW and FCAW semiautomatic arc welding is accomplished with welding equipment that controls the filler wire feed speed rate and the arc voltage. Travel speed and torch, or gun, angle is manually controlled.

143. Shielded Metal Arc Welding (SMAW)

Creating an arc between a consumable flux covered electrode and the surface of the base metal starts the weld with the shielded metal arc welding process. The molten filler metal fuses with the molten base metal in the arc area to form the weld.

Shielding is provided by the decomposition of the flux coating on the electrode. The flux provides both a shielding gas and a slag covering for the weld as it cools. The process name may be explained in this manner:

Shielded = the gas and slag residue from the flux.
Metal Arc = the consumable electrode.
Welding = arc welding.

The shielded metal arc welding process is the most portable of the arc welding processes. The initial welding equipment costs are less than for any other arc welding process.

144. Slope

The term "slope" has been used for a number of years by the welding industry. For the gas tungsten arc welding process (GTAW), the term "sloping off" is used to indicate a decrease in welding current at the end of a weld. This permits the welder to fill the crater at the weld end to eliminate possible crater cracking.

When discussing slope control for the gas metal arc welding process (GMAW), it is the shape of the static volt-ampere curve to which we refer. The higher the slope number, the steeper the angle of the volt-ampere curve and the lower the maximum short circuit current of the power source.

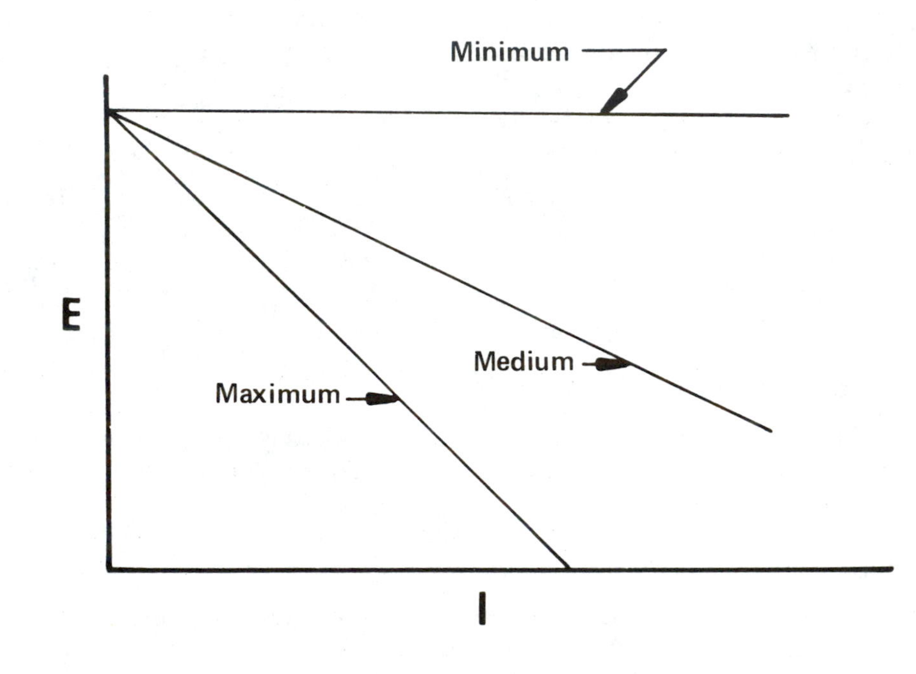

Figure 34. Typical Slope Curves.

145. Solid State

The term "solid state" refers to electronic control methods and systems used with welding equipment such as power sources and wire feeder/controls. Included in the solid state devices are silicon diodes, transistors, silicon controlled rectifiers (SCR), integrated circuits, and microprocessors. The use of solid state circuitry in welding equipment provides greater stability of output with many extra features not available with non-electronic control systems.

146. Spatter

Spatter is the metal expelled from the weld area during the actual welding operation. Spatter should be kept to a minimum since it reduces electrode efficiency, can cause burns to the welder and his clothing, and creates additional weld cleaning costs. Spatter consists of various sizes of metal globules, some of which may adhere to the surface of the base metal. Spatter indicates a poor welding condition or technique.

147. Spot Welding

A "spot weld" is normally made in materials having some type of overlapping joint design. Although the term is usually referred to in resistance welding, it is also applicable to the gas tungsten arc, and gas metal arc, welding processes. Note that resistance spot welds must have electrodes on both sides of the joint. The gas tungsten arc, and gas metal arc, spot welds are made from one side only.

Total metal thickness is limited with the resistance spot welding, and gas tungsten arc spot welding, processes. For the gas metal arc spot welding technique, only the top sheet or plate thickness is critical.

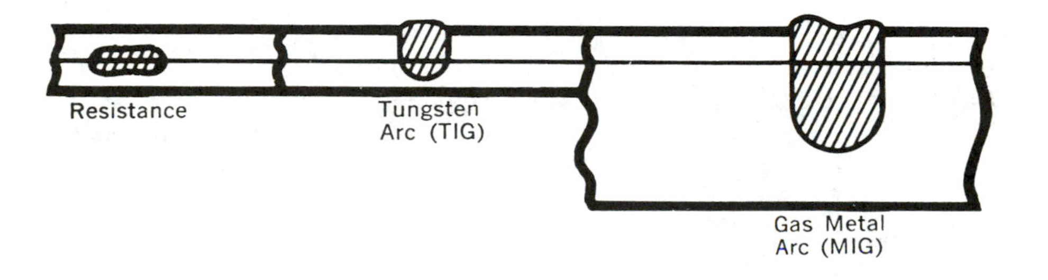

Figure 35. Typical Spot Welds.

148. Squeeze Time

Squeeze time is the time that elapses between the initial application of pressure and the start of current flow when resistance welding. It permits the positioning of the materials to be welded before the current begins flowing and the actual weld is made.

149. Static Electricity

Static electricity is electricity without movement. It is the opposite of dynamic electricity.

150. Straight Polarity (DCSP)

The term straight polarity is associated with DC welding. It is synonymous with the term "Direct Current Electrode Negative" (DCEN).

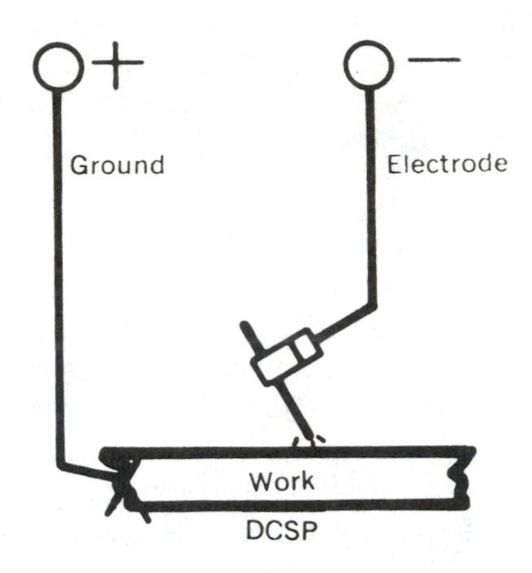

Figure 36. Connections for DC Straight Polarity—DCEN.

151. Stringer Bead

A stringer bead is a narrow, single pass bead made with little, if any, oscillation of the electrode. It is usually used for strength welds and for maintaining control of the fluid weld puddle. Stringer beads are preferred for most vertical-up welding. There is normally much less chance of slag inclusions, porosity, and other weld defects when a weld is made with stringer beads and proper welding procedures.

152. Tack Weld

A tack weld is a small size weld made to hold metal parts together during the fitting-up operation. The tack welds should be the same quality as the final weld requirement. In some applications, tack welds are ground out as the welder comes to them when making the root pass. This is to assure that there will be no lack of fusion at the root of the weld.

153. Thermal Conductivity

All of the metals have some measure of thermal conductivity. Thermal means "heat" and conductivity means "transfer". Thermal conductivity means heat transfer and, in metallurgy, it means heat transfer through metals and metal structures.

Carbon steel and stainless steels have relatively poor thermal conductivity. Copper, aluminum, and magnesium have relatively good thermal conductivity. Copper and aluminum are often used as "heat sinks" in weld tooling because of their ability to radiate heat to the atmosphere. Copper and aluminum also have good electrical conductivity.

154. Thermal Stress

Thermal, or heat, stresses in weldments are caused by uneven distribution of heat through the metal part. Such stresses will often cause distortion in the weldment.

155. TIG

The initials "TIG" are the abbreviation for Tungsten Inert Gas. This is a shop term for the Gas Tungsten Arc Welding (GTAW) process.

156. Timer

A timer is an electrical, or electronic, device that performs a pre-set function after a period of time has elapsed. It is also used to regulate and control welding time in many applications.

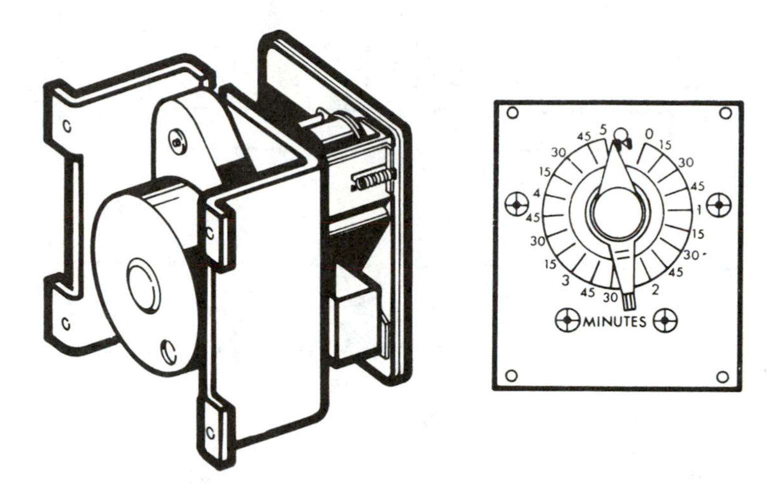

Figure 37. Front and Rear Views of a Timer.

157. Toe Crack

Toe cracks occur in the base metal at the toe of a weld.

158. Toe of Weld

The specific point at which the weld metal face intersects with the base metal is called the "toe" of the weld. Although applicable to all fusion welds, the term is most often used when discussing fillet welds.

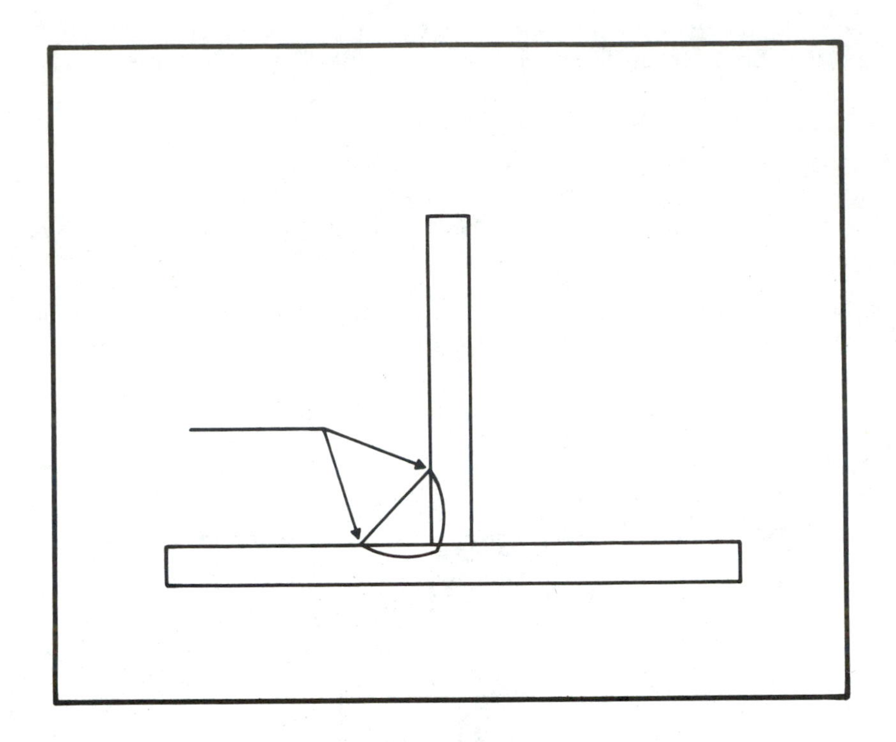

Figure 38. Toe of a Weld Joint.

159. Ultimate Strength

The ultimate strength of a material is the maximum stress it can withstand in tensile, shear, or compression loading. When a metal reaches its ultimate strength, it ruptures in a catastrophic manner.

160. Underbead Crack

A crack in the heat affected zone of the base metal, under the weld deposit, is called an ''underbead crack''. This form of crack normally does not propagate to the base metal surface.

161. Undercut

Undercut may be apparent at the underside of a root pass or at the toe of a weld. Surface undercut is a groove, or channel, melted in the base metal immediately adjacent to the weld. It is a defect because it decreases the cross sectional area of the base metal. Undercut also acts as a notch effect which could cause cracking in the metal.

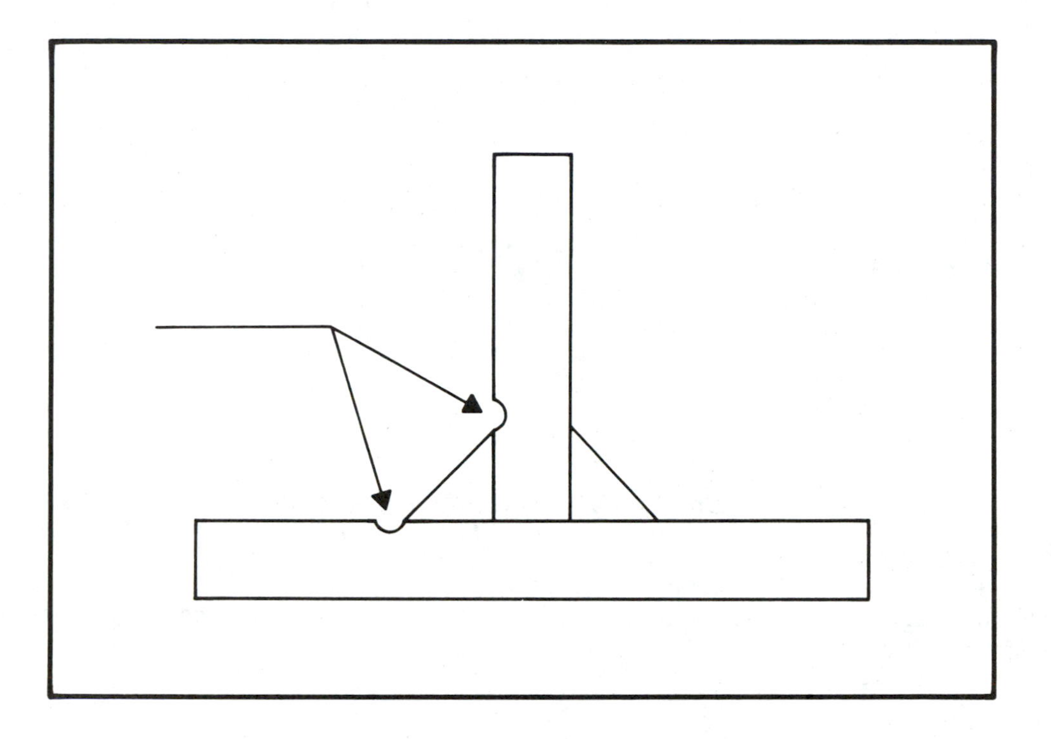

Figure 39. Typical Undercut in a Weld Joint.

162. Unit Charge

The electrical energy of one particle of matter is called a unit charge. For example, an electron is the fundamental unit of negative electricity and it has one negative unit charge. A proton has one positive unit charge.

163. Volt

The electrical term "volt" means electrical pressure, or force, that causes current to flow in an electrical conductor circuit. It is well to remember that voltage is a force that does not flow but which causes current to flow.

164. Volt-Ampere Curve

Volt-ampere curves are graphs which show the output characteristics of a welding power source. The vertical ordinate always indicates voltage. The horizontal ordinate indicates amperage. Volt-ampere curves are plotted under static conditions using a known resistive load. It is not possible to plot volt-ampere curves while welding due to the dynamic conditions of the welding arc.

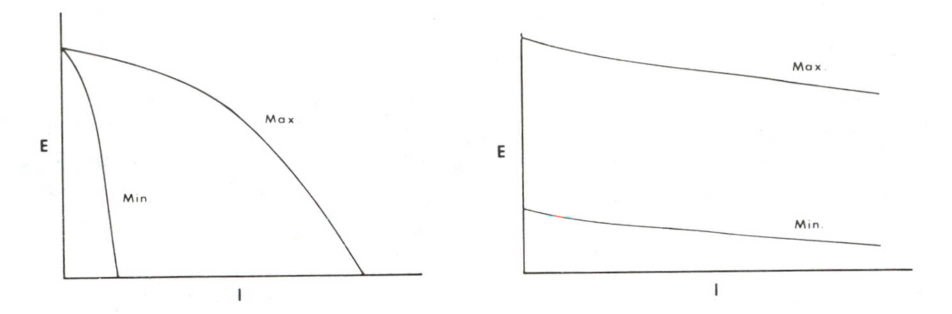

Figure 40. Typical Volt-Ampere Curves.

The point where the volt-ampere curve intersects the vertical ordinate is called "open circuit voltage". Where the curve intersects the horizontal ordinate the reading is "maximum short circuit current".

165. Watt

The watt is a unit of electrical power measurement. For DC power, watts are calculated by multiplying volts times amperes. A watt is the amount of total power required to maintain a current of one ampere flowing in a circuit where the electrical pressure is one volt. 746 watts are equal to one horsepower.

166. Watt-Hour

The watt-hour is the unit of electrical energy equal to the power of one watt being used continuously for one hour. It is watt-hours that are charged for on your electrical bill each month.

167. Weave Bead

Weave beads are made by oscillating the electrode tip from side to side of the weld transverse to the direction of travel. Weave beads are often used for the top pass of a multi-pass weld.

168. Weld

In metals, a weld is the uniting of two separate pieces by forging, melting and fusing, or through friction of the abutting surfaces.

169. Weldability

Weldability is the characteristic of a metal that permits it to be welded under a specific set of conditions. All of the common metals are weldable with one or more of the known welding processes.

170. Welder

A welder is a person qualified to perform a manual or semiautomatic welding operation. The term "welder" is often erroneously used to describe a welding power source.

171. Welder Certification

A "welder certification" means the welder satisfactorily performed a welding operation as set forth in a qualified welding procedure.

172. Welding Procedure

A welding procedure details the methods and practices used in making a completed weld. Welding procedures should be written for every welding application. Welding procedures should detail the joint design, number of weld passes, weld pass sequence, and the welding process used in addition to the mechanical and electrical requirements of the job.

173. Yield Strength

When testing metals, there is a definite proportion between the stress applied and the resulting strain. Yield strength is the value at which the metal shows a marked change from the normal proportionality of stress and strain. At the yield point, the metal permanently deforms with no increase in stress applied. This is the yield strength.

CHAPTER 2

Welding Metallurgy Fundamentals

Metallurgy is the study of metals in both their pure and alloyed forms. How metals react to other metals and chemical elements, to heat and cold, to both hot working and cold working, is all within the realm of metallurgy.

People who work with metals as engineers are called Metallurgists or Metallurgical Engineers. Associated with metallurgists are metallographers, metallurgical technicians, and testing laboratory personnel. Metallurgical personnel work right along with the welders to make sure that weld metal and base metal combine together in the best possible manner for strength and performance.

There are two broad classifications of Metallurgists in the metals industries. The **Extractive Metallurgist** is employed in the mining, smelting, and refining of mineral ores. The **Process Metallurgist** is concerned with the application of metals in the manufacturing industries. It is the process metallurgist who is most concerned with welding metallurgy. In this area, he may be the one to specify the welding procedure used, the welding process, electrode filler metal type and classification, any preheat or postheat required, and any other data necessary to a good weld joint.

Welding metallurgy is concerned with metals and how welding affects their physical and mechanical properties and chemical composition. Even slight changes in chemical composition can cause substantial changes in the physical and mechanical properties of metals.

Metals Welded

There are two basic types of metals in use today by the welding industry. They are the **ferrous metals** which contain substantial amounts of iron and the **non-ferrous metals** which have essentially no iron in their chemical content.

Some non-ferrous metals may be hardenable by cold working such as rolling, bending, hammering, etc. The cold working reduces ductility, and increases hardness, of the metals. When non-ferrous metals are heated substantially after cold working, they tend to soften. This will increase ductility but reduces hardness and strength.

Some of the non-heat-treatable aluminum alloys include the following series: 1xxx, 3xxx, and 5xxx.

Some alloys of non-ferrous metals may be hardened by heat treatment. For example, certain aluminum alloys are classed as heat-treatable. Some of the heat-treatable aluminum alloys include the following series or alloys: 2xxx, 4032, 6xxx, and 7xxx.

Welding heat-treatable aluminum alloys after they have been heat treated will cause some loss of strength in the heat-affected zone (HAZ) due to overaging of the metal. There is no known fusion welding process that can weld heat treated and aged aluminum alloy parts and obtain an "as welded" part that has 100% joint efficiency. It is preferable to weld heat-treatable aluminum alloys in the annealed condition and then heat treat them subsequent to welding.

There is a substantial loss of ductility when welding heat-treatable aluminum alloys. Heat treatment after welding will restore much of the strength of the metal but will not improve the ductility lost to welding.

Ferrous metals are iron based alloys. An alloy is a combination of two or more elements which may make a metal with better physical or chemical properties than either of the two, or more, elements basic to the alloy. In steel it is carbon that is the most important primary alloying element.

Most heat treating of iron and steel is done between the temperatures of 1400°-1850° F. Of course, when welding iron or steel, the welder actually melts the base metal in the localized area of the weld joint. Since iron and steel have a nominal melting temperature range of 2500°-2600° F. in the commercial grades, it is logical that a temperature gradient exists between the molten weld metal deposit and the unheated base metal. To define:

gradient = "change in value of a quantity over a specific distance."

In this case, the temperature gradient is from a high temperature at the melting point of the weld metal, through the heat-affected zone, to the base metal still at room temperature.

Somewhere between the melting point of the weld metal and the room temperature of the base metal, part of the heat-affected zone is in the range of 1400°-1850° F. heat treating temperature. It becomes apparent that the welder "heat treats" the heat-affected zone each time he makes a weld simply because the heat from the weld travels through the heat-affected zone on its way to dissipating in the mass of the base metal. A very good rule to know is that heat will transfer much more rapidly through cold metal than it will through hot metal. This means that the heat of the weld will flow quickly to the colder base metal being welded. That is the reason a weld will lose its red heat color very fast as the welder moves along the weld joint.

All of the carbon steels have some alloying elements in their chemical composition. Not all of these elements are really desirable. For example, all plain carbon steels contain iron (Fe), carbon (C), manganese (Mn), sulphur (S), silicon (Si), and phosphorous (P). Both the sulphur and phosphorous content is held to a minimum fractional percentage, usually not more than 0.05% maximum.

In addition to the plain carbon steels, there are a group of steels called "alloy steels". It may seem a little odd that certain steels are called alloy steels, and carbon steels are not included in that group, when in fact **all steels are an alloy of two or more elements.** "Alloy steels" are so named because one or more elements are added for a specific purpose. In many cases, alloying elements are added to steel to increase physical properties such as ductility, strength, and toughness. Some of the alloying elements in low alloy steels include nickel (Ni), molybdenum (Mo), chromium (Cr), tungsten (W), vanadium (V), titanium (Ti), and columbium (Cb).

Some elements, including non-metallic ones, are added to steel for other reasons. Silicon (Si), for example, may be added to promote fluidity in the weld puddle and to achieve better wetting action at the toes of the weld. Silicon is also used as a deoxidizer — an oxygen scavenger — in steel manufacturing. This means that the silicon scavenges oxygen from the steel when it is being manufactured at the steel mill. Such steels are called "killed", or "semi-killed" steels. They may be easily welded with most arc processes without porosity problems. Some welding electrodes have deoxidizers such as silicon and aluminum as part of their chemistry for the purpose of removing oxygen from the weld puddle. This action is similar to the steel makers use of silicon except on a much smaller scale.

Basic Metal Structures

Any substance that occupies space is called **matter.** Everything is some kind of matter, even the air we breathe. There are presently three known **states of matter.** They have been defined as **solid, liquid, and gaseous.** There has been some thought that superheated gas plasma might be a fourth state of matter but no conclusion has been reached on this concept by the academic community as a whole. In welding, we are basically concerned with metals in either the solid or liquid state.

Metals are normally solid state materials at room temperature. When a metal is heated to its specific melting point it changes from the solid to liquid form. If it is a pure, unalloyed metal, the solid-liquid transformation occurs at a specific temperature.

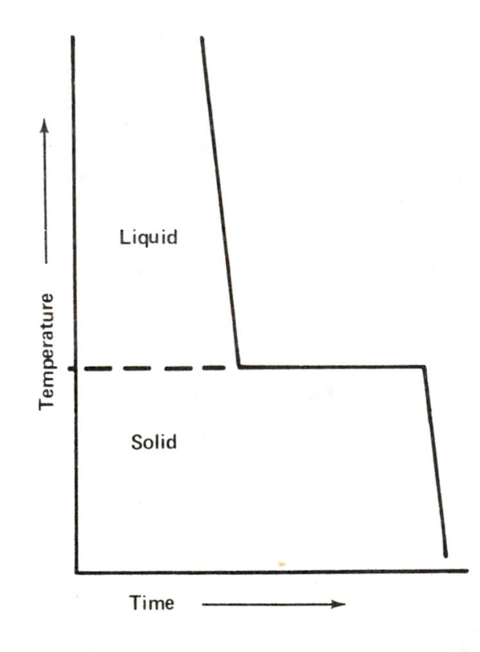

Figure 41. Solid-Liquid Transformation, Pure Metal.

If the metal is an alloy of two or more elements the solid-liquid transformation usually occurs over a range of temperatures. This is illustrated in Figure 42 where the transformation range is shown as both solid and liquid (S&L).

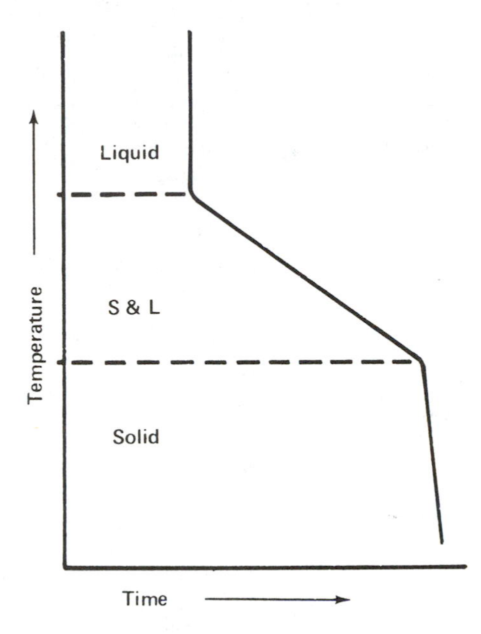

Figure 42. Solid-Liquid Transformation, Alloy Metal.

When weld metal is liquid it is a collection of atoms having no definite atomic pattern or arrangement. As a liquid, metal has no crystalline shape or form. When the liquid metal cools to a temperature at, or below, its solidification point, it solidifies into crystals (or grains). As used here, the terms "crystal" and "grain" have the same meaning. A grain is a crystal formation in metal that has an irregular shape and, therefore, irregular boundaries.

As molten (liquid) metal cools it reaches a temperature where it begins to solidify. Particles of matter called **grain, or crystal, nuclei** begin to form. Naturally, the grain nuclei forms at the coolest spot in the weld joint. This would normally be at the interface between the molten weld deposit and the unmelted base metal. The grain nuclei tend to attach themselves to existing grains of the base metal at the weld interface. Figure 43 shows the solidification concept in weld deposits.

As the molten weld metal continues to cool, the solidification of metal progresses with the nuclei growing into full size, irregularly shaped grains. As illustrated in Figure 43, the grains grow unimpeded until they meet the abutting surface of another grain. The grain, or crystal, shapes are of random pattern and design.

In Figure 43-A, the crystal nuclei are just beginning to form and attach themselves to existing grains in the base metal. Figure 43-B indicates the initial grains have formed. In Figure 43-C, the solidification of the weld metal has been completed. Note that the grains

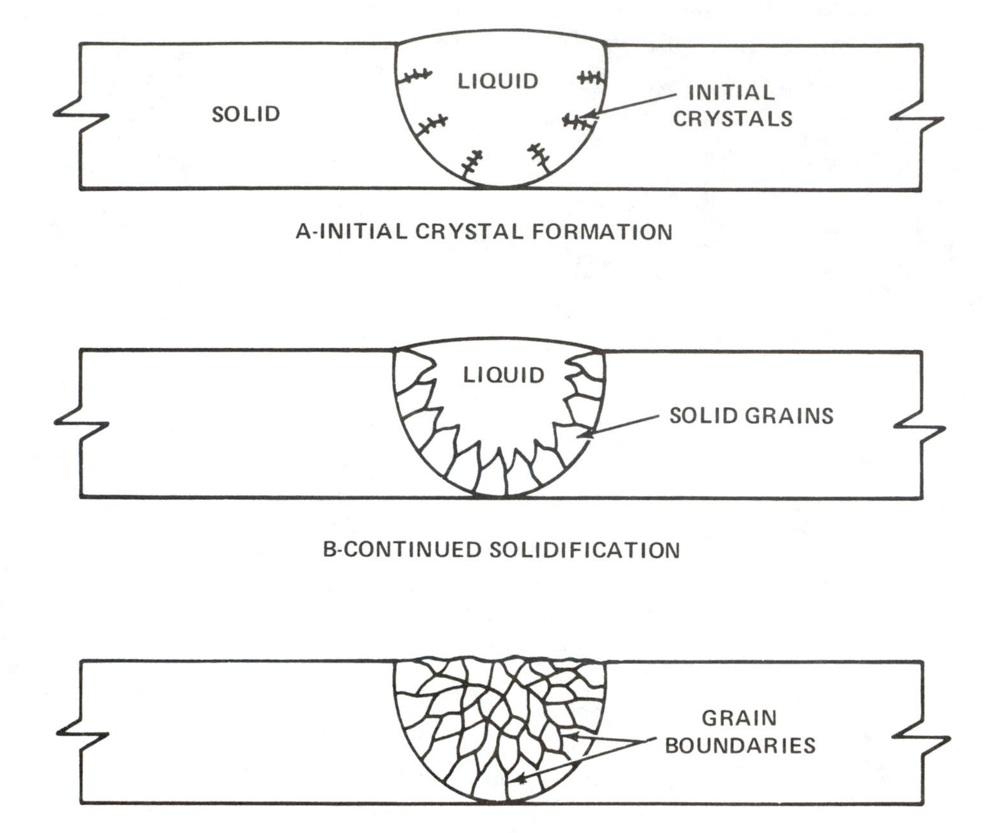

Figure 43. Nuclei and Crystal Formation In Welds.

in the weld deposit center are smaller and finer in texture than the grains at the weld and heat-affected zone (HAZ) interface. This may be explained by the fact that the grains at the HAZ-weld interface solidified first and were, therefore, at higher temperatures for a longer time while in the solid state. This provided more time at elevated temperatures for grain growth.

Remember: As the weld metal cools from liquid to solid form, the heat flow out of the liquid metal must go through the grains that solidified first, then through the heat-affected zone (HAZ) of the base metal until the heat dissipates into the mass of the base metal.

Formation of Crystal Structures in Metals.

We have found that liquid metal has no crystal (grain) structure since the atoms are in random location and arrangement. As the liquid metal cools and solidifies, crystal solids form irregular grain shapes. The **solid grains (crystals)** have specific atomic patterns or arrangements which are typical of the metal being welded. It is the atomic arrangement in the crystals that determines the type of metal.

The crystal structures that form in metals may be classed in three main categories. They are: **body centered cubic (BCC), face centered cubic (FCC), and hexagonal close packed (HCP).** These structures are found, in one form or another, in most metals. Some metals retain the same crystal structure from ambient (room) temperature to their melting point. Other metals have specific transformation temperatures where they change from one crystal structure to another. Diagrammatic views of the three crystal structures are illustrated in Figure 44.

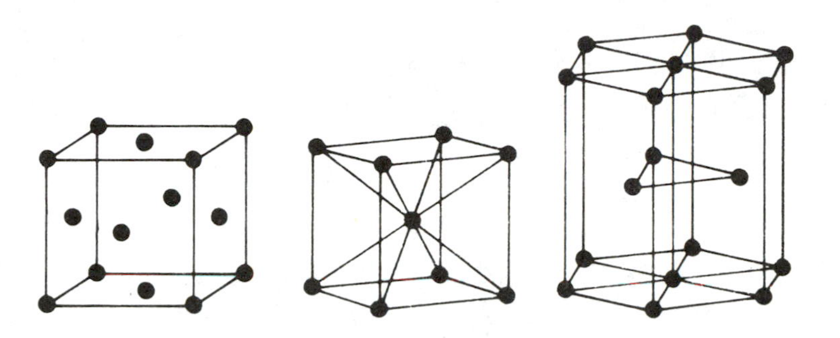

Figure 44. Some Crystal Forms.

The chart in Figure 45 shows some of the metals and their crystal structures at room temperatures unless otherwise noted.

STRUCTURE	METALS
Body Centered Cubic (BCC)	Iron (Alpha & Delta phases); Chromium; Columbium; Tungsten; Vanadium; Molybdenum.
Face Centered Cubic (FCC)	Iron (Gamma phase); Copper; Gold; Lead; Nickel; Silver; Aluminum.
Hexagonal Close Packed (HCP)	Cobalt; Magnesium; Tin; Zinc; Titanium; Zirconium.

Figure 45. Crystal Structures in Some Metals.

Phase Transformation

There are many different phase diagrams presently available to the metal working public. The American Society for Metals (ASM) publishes the *Metals Handbook, Sixth Edition,* which shows the various metal alloys and the phase diagrams that are pertinent to them. In this brief discussion of Welding Metallurgy Fundamentals, however, we will concern ourselves basically with iron-carbon alloys and the iron-iron carbide phase diagram. Additional information can also be obtained from the American Welding Society (AWS).

The interpretation of phase diagrams is based on several concepts. They include: (1) the temperature of the metal, (2) the pressure of the metal, and (3) the "state", or "phase", of the metal. The three **states** of matter correspond exactly to the three **phases** of metal. They are solid, liquid, and gaseous. It has been truly said in many classrooms that it is only the "**points** in the phase diagram that have physical significance". Any given point on a phase diagram for any metal or alloy is determined by a specific temperature and a specific pressure.

A **phase transformation** occurs in some metals when they are subjected to temperatures elevated above room temperature. By "phase transformation" we mean the changing of a metal from one crystalline structure to another. For example, in pure iron the liquid metal solidifies at 2,795° F. The crystal structure that forms is body centered cubic (BCC) and is called **delta iron** (Figure 46).

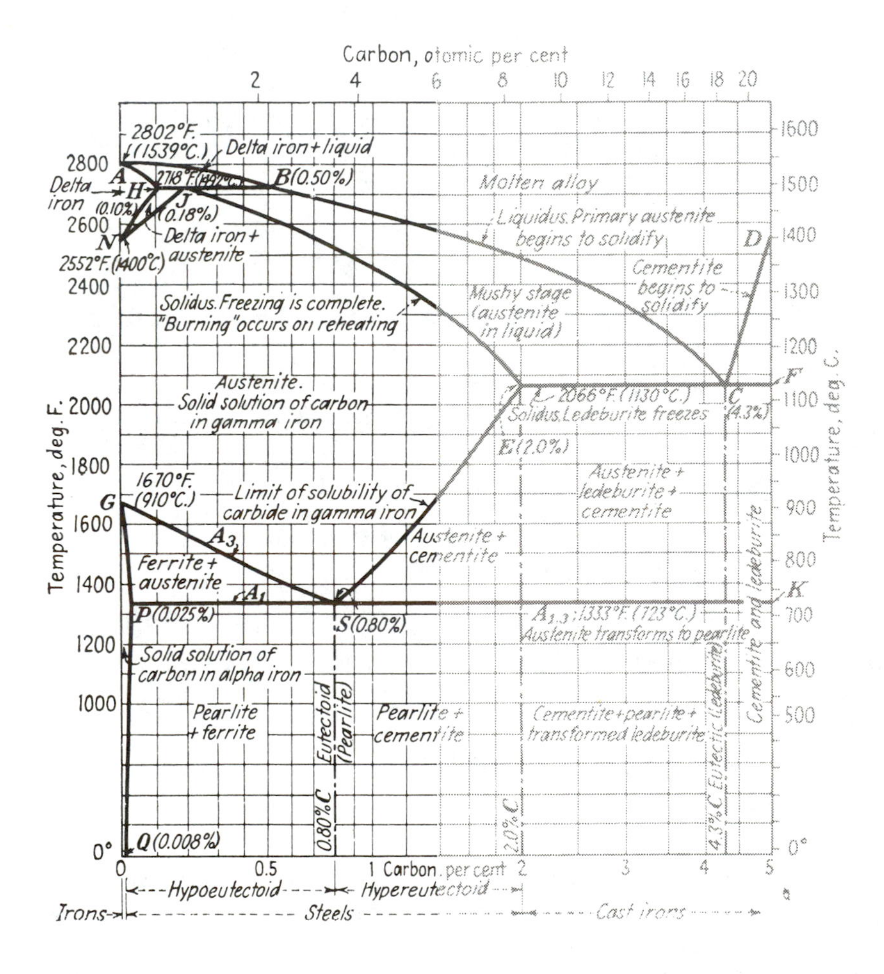

Figure 46. Phase Diagram for Pure Iron.

As the iron slowly cools, the body centered cubic delta structure changes (transforms) to face centered cubic (FCC) at a temperature of 2,535° F. The face centered cubic structure is called **gamma iron.** Gamma iron is also known as "austenite". Iron in this form and crystalline structure is non-magnetic. The face centered cubic austenite structure is retained by the pure iron as it continues to cool. Figure 46 shows the left side of the iron-iron carbide phase diagram. The various temperatures and structures of **pure iron** are shown in this diagram. The full iron-iron carbide phase diagram is illustrated in Figure 47.

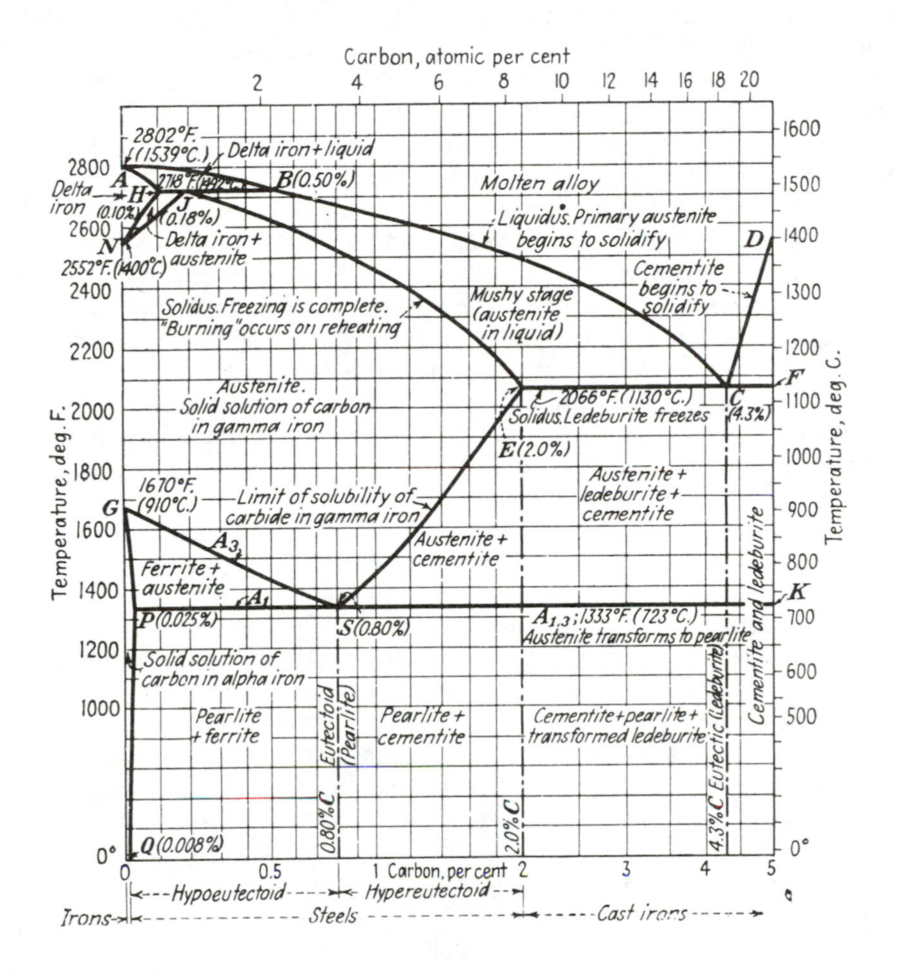

Figure 47. The Iron-Iron Carbide Phase Diagram.

When the pure iron reaches a temperature of 1,670° F., the face centered cubic structure transforms back to the body centered cubic crystalline form. Below this temperature the metal is known as "alpha iron". It is interesting to note that both "alpha" and "delta" iron have the body centered cubic crystalline structure. They are given different names to identify the **high temperature phase (delta)** and the **low temperature**

phase (alpha). The ability of a metal to change phase through transformation is called **allotropic.** To define:

> **Allotropic** = ''The ability of atoms to change their orderly arrangement in metal and exist in two or more crystalline structures at **different temperatures''.**

Iron and steels are allotropic metals. Other metals that undergo allotropic transformation are titanium, cobalt and zirconium.

In looking at the iron-iron carbide phase diagram it is easy to see that, **as the carbon content increases,** there is a marked change in the type of iron or steel present. Careful study of the phase diagram will show that, as carbon content varies in the iron or steel, phase transformation temperatures will increase or decrease.

Note that the alloy ''steel'' may have carbon content up to 2% of the chemical analysis. It is very difficult to weld steels with over 1% carbon content. Special techniques which include preheat and postheat are required. The iron-iron carbide phase diagram shows that metals with over 2% carbon content are classed as cast iron.

Commercial carbon steels are normally classed by carbon content. For example, **low carbon steel** is considered to have a maximum of **0.15% carbon content.** The **mild steels** range up to a maximum of about **0.30% carbon content.** Above 0.30% carbon, and up to about **0.60% carbon content,** the steel is called **medium carbon steel.** The medium carbon steels are heat treatable for hardness.

Steels with a range of carbon content from **0.60% up to approximately 1.5%** carbon are classed as **high carbon steels.** Such steels are normally used for spring steel and tool steel applications. These materials are very hard and heat treat easily but they have poor weldability. Special welding techniques must always be used with medium and high carbon steel. Cooling rates for medium and high carbon steels must be considerably slower than the cooling rates used for low carbon and mild steels.

Fast cooling rates for medium and high carbon steels will often cause the microstructure ''martensite'' to form in the heat-affected zone (HAZ) of the weld. Martensite is a very brittle, hard structure that is not normally used in engineering applications. If not properly heat treated, the martensite will cause the weldment to fail under stress loading.

High carbon steel is often oil quenched to retard the cooling rate to prevent the formation of martensite. Low carbon and mild steels may normally be quenched in water since they do not have sufficient carbon content to cause formation of martensite. Water, of course, will cool steel at a rate 6-7 times faster than oil.

Another method of slowing the cooling rate of steels is with preheat. Usually 300°-400° F. is sufficient to protect most steel weldments. Preheat will slow the cooling rate because heat will move more slowly through hot metal than it will through cold metal. The heat in the weld metal is, therefore, not as abruptly quenched as it would be with cold base metal.

Cast Iron

In the iron-iron carbide phase diagram you see that, above 2% carbon content, the metal is considered to be cast iron. There is a marked difference between the carbon in steels and cast iron. True, the percentage of carbon in cast iron is higher but that is not the key factor. It is the **form of the carbon** that makes such a difference in the metals.

For example, most carbon steels have carbon in either dissolved form solid solution or as iron carbides. Not so with cast iron. Although some carbon is in the dissolved form in cast iron, most of it is in the form of elemental carbon or graphite. There is no particular arrangement of the graphite flakes in the cast iron. Rather, it is in random disarray at all angles. When stressed, it isn't the cast iron metal that breaks; it is the flakes of graphite which propogate the crack and ultimate fracture. It is logical that, the smaller the graphite flakes, the less opportunity for fracture and the stronger the cast iron.

Grey and nodular cast iron may be welded by several welding processes. One of the oldest, and best, is with preheat, an oxy-acetylene torch and cast iron rod. It is hot, dirty work because the cast iron filler metal must be stirred vigorously into the weld.

The use of flux covered nickel electrodes of various alloys with the shielded metal arc welding (SMAW) process has been very successful in welding cast irons. In recent applications, the flux cored arc welding (FCAW) process has been used with a special flux cored nickel electrode material for cast iron. It has been quite successful in large applications.

Arc welding of cast iron may be done with a low preheat of about 200° F. A very important factor in welding cast iron is to remove all fatigued metal and any old weld or braze material prior to welding. The area to be welded must be prepared correctly and the entire weld area cleaned.

If the part is broken into several pieces, each piece must be cleaned and degreased. The part should be prepared with the correct edge preparation for the joint.

Carefully tack weld the individual pieces into the weldment. Align all pieces carefully so the completed weldment will have correct dimensions. When making the weld use qualified welding procedures. This will minimize the possibility of locked in stresses in the weldment.

Carbon Steel

The chemical combining of iron and carbon in carbon steels produces an iron-carbide molecule having three atoms of iron and one atom of carbon in the molecule (Fe_3C). It is carbon that primarily controls the hardenability of carbon steel. The higher the carbon content, the easier it is to harden steel.

Some of the other elements found in low carbon steels include manganese, phosphorous, sulphur and silicon. Some of the effects of these elements on carbon steel are examined in the following paragraphs.

Manganese (Mn) is used in steels basically to combine with sulphur to form *manganese sulphides.* The intent is to prevent, or minimize, the formation of iron sulphides which tend to cause "hot shortness", or crack sensitivity, in the steel when it is heated. Iron sulphides form coatings around the individual grains of metal. Since they have a low melting point, iron sulphides are in the liquid form at the rolling temperatures of the steel. If they are allowed to form, due to insufficient manganese in the steel composition, iron sulphides cause loss of intergranular adhesion during the rolling process at the steel mill. The addition of manganese at 2-3 times the sulphur weight creates the **manganese sulphides,** which are spherical in form, within the grains of metal. The metal may be easily rolled with no concern for hot shortness due to iron sulphide formation.

Manganese is also added to some low alloy steels for its effect of increasing hardenability of the steel. The normal maximum percentage of manganese is 1.5%. Higher percentages of manganese may change the steel from a water hardening material to an oil hardening material. This would indicate that a much slower cooling rate is necessary for steels with more than 1.5% manganese content. Special welding procedures would also be required.

Phosphorous (P) is present in all steels but not by choice. Phosphorous cannot be totally eliminated from steel in smelting and refining operations. It's level can, however, be controlled to very fractional percentages. Usually carbon steel has a maximum of 0.05% phosphorous content. This element has a tendency to increase the brittleness of steel thereby making the metal crack-sensitive. Phosphorous does add strength to some steel but the added brittleness caused by the phosphorous makes welding of the steel difficult due to its crack sensitivity.

Some low alloy, high strength steels have no phosphorous limitations in their compositions. They are structural steels which meet the ASTM A-242 specification. Steel makers are using phosphorous as a low cost strengthener to meet this specification. Welding people should check the steel specifications carefully to make sure they do not try to weld the unlimited phosphorous version of ASTM A-242 low alloy steel.

Sulphur is another undesirable element in steel. The effect of sulphur in combination with manganese has been discussed. However, certain free machining steels use sulphur in the form of manganese sulphides to improve machinability of the metal. This is most evident in certain bar steels. Although the bar stock is often referred to as "cold rolled", it is actually "cold drawn" through a die to obtain a clean finished surface and precise dimensional sizing.

During the drawing operation the manganese sulphides, which are present in the steel grains as spherical shapes, are elongated to needle-like inclusions for better machinability. Such high sulphur steels are difficult to weld successfully. The best technique would be a braze type weld with **minimum dilution** of the base metal with the filler metal. The preferred electrode would probably be an E7018 low hydrogen electrode. The amperage used should be on the low side of the manufacturers recommended amperage range.

One fallacy that should be dispelled at this time is the statement, "You can't weld cold rolled steel successfully". The statement is not true and may be disproved very easily. As a matter of fact, the finished surface condition has nothing to do with the weldability of the metal. As long as the metal surfaces are clean of dirt, oxide and other contamination, only the composition of the steel will control its weldability. Certainly the external finish, such as "cold rolling", has nothing to do with its weldability.

All metals are granular; that is, all metals have grains or crystals as their basic structure. The grain size may vary from very small to quite large in various metals. Grain size does have some effect on the weldability of steel. For example, steels having small grain size are tougher, stronger, and normally easier to weld.

Large grain size may make welding some metals more difficult. The large grains promote crack sensitivity because they have less ductility and less toughness. In addition, large grains are susceptible to cross-grain fracture. As an example, it is much easier to break a peppermint candy stick than it is to break a peppermint candy ball. The analogy relates, of course, to large elongated metal grains as compared to small, spherical grains of metal.

Grain size can be controlled to some degree at the steel rolling mill. This is done by making the last rolling pass on the material at as low a temperature as possible to produce the finest grain size.

Another method of controlling grain size is "normalizing" the steel. Normalizing is accomplished by re-heating the steel to a temperature above its upper critical temperature (A_3) of 1,333° F. minimum, holding it at that temperature for one hour for each inch of cross section of the metal, or portion thereof, and then cooling as evenly as possible in still air. The result is smaller grain size, better physical properties in the steel, and better weldability.

Any heat treatment or conditioning of the base metal must always become a part of the history of the metal part(s). Such a history is necessary to provide a measure of predictability to subsequent working of the metal.

Alloying Elements in Carbon Steel

Some of the alloying elements in carbon steel have been mentioned and some of their functions described. There are other alloying elements that are used in carbon steels and low alloy steels that have not been discussed and this is a good time to consider them. The following list names some of the elements and describes their principal functions.

Element		Function
Carbon	=	The most important alloying element in steel.
Sulphur	=	An undesirable element in steel; causes brittleness.
Phosphorous	=	An undesirable impurity in steel.
Silicon	=	Used mainly as a deoxidizer in steel; promotes fluidity in the weld puddle.
Manganese	=	Increases hardenability and strength of steel; combines with sulphur as manganese sulphides.
Chromium	=	Soluble in iron, chromium (Cr) is added in amounts to 9% to increase oxidation resistance, hardenability, and elevated temperature strength.
Molybdenum	=	A carbide former, molybdenum (Mo) is normally added in volumes of less than 1%. Mo increases the hardenability of steel as well as its strength at elevated temperatures.
Nickel	=	Nickel (Ni), in amounts to 3.5%, is used in low alloy steel to increase toughness and hardenability. Up to 35% Ni may be used for certain high alloy steels and stainless steels.
Aluminum	=	Aluminum (Al) is often added to steel in fractional percentages as a deoxidizer and grain refiner.
Dissolved Gases	=	Hydrogen, oxygen and nitrogen are all soluble in steel. These dissolved gases must be removed because they will embrittle the steel.

Anytime alloying elements are added to carbon steel they tend to make the metal more difficult to weld. More care must be taken with the welding procedure development, the selection of welding process and filler metal composition and classification to insure strong, sound welds at proper strength.

Small changes in chemical composition in steels will possibly cause considerable change in the physical properties of the metal. The result is that the heat of welding can have significant effects on element loss through the arc as well as the metallurgical properties of the weldment.

The Effect of Cooling Rate on Carbon Steel

The ability of austenite to transform to certain desirable lower temperature microstructures in steel is essentially the ability of carbon steels to be hardened by heat treatment. The transformation temperatures will vary with the composition of the carbon steel.

The effect of the cooling rate on austenite and the resulting lower temperature microstructures of carbon steel is illustrated in Figure 48 and described in the following paragraphs.

a. Pearlite

When austenite is slow cooled there is sufficient time for the extensive diffusion of metal atoms to form ferrite and carbide layers. This is typical coarse pearlite and the resulting carbon steel microstructure is soft and ductile.

b. Fine Pearlite

Fine pearlite is formed by using a slightly faster cooling rate than that used for coarse pearlite. The resulting steel is less ductile and is harder than coarse pearlite.

c. Bainite

When even faster cooling rates are applied, no pearlite appears. Bainite appears at lower transformation temperatures, with a feathery arrangement of fine carbide needles in a ferrite matrix. This structure has high strength, low ductility, and fairly high hardness characteristics.

d. Martensite

Martensite is achieved by extremely fast quenching of the austenite. It is a hard, relatively brittle phase of steel. The martensitic structure has an acicular, needle-like appearance. Quenched martensite is normally too brittle for industrial use and must be tempered for ductility. Tempering martensite provides a tough, more ductile steel with slightly decreased hardness.

Certain commonly used heat treating terms refer to the **rate of cooling** of austenite to room temperature. In the table following, specific cooling methods are listed in the order of increased (faster) cooling rates from above the A_3 temperature as shown on the iron-iron carbide phase diagram.

1. Furnace annealing : Slow furnace cooling (slowest method).
2. Normalizing : Cooling in still air (slightly faster).
3. Oil quench : Quenching in an oil bath (medium fast).
4. Water quench : Quenching in a water bath (fast).
5. Brine quench : Quenching in a salt brine bath (very fast).

If a heat treatable steel of specific type were heat treated and quenched with materials and methods in the order named above, beginning with furnace annealing, the metal would increase in hardness if quenched from a specific temperature using each of the quenching mediums in turn. For example, the metal would be softest after furnace annealing and hardest after brine quenching. Note that rate of cooling is very important to the hardening of steels. The faster a heat treatable steel is cooled, the harder it will be.

The Isothermal Transformation (TTT) Diagram

In metallurgy, the TTT diagram is a method of showing graphically the **Time, Temperature** and **Transformation** at which pearlite, bainite, and martensite form when austenite is cooled. It is important to know that every steel composition has its own TTT diagrams.

The TTT diagram shown in Figure 48 is for a 0.80% eutectoid steel. Creating the TTT curve was accomplished by heating samples of the 0.80% carbon steel to their austenitizing temperature of approximately 1550° F., then quenching them to a series of temperatures between 1300° F. and room temperature. The quenching mediums were hot salt pots, hot lead pots, oil quenches and water quenches, all at the proper temperatures.

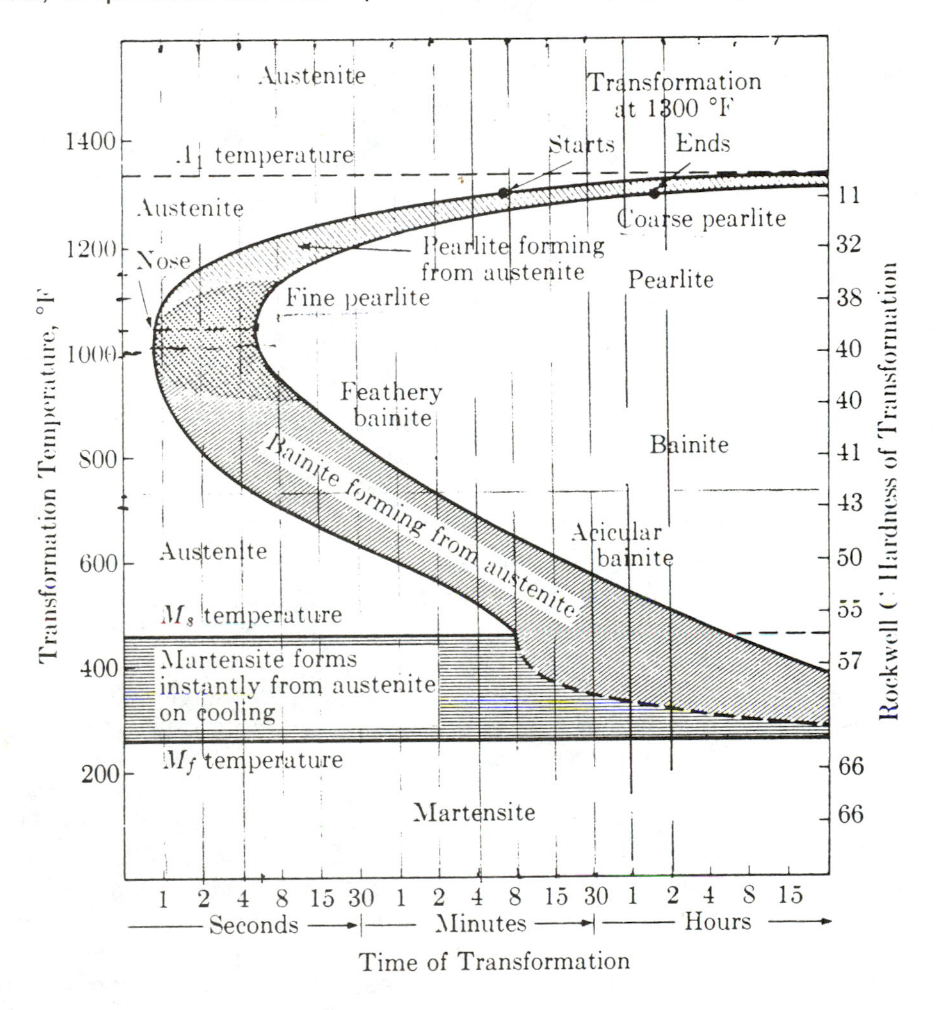

Figure 48. TTT Diagram for 0.80% C Eutectoid Steel.

The curve near the top of the diagram shows that the specimen held at 1300° F. did not begin to transform for about 500 seconds (about 8 minutes) and did not complete transformation until about 4,000 seconds (about 1.1 hours). The microstructure formed was coarse pearlite and was fairly soft (about Rockwell "C" 15).

When the quench temperature was reduced to 1050° F. the transformation action speeded up. The transformation began in one second and was fully completed in five seconds. This is shown as the "nose" of the TTT curve. The microstructure of the transformed steel was fine pearlite with an increased hardness to Rockwell "C" 41. (**NOTE:** Take the time to examine the isothermal TTT diagram and plot the difference in quenching temperature and time to find the transformation characteristics).

As shown in the TTT diagram, further decreasing the quenching temperature increases the time before transformation begins. The resulting microstructure becomes bainite with a corresponding hardness in the steel specimen.

If the specimen is cooled rapidly enough to get past the "nose" of the TTT curve, which would be cooling to 450° F. or less instantaneously, the microstructure formation would be martensite. Martensite only forms at temperatures below the M_s point and is substantially completed at the M_f point. The slowest cooling rate required to form 100% martensite may easily be determined from the TTT curve. It is called the critical cooling rate which just misses the nose of the TTT curve. As you can see, the time permitted is less than one second.

As carbon content and alloying content increase in steel, the nose of the TTT curve moves to the right. In this situation, the heat treating conditions which produce less rapid cooling can be used and martensite is easily formed. When there is lower carbon content and alloy content in the carbon steel, the nose of the TTT curve moves to the left. In this situation, it is almost impossible to cool the metal fast enough to form martensite. Low carbon steel and mild steel below 0.30% carbon fall into this group of steels that are considered non-heat-treatable for hardness.

Hardenability and Tempering Martensite

Hardenability of steel is a measure of the ease with which the formation of non-martensitic microstructures can be prevented and 100% martensitic microstructures obtained. In general, the steels with higher carbon content and/or alloy content have higher hardenability characteristics.

To harden steels properly, it is necessary to produce martensite in the base metal by proper chemistry, heating and quenching. As we know, martensite is very hard and brittle and totally unsuitable for engineering purposes and applications of steel. However, by tempering martensite it is possible to achieve a microstructure in the metals that is more ductile, increases toughness, and decreases brittleness without significantly decreasing the metal strength.

The heat treatment for tempering martensite is relatively simple. The martensite must be heated to some temperature below the A_1 temperature which is 1333° F. approximately. This permits the unstable martensite to change to tempered martensite by allowing the carbon to precipitate in the form of tiny carbide particles of Fe_3C. (Fe_3C is iron carbide, you may recall).

The desired strength and ductility can be controlled by selecting the proper tempering temperature and time. The use of higher tempering temperatures results in softer metal with more ductile properties and less strength.

Metallurgy and the Weld Joint

In order to intelligently consider the welding of a steel material it is necessary to know the chemistry of the metal and any previous heat treatment that it may have had. This is called the metallurgical and heat treatment history of the material.

Weld joints have two basic areas that are of interest and concern to the metallurgist and welder. The two areas are the **heat-affected zone (HAZ)** and the **deposited weld metal.** The weld deposit is a dilution of the deposited filler metal and the base metal. This metal has been in the liquid (molten) form and is a cast structure when it solidifies. The HAZ is unmelted base metal which has undergone physical and chemical changes due to the heat of the welding operation.

We have discussed the fact that the heat treating range for carbon steels is approximately 1400° F.-1850° F. The heat-affected zone of a weld is in the temperature range from about 1300° F. in the base metal to a temperature just below the melting point of the steel at the weld-HAZ interface. This means that some portion of the HAZ is within the heat treatment range of 1400° F-1850° F. and it will be transformed into the austenitic microstructure by the heat of the weld. It is apparent that the properties of the heat-affected zone (HAZ), for a particular steel, will be determined to a major extent by the iron-iron carbide phase diagram and the TTT diagram and curve.

The illustration in Figure 49 shows the temperature profile of the maximum temperatures achieved at various points (1-5) in the heat-affected zone of a typical welded mild steel having 0.30% carbon. This has been related to the iron-iron carbide phase diagram to show the microstructure of the steel at those given temperatures. As you can see by referring to the iron-iron carbide phase diagram, much of the profiled area has been heated to points above the A_1 temperature.

Point 1—has been heated in excess if 2400° F. The austenite that forms will be coarse grained because of the grain growth that occurs at this temperature.

Point 2—has been heated to about 1800° F. and the material is fully austenitized. Grain growth has not occurred; some grain refinement may occur.

Point 3—has been heated to just above the A_3 temperature. This is not enough temperature to fully homogenize the austenite.

Point 4—this area has been heated to approximately 1400° F. which is between the A_3 and the A_1 temperatures. Part of the structure is converted to austenite and the balance is ferrite. The resulting mixture of products that occurs during cooling can result in poor notch toughness.

Point 5—This area has been heated to about 1200° F. which is below the A_1 temperature so no austenite is formed. The base metal may be spherodized and softened.

The effects of welding on the cooling rate of the weldment depends on the welding conditions such as the base metal type and thickness, the welding process used, the amount of heat input in joules per linear inch of weld, the amount of preheat in the base metal and the speed of welding.

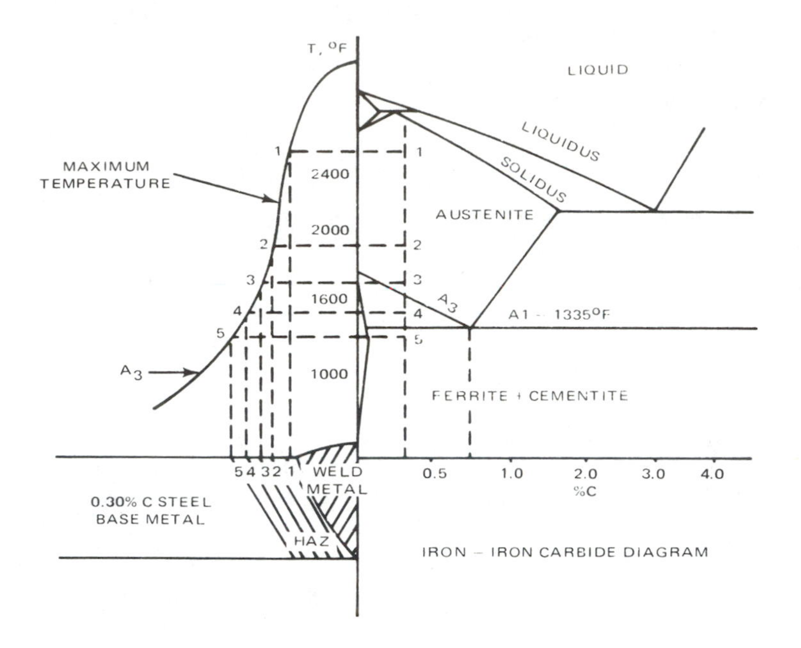

Figure 49. Weld and HAZ Profile With Iron-Iron Carbide Diagram.

The greater the overall amount of heat input to the weld area, the slower will be the cooling rate of the weldment. Heat energy input in welds is usually measured in joules per linear inch of weld. (See JOULES, Chapter 1). Base metal thickness is important to the cooling rate since thicker metals will act as a "heat sink". This means that the heat from the weld will dissipate faster through thicker metals than it will through thinner metals. The thicker metal parts have more mass to absorb the heat. This can actually cause a quenching action in the weld area. If there is too much quenching action, there can be defects in the weld such as lack of side wall fusion in the joint. Remember: heat will flow faster through cold metal than it will through hot metal. The hotter the base metal, the slower the cooling rate will be.

Stress Relieving

Welding often creates localized stresses in the weldment due to uneven heating and cooling. These stresses can, and do, cause distortion of the part in many instances. Although there are several methods to relieve stresses in weldments, probably the most used method is thermal stress relief.

Thermal stress relief can occur at different temperatures, depending on the base metal being worked. Steels will often stress relieve at low temperatures in the 400°-500° F. range. The stress relief is not usually complete but it is sufficient to minimize distortion in the part.

Certain carbon steels with over 0.40% carbon, or those with low alloy additions of some magnitude, may have both bainite and martensite in the heat-affected zone of the welds. As welded, these structures would cause a problem because they would tend to make the heat-affected zone brittle and prone to cracking. Stress relieving such welds at temperatures in the range 900°-1200° F. will temper the microstructure and make it more ductile. Unfortunately, it will not remove any cracks that may have occurred due to the stressing of the material while welding and cooling.

Deposited Weld Metal For Carbon And Low Alloy Steel

The deposited weld metal used in making a joint must meet the same minimum structural strength requirements as the base metal. There are four main factors that govern the metallurgical characteristics of weld metal deposits. They are:

1. The chemical composition of the consumable filler metal.
2. The chemical composition of the base metal.
3. The chemical reaction of (1) and (2) when the base metal is molten and mixed together.
4. The cooling cycle of the solid weld metal deposit.

A brief examination of these factors will provide some idea of what is being done to protect the integrity of the weld metallurgical characteristics.

Most **filler metals** have low carbon content. This will not permit the formation of martensite and minimizes crack sensitivity.

Knowing the **base metal** chemistry lets the welder select the proper filler metal material for that metal.

With the correct filler metal matched to the base metal chemical composition, the **admixture** will provide the minimum guaranteed strength of the base metal to the weld metal strength.

The cooling cycle of the weldment must be such that it does not permit the formation of martensite in either the weld or the heat-affected zone. All four of these factors will normally be included in a properly developed welding procedure.

The Fundamental Metallurgy Of Stainless Steel

The "stainless steels" are iron-based alloy materials that are available in two basic types, or series. The two types are the **300 series** and the **400 series.** Each type will be explained as we progress.

The two main elements used as alloys to make stainless steels are chromium (Cr) and nickel (Ni). Chromium is an essential part of all stainless steel because it is the chromium oxide that coats the iron-based metal and keeps it from "rusting". The minimum amount of chromium needed to make a true stainless, or rustless, steel alloy is about 12%. The chromium-iron (Cr-Fe) stainless steels are the 400 series materials which have the same magnetic qualities as carbon steels.

Adding both chromium and nickel to the iron produces the 300 series stainless steel alloys which are non-magnetic.

In general, the stainless steel alloys have excellent resistance to oxidation and corrosion.

Some of them will have improved physical properties at elevated temperatures due to the addition of other alloying elements. Chromium is the principal alloying element in all stainless steel. The amount of chromium may vary from about 12% to about 30%.

There are three basic types of "stainless steels". They are:

1. Martensitic stainless steel
2. Ferritic stainless steel
3. Austenitic stainless steel

Both martensitic and ferritic stainless steels are alloys with varying amounts of iron, carbon and chromium. If nickel is present it is usually in amounts of less than 1.5%-4% by volume.

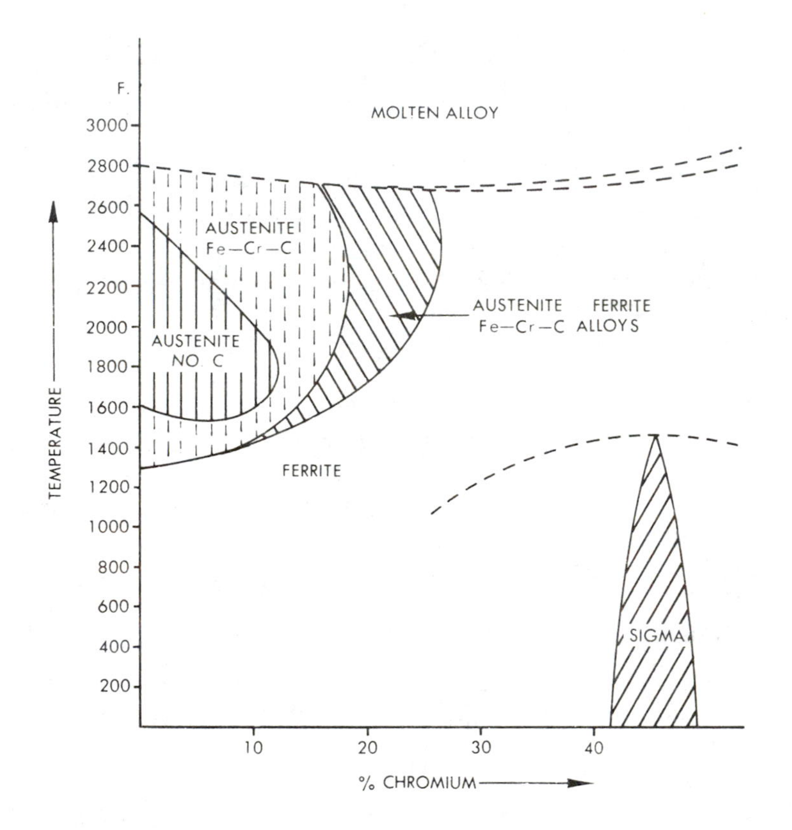

Figure 50. The Iron-Chromium-Carbon Phase Diagram.

Please note the iron-chromium-carbon phase diagram, Figure 50. The loop at the extreme left side of the phase diagram shows that austenite can only exist in **carbon-free** iron with up to about 12% chromium. With carbon additions, and depending on the amount of carbon added, the austenite may extend to approximately 18% chromium in iron.

There are certain high chromium, low carbon, iron alloys which are called **ferritic stainless steels.** The ferrite phase of these alloys is stable up to the melting point of the alloy. There is no phase transformation in this type of stainless steel.

Martensitic Stainless Steel

Martensitic stainless steels are part of the 400 series alloys. This group of alloys is hardenable by heat treatment. Therefore, extreme care should be exercised when welding martensitic stainless steels. Both preheat and postheat should be used with the welding procedure. This will eliminate, or minimize, the formation of martensite in the heat-affected zone of the weldment.

The percentages of chromium in the martensitic stainless steels will range from about 12% to about 17%, depending on the specific alloy. Some of the uses for martensitic stainless steels include cutlery blades, razor blades, and other applications where the ability to hold a high degree of hardness is required.

Ferritic Stainless Steel

The ferritic stainless steels contain a range of chromium from about 17% to about 27% without significant amounts of nickel, carbon, or other austenite-forming elements. Ferritic stainless steels are non-hardenable and will retain the ferrite phase right up to the melting point of the alloy. This is illustrated in the iron-chromium-carbon phase diagram shown in Figure 50. Due to the fact of retaining the ferrite phase to the melting point of the alloy, there is essentially no austenite formed as the metal is heated. Although not normally considered hardenable metal alloys, some ferritic stainless steels can develop martensite at the grain boundaries. This can change to martensite in the heat-affected zone when the metal is cooled.

Ferritic stainless steels are subject to grain growth at temperatures above 2,000° F. Rather large, coarse grains will probably be evident in the heat-affected zone of the base metal when the metal is cooled from welding temperatures. Fortunately, the grain size can be refined (reduced in size) by post-weld heat treatment. This treatment will also usually increase the notch toughness.

Ferritic stainless steel filler metal may be used for welding ferritic stainless steel alloys. In actual practice, however, the austenitic stainless steel alloys 312, 310, or 309 are most often used as welding filler metal for the ferritic stainless steel alloys.

Austenitic Stainless Steel

The austenitic stainless steels are basically an iron-chromium-nickel (Fe-Cr-Ni) group. Varying amounts of carbon and other alloying elements are added to provide special properties such as oxidation and corrosion resistance, high temperature strength and improved weldability.

The austenitic microstructure is stabilized at all temperatures by the addition of nickel and lesser amounts of carbon and manganese. By stabilizing the austenite at all temperatures the austenite-ferrite transformation is eliminated.

The austenitic stainless steels cannot be hardened by heat treatment since there is no transformation to ferrite. This is beneficial because there can be no hardened areas in the base metal. The austenitic stainless steels have excellent weldability in all the 300 series alloys.

It is not necessary to preheat any of the 300 series austenitic stainless steels before welding. In fact, these alloys should never be preheated. Interpass temperatures should be held to less than 200° F.

Stainless Steel Filler Metals

The most important single factor in the selection of filler metals is to match the base metal as closely as possible in chemical content. Strength requirements, corrosion resistance and oxidation resistance are often the factors that most influence weld metal electrode selection. The best method is to use the American Welding Society Filler Metal Standards as the basis of selection.

Although there is no alpha ferrite in austenitic stainless steels, there is usually a small amount of delta ferrite in the metal. This is advantageous since it will help to minimize hot-cracking in the weld during the cooling cycle. Ferrite control is highly desirable in austenitic stainless steels. The addition of small amounts of delta ferrite minimizes the welding problems of hot-cracking in the weld metal.

Sigma Phase

Although ferrite in the weld metal aids in the actual welding process it can cause physical property problems when subjected to high temperature service. Some austenitic stainless steel alloys, particularly those containing ferrite, can develop a brittle phase called the **sigma phase** when held in a temperature range between 900°-1775° F. At specific temperatures within this temperature range sigma phase formation is very rapid. For example, sigma phase will form in one hour at a temperature of 1550° F.

When sigma phase formation occurs in the microstructure of the alloy, corrosion resistance and ductility decrease. Hardness increases markedly while notch toughness is reduced considerably.

While the sigma phase may be removed from stainless steel by annealing at temperatures above 1800° F., it is important that careful choice of the alloys be made, especially when the alloys will be used for high temperature service above 1,000° F. The amount of delta ferrite present in the alloy selected should be held to a minimum. Another area that deserves careful consideration is the selection of the electrode alloy used for specific high temperature applications. The reason, of course, is to minimize the amount of delta ferrite in the weld deposit.

Carbide Precipitation

There is a critical range of temperatures to consider when austenitic stainless steels are welded or otherwise heated. In the temperature range between 800°-1500° F., chromium precipitates from the solid metal grains and combines with carbon at the grain boundaries to form chromium carbides. The chromium carbides have no corrosion resistance and will cause corrosion attacks to occur preferentially at the grain boundaries as well as in the base metal grains which have been depleted of chromium.

Carbide precipitation occurs within those areas of the heat-affected zone of the base metal which have been heated to within the critical temperature range. Of course, the time the alloy is within the critical temperature range is important. The longer the time at temperature, the greater the carbide precipitation danger.

There are several methods by which carbide precipitation can be reduced or eliminated in the austenitic stainless steels. These include the following:

1. Heat the metal to above 1800° F. and quench cool rapidly. This will dissolve the chromium carbides and put the chromium back into solution in the base metal grains. Rapid cooling prevents carbide precipitation by cooling the metal through the critical range in minimum time.
2. Decrease the carbon content of the stainless steel to a low enough level that there is not sufficient carbon to form carbides. Extra low carbon grades of filler metals and base metals have the suffix "L" or "ELC".
3. The third method is to add a "stabilizer" element such as columbium (Cb) or titanium (Ti) to the alloy steel or filler metals. Both of these elements have a greater affinity for carbon than chromium. The formation of titanium carbides or columbium carbides uses up the available carbon and there is very little carbon left to combine with the chromium. This type of stabilized stainless steel is often used where control of temperatures in the critical range is not possible.

Summary

1. Strength and ductility of steels, both carbon steel and alloy steels, will vary widely with either, or both, of the following:

   a. Comparatively small changes in chemical composition.
   b. Changes in the heat treatment cycle.

2. Preheat, slower cooling rates, and post heat treatment will affect the strength and ductility of the welded joint.

CHAPTER 3

Oxy-Acetylene Welding and Cutting Process Fundamentals

Flame! A mixture of oxygen and some type of combustible material which, when ignited, produces heat and light. A controlled flame for welding metals is what this chapter is all about.

Since before recorded history, man has known fire—for warmth, for food preparation and later for heating and shaping metals. A basic tool handed down from antiquity is the blacksmith's forge. The forge is the container in which the fire is shaped and controlled by the blacksmith. The welder of today, with his relatively sophisticated welding equipment, has a rich heritage from the blacksmiths of yesteryear. The very term "blacksmith" comes from the work he did and the metals he used. The black iron (wrought iron) he worked with provides part of the name. His actions in "smiting" (striking) the iron with a hammer gives us the rest of the name. A blacksmith smites black iron to shape it into usable objects.

It was not uncommon for blacksmiths to manufacture all of their own tools including their anvil. Specially shaped tools were devised for each special application. Unfortunately, the art and craft of blacksmithing has almost disappeared from the American scene. It is still an honored profession in many parts of the world, however, including Europe, Africa and the Far East.

It was about 1900 when the first practical methods were developed for manufacturing both acetylene and oxygen. Shortly thereafter, gas welding and cutting torches were developed and a new welding process was born. The manufacture and use of carbon steel has progressed very greatly in recent years. As a matter of fact, it is considered that about 78% of all welding accomplished with all welding processes is on some type of iron or steel.

In the oxygen-fuel gas welding processes, the flame heats the metal to be welded where the two pieces join or abut. This area is called the **weld joint.** When the weld joint reaches the proper temperature the edges of the base metal melt. The molten metal of the edges is caused to flow together by proper manipulation of the welding torch.

As the welder moves the welding torch and flame along the metal joint the molten weld metal cools and solidifies. The result is a weld joining two pieces of metal into one solid homogenous mass through the use of controlled heat input. The essential ingredient for fusion welding is **controlled heat input to the weld.**

Most gas welding is done with a combination of oxygen and acetylene gases. The process name is normally referred to as **oxy-acetylene welding.** As a matter of fact, the fuel gas for welding is selected because of its ability to produce high heat output when mixed with oxygen. For example, oxy-acetylene mixtures are normally used for metals having relatively high melting points. These metals include copper, carbon steels, cast steels, cast iron, and stainless steel.

Other fuel gases with lower flame temperatures may be used for lower melting metals and alloys such as aluminum and magnesium. For example, such gases as hydrogen, natural gas, propane or methylacetylene propadiene, stabilized (MAPS) may be combined with oxygen to produce flames hot enough to weld aluminum, magnesium, lead, zinc, and some of the precious metals.

The hottest flame produced by any of the oxygen-fuel gas mixtures is the oxy-acetylene flame. The three basic types of flames used for oxy-acetylene gas welding are illustrated in Figure 51. Each flame type has its specific use for different metals and alloys. For example, a carburizing flame is normally used when gas welding aluminum and aluminum alloys; a neutral flame is used for carbon steels; and an oxidizing flame is preferred for brazing with bronze and brass filler metals.

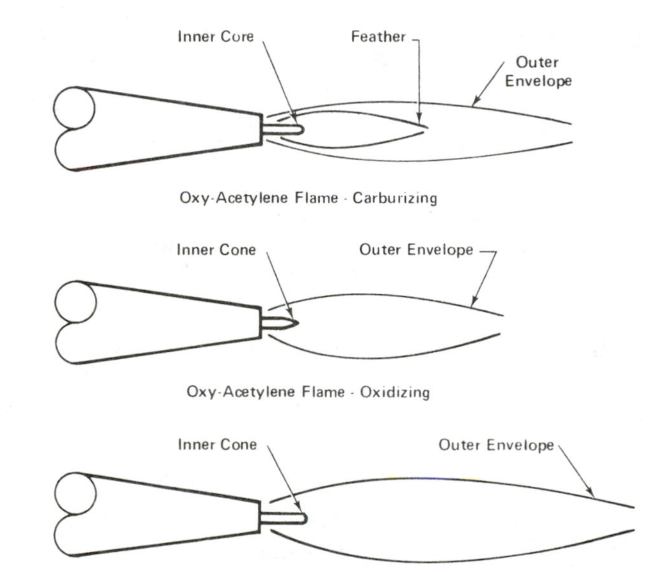

Figure 51. Typical Oxy-Acetylene Flame Types.

Safety With Oxy-Acetylene Equipment

In the welding business there are basically two kinds of people—the "quick" and the "dead."

The term "quick" means the live, SAFE workers who are concerned about safety for themselves and others. Only a fool takes chances with welding equipment.

The term "dead" as used here means the UNSAFE workers who are often injured or killed through careless and ignorant use of welding equipment.

Welding is a safe occupation for anyone who practices a few common sense rules for safe operation. This part of the safety discussion will be concerned only with the gas welding safety practices. Electric welding safety will be considered under the proper chapter title.

The governing Standard for welding safety is the AWS/ANSI Standard Z49.1 "Safety in Welding and Cutting". It is available through the American Welding Society (AWS), P.O. Box 351040, Miami, FL 33135. Every welding shop should have a copy of this safety Standard available to welders and others involved with welding safety.

The safety rules presented here are common sense directions that can save lives, prevent accidents and decrease equipment malfunctions. Specific safety rules for storage of gases, preparation of equipment, and personal safety equipment for welders, are written in the AWS/ANSI Z49.1 Standard. Read them, understand them and heed them!

Rule 1

Oxygen is a tasteless, odorless, and colorless gas. It will not burn or explode by itself but it will support combustion in all materials. NEVER USE OXYGEN TO BLOW DUST OR DIRT FROM CLOTHING, WORKBENCHES, OR OTHER AREAS! Even so-called non-flammable and fire retardant clothing and materials will burn if saturated by oxygen. All that is needed is a slight spark to set it off.

Rule 2

Oxygen must never be used with, or around, grease and oil. Even the slightest trace of grease mixed with oxygen will cause violent explosions to occur. NEVER USE GREASE OR OIL ON ANY OXYGEN OR ACETYLENE WELDING EQUIPMENT INCLUDING TORCHES, VALVES, REGULATORS, CHECK VALVES AND OTHER HOSE CONNECTIONS. Oxygen is so sensitive to grease that even a slightly oily rag, which might be used to wipe off cylinder valve threads, would leave enough oil to cause an explosion.

Rule 3

Acetylene is a fuel gas that becomes unstable at pressures above 15 psig. (**psig** = pounds of force per square inch, gage pressure). All acetylene regulators are marked with 15 psig maximum on the low pressure (working pressure) gage. Acetylene is dissolved in acetone in the acetylene cylinder. The maximum withdrawal rate for acetylene from a cylinder is 1/7 of the gas volume per hour. Attempting to obtain greater acetylene flow rates from a single cylinder will cause acetone to be drawn from the cylinder in liquid form. ACETYLENE SHOULD NEVER BE USED AT OPERATING PRESSURES ABOVE 15 PSIG.

Rule 4

Acetylene cylinders may be manifolded for greater volume flow to shop work stations. ACETYLENE PIPING SYSTEMS SHALL NEVER BE CONSTRUCTED OF COPPER PIPE OR TUBING. Copper, and copper alloys with more than approximately 67% copper content, will form acetylides with acetylene gas. Acetylides are violently explosive and can be set off by a slight shock or an increase in temperature.

Rule 5

AWS/ANSI Standard Z49.1 provides detailed instructions for storage of all types of compressed gases. Oxygen must be stored separate from fuel gases. All cylinders MUST be stored in the upright position and secured by chains or cables. Both oxygen and fuel gases should be stored in clean, dry areas away from any petroleum products such as oil and grease. NEVER TRANSPORT, OR USE, ACETYLENE CYLINDERS IN ANY POSITION OTHER THAN VERTICAL. Severe damage to equipment and personnel could result.

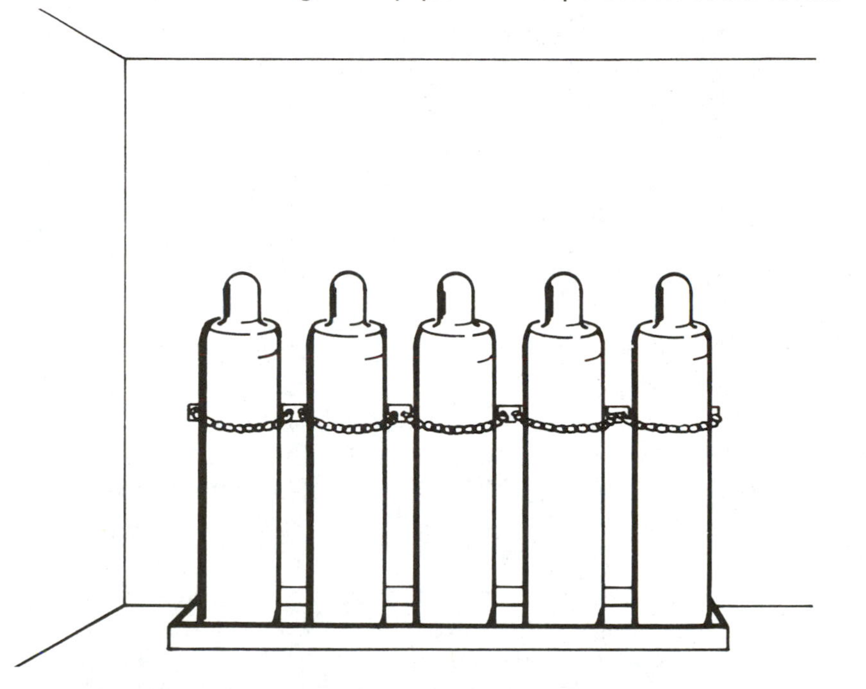

Figure 52. Safe Storage of Compressed Gas Cylinders.

Rule 6

Protective hardened safety glasses with properly shaded lenses must be worn when using oxy-acetylene processes. Vented side shields are required for maximum protection. Molten weld metal and other spatter can seriously damage, or destroy, eyesight. Regular hardened safety glasses with side shields should always be worn when working in any metal shop. You only get one pair of eyes. There is no replacement if they are lost.

Rule 7

THERE SHALL BE NO HORSEPLAY OR SKYLARKING AROUND IN ANY WORKSHOP AREA. Serious injury can result because people were "just fooling around" Don't do it!

Rule 8

Never pick up a piece of "cold metal" without testing it first to be sure it IS cold. The blacksmith who was making horseshoes and tossing them on the ground to cool as he finished each one watched in amusement as a young apprentice walked over to the pile of 'shoes, picked one up and immediately dropped it. "Whats the matter, son? Was it hot?" he asked. The young fellow looked ruefully at his burned hand and said in reply, "Not particularly so. It just doesn't take me long to inspect a horseshoe!" Save your hands by using your head.

Rule 9

Never play practical jokes on people in the work area. Putting a mixture of acetylene and oxygen into a paper bag and exploding it under or around someone could cause serious injury or death, including yours.

Rule 10

Keep the shop working area as clean as possible at all times. Scrap metal and electrode stubs carelessly left on the floor can cause injuries to other workman. SHOP SAFETY IS EVERYBODY'S RESPONSIBILITY!!

Oxy-Acetylene Welding Equipment

Oxy-acetylene torches, regulators, hoses and connectors are all important tools of your trade as a welder. If these tools are damaged through improper use, or through poor maintenance, they cannot function properly. In some instances, damaged tools can be a serious hazard to safety. A skilled craftsman takes care of his tools as an investment.

Torch Bodies

There are two basic types of oxy-acetylene torches for welding and brazing. They are the **injector** type (low pressure) and the **equal pressure** type (medium pressure).

The **injector,** or low pressure, torch operates with low acetylene pressure (usually 1-3 psig) but with relatively high oxygen pressure (usually 10-35 psig). By design, the flow of oxygen through the torch pulls acetylene gas into the mixing chamber by a venturi action.

The **equal pressure,** or medium pressure, torch operates with acetylene pressures over one (1) psig but not greater than 15 psig. Oxygen pressures are normally equal to acetylene pressures when welding with this style of torch. Actual flow rates are balanced for both gases. The specific gas pressures used will depend on the torch type and the tip size used for welding. Detailed instructions on how to set up and prepare an equal pressure torch for welding are provided in another part of this chapter.

The welding torch body is equipped with two needle valves at the gas input end. One valve is for oxygen and the other is for acetylene. The gases may be supplied from cylinders or from a manifolded and piped system.

Both torch needle valves have packing glands and packing nuts to seal gas leaks. The packing nuts and packing glands should always be checked prior to beginning welding or brazing operations with the torch. Gas leaks are dangerous!

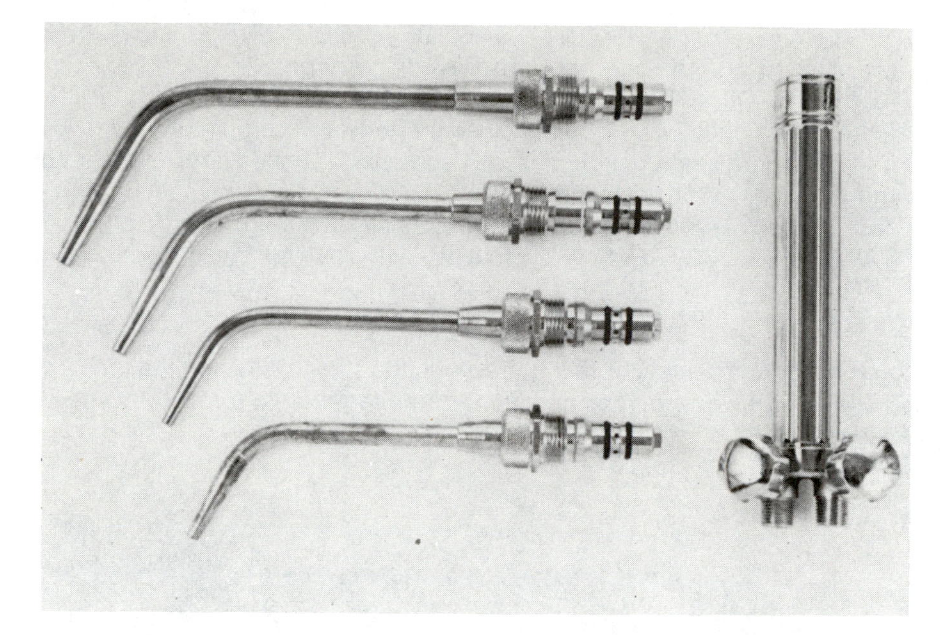

Figure 53. Typical Oxy-Acetylene Welding Torch With Tips.

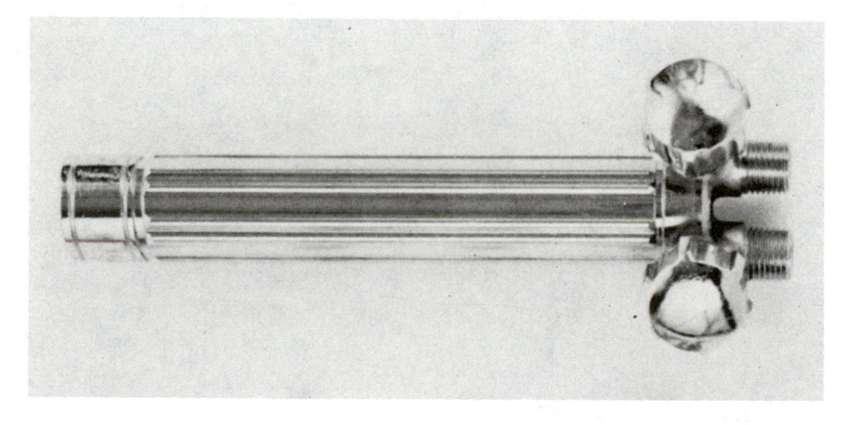

Figure 54. Oxy-Acetylene Torch Needle Valves.

Some common sense facts concerning torch body care and maintenance are listed for your information.

1. USE WRENCHES GENTLY TO TIGHTEN BRASS FITTINGS ON ANY WELDING EQUIPMENT. Excessive wrench pressure will possibly strip the threads or otherwise

damage the fittings. Some torch tips and mixers are so designed that only hand pressure is needed for adequate tightening of the threaded joint.

2. NEVER TIGHTEN A VISE ON A GAS TORCH BODY. The tubing of the body could be crushed and the internal parts severly damaged.
3. KEEP ALL OIL AND GREASE AWAY FROM TORCH FITTINGS AND VALVES.
4. If the torch needle valves do not close properly and completely, remove the valves and clean the seat connections and surfaces. If necessary, remove and replace the packing in the packing glands.
5. NEVER USE ANY OXY-ACETYLENE WELDING TORCH UNLESS YOU ARE CERTAIN IT IS IN PERFECT WORKING CONDITION. It could save your life.

Torch Mixers

The torch mixer is a device, or fitting, between the torch body and the torch welding tip. It is in the mixer that the oxygen and acetylene come together to form a gas compound usable for welding.

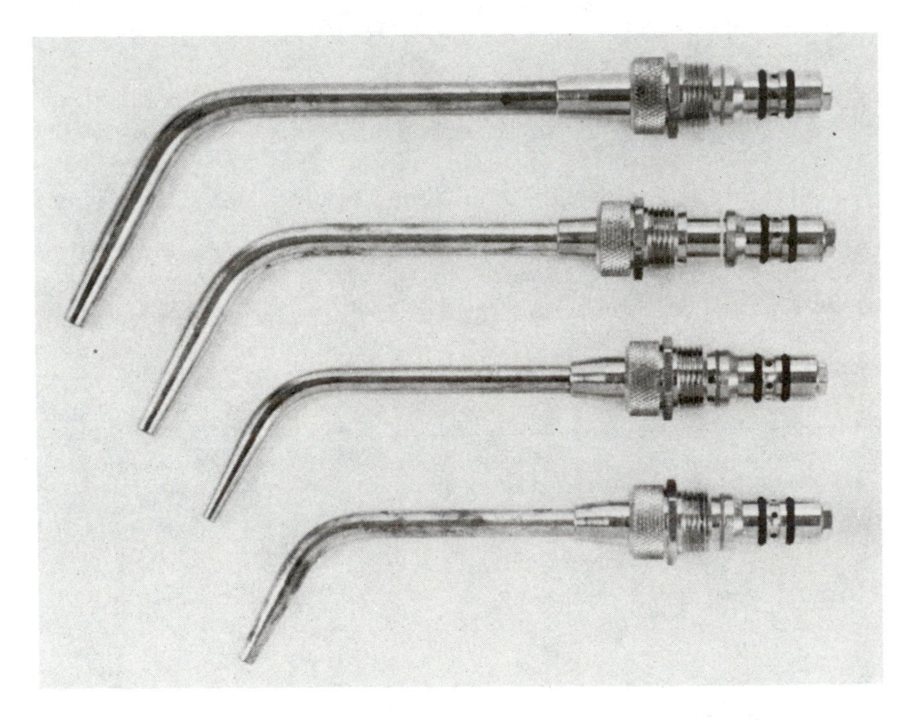

Figure 55. Welding Torch Tips With Mixers.

Some torch designs have a mixer that accommodates a range of welding tip sizes. The actual torch tip is threaded into the mixer. Other torch designs employ a separate gas mixer for each torch tip size. In this design, the mixers are an integral part of the welding torch tip structure.

The basic function of a gas mixer is to combine the oxygen and acetylene thoroughly for proper combustion at the torch tip. For equal pressure torches, the gas mixture is normally a 1:1 ratio. This ratio prevents either oxygen or acetylene gas "creep" from changing the torch flame while welding.

Torch Tips

Welding torch tips are manufactured in a variety of orifice (output hole) sizes for industrial use. For most commercial torches, single flame tips are available from size "000" to size "12". Multiflame tips are also available for heating purposes.

For light, delicate work, there are small "aircrafter" or "jewelers" torches which use only very small tip sizes. Such torches may be used for light sheet metal work, jewelry manufacturing and repair, and other low heat input applications.

Torch tips are usually made of copper or high copper alloys. Care should be taken when fitting a torch tip to a torch mixer. Cross threading is very easy in this soft metal. This would seriously damage the tip threads. Firm hand tight connection between the tip and mixer is all that is necessary.

Gas Pressure Regulators

Gas pressure regulators, as the name implies, control and regulate the flow of gas from the cylinder or manifold piping system. Separate oxygen and acetylene regulators are used for the two gases.

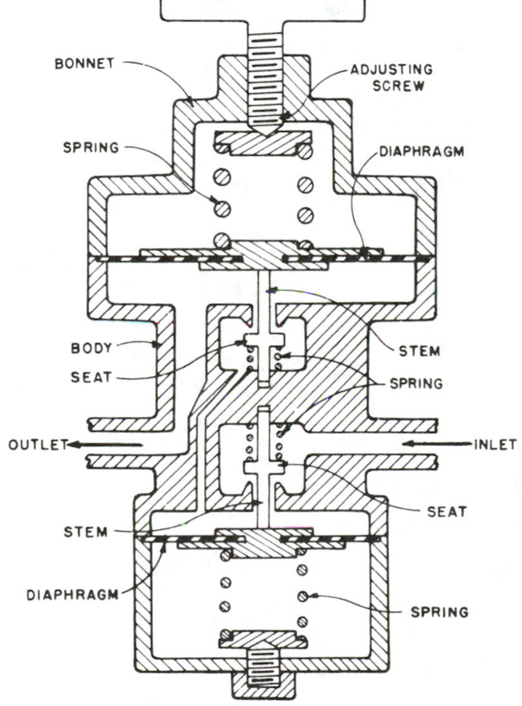

Figure 56. Typical Single Stage Gas Regulator.

All **fuel gas** pressure regulators, and other fuel gas fittings, have left hand threads. They also have a groove cut into the outer periphery of the fittings to identify them as fuel gas fittings.

Oxygen gas pressure regulators and fittings will always have a right hand threaded connection nut and fittings. The outer surfaces of oxygen connection nuts and fittings are smooth.

The two common types of gas regulators used for oxygen and acetylene are the **single stage** and **two stage** regulators. Single stage regulators are less expensive than two stage regulators but they do not provide constant regulation of the working gage pressure as the cylinder empties. Two stage regulators provide constant working pressure until the cylinder is virtually empty.

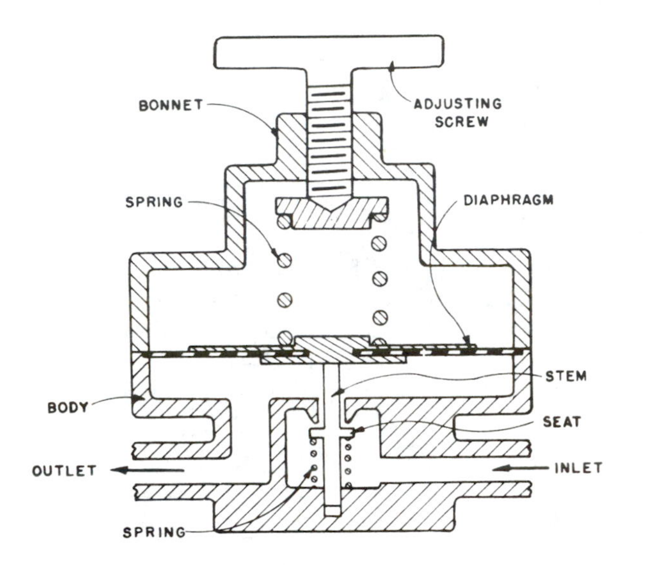

Figure 57. Typical Two Stage Gas Regulator.

Both single stage, and two stage, gas regulators have two gages. One gage meters the cylinder pressure at all times. The second gage meters the actual working pressure of the gas at the torch. In all cases, the covering cases of regulator gages must be tight and leakproof. Gage glass covers must be clean and unbroken.

Connecting Regulators to Cylinders

Some basic common sense shop practices have been found useful in connecting and using gas pressure regulators. By following the safety guidelines listed below, safe and sensible operations may be expected.

1. Remove the cylinder cap. Do not use a pry bar or other tool to loosen the cylinder cap. Use only your hands. If the cap will not come loose, do not try to force it. Get another cylinder. Remove the defective cylinder, tag it, and call the industrial gas supplier to pick it up and remove it from the shop area.

2. After removing the cylinder cap, stand to one side of the cylinder valve away from the valve output orifice. "Crack" the cylinder valve momentarily to blow out any dust or dirt in the valve. ("Cracking" the cylinder valve means to open and close the cylinder valve quickly to let some gas escape. Standing to one side while cracking the cylinder valve is a safety precaution.)
3. Attach the correct gas pressure regulator to the cylinder valve. Be careful that the threads of the connection nut are not cross-threaded. Tighten the cylinder connection nut gently but firmly with a cylinder wrench.
4. Release the pressure regulator adjusting screw by backing it out to the no-pressure position BEFORE opening the cylinder valve. Opening the cylinder valve with pressure on the adjusting screw can damage, or possibly destroy, the regulator diaphragm.
5. STAND TO ONE SIDE OF THE REGULATOR AND GAGES when opening the cylinder valve on any compressed gas cylinder. The sudden pressure could cause the faces of the gages to blow off and severely injure anyone standing in front of them.
6. Open the cylinder valve slowly. An acetylene cylinder valve should be opened not more than 1/4 turn. If a cylinder wrench is used, leave it in place so that a fast shut down can be accomplished in an emergency. The oxygen cylinder valve is opened all the way since it is a double seated valve. It must be seated at the top, or full open, position.
7. At this point the cylinder and regulator are ready to have the hose and torch attached. It is recommended that check valves be installed on the output fittings of the regulators and at the input fittings of the torch to prevent any possibility of backfires or flashbacks reaching the cylinders.
8. The hose must have an adequate ID to convey the required volume of gas. Green hose is always used for the oxygen gas lines. Red hose is commonly used for the acetylene, and other fuel gas, lines. All oxygen fittings are right hand threads; all acetylene, and other fuel gas, fittings are left hand threads and have a groove cut into the outer periphery of the fittings.

After attaching all hose connections, tighten them firmly but gently with a cylinder wrench. To assure proper gas tightness in the system, it is wise to conduct a soap and water test on all gas connections from the regulator to the torch after gas has been supplied to the system.

Torch Tip Selection

Before any welding or brazing can be done with the gas welding process the proper torch tip size must be selected. The tip size will depend on the type of metal to be welded and the thickness of the metal. Thin gauge metals require less heat volume, and therefore a smaller welding tip size, than heavier thicknesses of metal.

The thermal, or heat, conductivity of the metal being welded must also be considered. For example, aluminum will require more heat input for a given thickness than steel because aluminum will transfer heat much more quickly through its entire mass than will steel. Heat moves relatively slowly through steel.

The **volume** of flame from various sizes of welding tips will vary significantly. Keep in mind that although there is a greater volume of flame with a larger tip size, the flame temperature will remain the same for a given type of welding flame. For example, a neutral flame has an approximate temperature of 5800° F. regardless of the tip size used for welding.

After selecting the correct welding tip size, and having properly connected all fittings and other equipment, it is time to bring the gases to the welding torch. Check all safety factors including proper restraint of the gas cylinders, correct arrangement of hoses so they do not get damaged, and the needle valves on the torch.

Setting the Equal Pressure Torch for Welding

The following exercise in setting an equal pessure torch for welding is a positive method of balancing the flow of oxygen and acetylene to the torch tip. If done correctly, there is no possible way that one gas or the other can increase in flow rate (creep) and cause a distortion in the gas flame desired. Each step must be done in sequence to achieve the proper gas flow rates for the welding tip size selected.

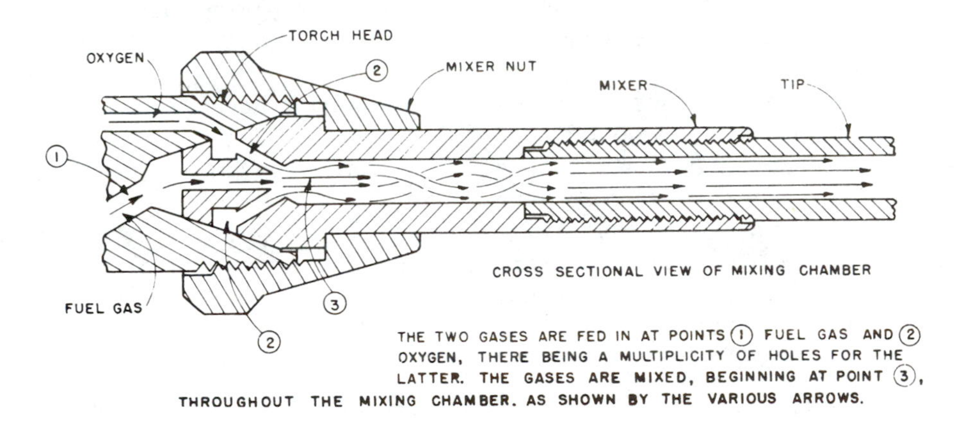

Figure 58. Typical Equal Pressure Torch.

1. The gas regulators are installed with the adjusting screws backed out to the "no-pressure" position.
2. The cylinder valves are opened as previously described. The acetylene valve is opened only 1/4 turn; the oxygen cylinder valve is opened fully and seated at the top valve position. There is no gas pressure on the hoses at this time.
3. Hold the torch in one hand and open BOTH needle valves on the torch to full on. Provide a spark lighter designed for torch flame ignition.
4. Slowly turn the acetylene regulator adjusting screw in until the sound of escaping acetylene gas can be heard flowing from the torch tip. The acrid odor of acetylene will certainly be noticeable.

Ignite the acetylene with the spark lighter designed for the job. Do NOT use matches or a cigarette lighter as these devices may cause severe burns when the acetylene ignites. Always be careful where the torch tip is pointed. Never point the torch tip in the direction of a fellow workman.

5. Increase the acetylene regulator pressure until the flame is dancing about 1/2'' off the end of the torch tip. That is the correct amount of acetylene for the welding tip size. All further regulation of the torch will be done at the needle valves on the torch.
6. Using the torch acetylene needle valve, reduce the acetylene gas flow until the flame returns to the end of the torch tip. The flame should be yellow-orange in color with little or no black smoke.
7. Very slowly turn in the oxygen regulator adjusting screw. The acetylene flame will turn white as oxygen is added. As more oxygen is slowly added, the white ''feather'' will decrease in size and a small blue inner cone will appear. This is the actual welding flame. Slowly increase oxygen until the white feather disappears.
8. Use the torch acetylene needle valve to increase the flow of acetylene gas. The white ''feather'' will appear again, indicating an excess of acetylene gas. Increase acetylene flow until the white feather is approximately 10'' in length.
9. Using the oxygen regulator adjustment screw, increase oxygen until the white feather disappears. The inner cone should be a medium blue color with a slightly rounded end on the cone.
10. Use the acetylene needle valve on the torch to increase the flow of acetylene gas. At this time, you can probably turn the acetylene needle valve full on.
11. Slowly increase the oxygen flow rate with the oxygen regulator adjusting screw. The white feather will disappear and the blue cone should be a medium blue color. Both welding torch needle valves are now full open and there is a neutral flame at the torch tip.

    Observe the flame for about a minute to permit natural adjustment of the gas pressures in the hoses and torch. Some slight adjustments may be necessary as the flame changes slightly to reflect the gas pressures. Any changes must be made at the oxygen regulator adjustment screw.
12. At this time the torch is now ''balanced'' for the type and size of welding tip to be used. There is equal pressure on both the oxygen and acetylene regulators despite the fact that the working pressure gages of the two regulators have different output readings. It is well to know that regulator gages are probably no more accurate than plus or minus about 5%. Note that the working gage pressures are very low. In fact, it is probably difficult to read the oxygen pressure at all! Do not increase either the oxygen or acetylene working pressures at the cylinder regulators. The pressures set by the steps listed above are the maximum that may be properly used with the torch tip size selected.
13. The blue inner cone of the flame is where most of the welding and brazing heat is developed. It should never come in contact with the part being welded or brazed. The outer flame, properly called the ''outer envelope'', is a combination of almost transparent blue and white flames with some yellow and orange mixed in. The outer envelope acts as a shielding gas to exclude the atmosphere from the weld area.

14. When shutting down the equal pressure torch, the torch needle valves are used. It is strongly recommended that the **oxygen gas** be shut off first by closing the oxygen needle valve on the torch. When the oxygen needle valve is completely closed, shut off (close) the acetylene torch needle valve quickly. This will prevent soot formation in the air. At this point, there should be no flame and no gas issuing from the torch. Safety is the issue here. It DOES make a difference which needle valve is closed first. The method described is considered safest because acetylene without pure oxygen has a much lower flame temperature plus it must then depend on atmospheric oxygen for combustion. The positive pressure from the acetylene cylinder, through the gas hose, will prevent atmospheric oxygen from coming into the torch and will virtually eliminate what is known as "flashback" in the torch and hose assembly. If the acetylene needle valve is shut off first, there is the possibility of a flashback in the torch and hose assembly since oxygen supports combustion in even very small amounts of flammable material. The volume of pure oxygen could literally suck the flame back into the torch and possibly into the acetylene hose. An explosion could be the result.

The rule is: **SHUT OFF THE OXYGEN NEEDLE VALVE FIRST!**

The reason for setting the pressure regulators, adjusting screws and torch needle valves as described is very simple. The two gases used for welding, acetylene and oxygen, are at exactly the same working pressures. Neither gas has an excess pressure that would permit forcing itself into the other gas hose and possibly causing pre-combustion. In this situation, it is virtually impossible to create a flame—that is, a flashback—beyond the torch mixer into the torch body and hoses.

In all cases the torch flame may now be set at the torch needle valves without further adjustment of the regulator adjusting screws. Remember: the reason for balancing the welding torch flame as described is to insure the inability of either gas, oxygen or acetylene, to cause "creep" at the welding flame. "Creep" may be defined as the slow increase of one gas pressure at the welding flame.

Excess acetylene would be apparent by the white feather extending through the outer envelope of protective flame. *Excess oxygen* would be apparent as a darker blue inner cone having a rather pointed end and a harsh, turbulent affect on the weld puddle.

Changing the welding torch tip size requires a re-adjustment of the oxygen and acetylene pressures at the regulators. For each welding tip size, the regulator adjusting screws must be adjusted as previously outlined. It is very important that the gas pressures be rebalanced each time the torch tip is changed.

Welding is essentially a common sense craft and skill. When the equal pressure gas torch is balanced as directed the gas welder should have no problem with gases or welding equipment.

Setting the Injector Torch for Welding

Injector, or low pressure, torches operate on a different principle than equal pressure torches. They are constructed so that oxygen, flowing through a center orifice in the torch body, pulls in the acetylene (or other fuel gas) by suction. The action is called the "venturi"

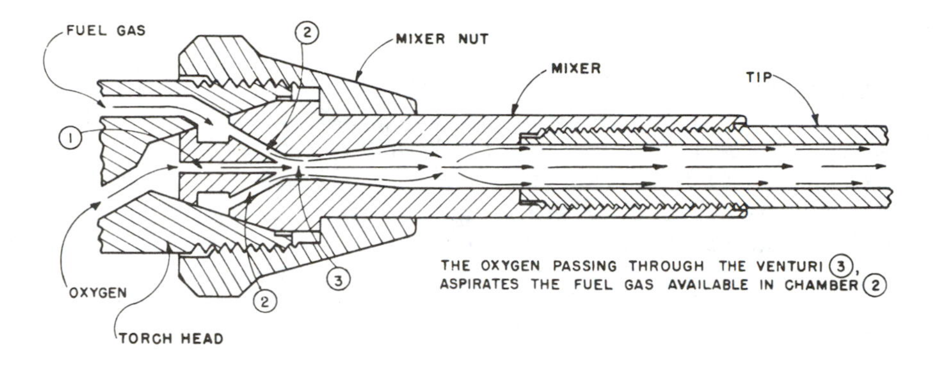

Figure 59. Typical Injector Low Pressure Welding Torch.

principle. The movement of relatively high pressure oxygen causes the acetylene to flow through a metered orifice in the welding torch body in proper amounts to create a good welding flame condition. (A **metered orifice** is a hole of specific diameter which will allow only so much gas to flow through in a given condition).

To set up prior to initiating the injector type torch flame, the welding torch needle valves are left closed. All pre-flame regulation is done at the pressure regulating adjustment screws.

The adjusting screws of both the acetylene and oxygen regulators are turned out to the "no-pressure" position. The cylinder valves are opened as previously described with the oxygen valve full open and seated. The acetylene cylinder valve is opened only 1/4 turn.

The acetylene regulator adjusting screw is turned in until about two (2) psig pressure shows on the acetylene working gage. The oxygen regulator adjusting screw is turned in until about 15-30 psig pressure shows on the working gage. The exact working pressure of both gases will depend on the torch tip size used.

With a spark lighter ready, turn on the acetylene and ignite the fuel gas. BE CAREFUL WHERE THE TORCH IS POINTED SO NO ONE IS BURNED BY THE FLAME.

Turn on the oxygen torch needle valve and adjust it for the type of welding flame required. The movement of the oxygen through the torch body will pull the required amount of acetylene gas into the mixer and torch tip.

In shutting down the injector torch, shut off the oxygen first for the safety reasons previously explained. Then shut off the acetylene torch needle valve. Be sure both torch needle valves are completely closed. The injector type torch is used a great deal where natural gas is the fuel gas. For oxy-acetylene welding, the equal pressure torch is preferred by professional welders.

Filler Rods for Gas Welding

The American Welding Society (AWS) develops and publishes all specifications for filler metals for welding applications. For oxy-acetylene welding of iron and steel, the filler rods are listed in AWS A5.2 "Specification for Iron and Steel Oxyfuel Gas Welding Rods". Filler metals for other base metals may be located in the appropriate AWS Specifications.

Fluxes for Gas Welding

Gas welding of non-ferrous metals normally requires some type of chemical flux to dissolve and remove surface oxides. (Non-ferrous means "without iron"). Exceptions to this rule include lead, zinc and most precious metals. Chemical fluxes may be furnished as powders, liquids, or in paste form.

Chemical fluxes are not required for welding carbon steels. Gas welding of stainless steels, however, does require some type of chemical flux for good results. Aluminum and magnesium require a fluoride flux for gas welding applications. All types of brazing operations require some form of flux.

Practical Data for Gas Welding

The most important consideration in the oxy-acetylene welding process is control of heat input to the base metal being welded. Heat energy input causes the base metal to expand where the heat is applied. When the metal cools after welding it will contract, often with severe distortion of the welded part.

The distortion of the base metal in gas welded structures is usually caused by the difference in temperature in the part. For example, carbon steel has a melting temperature of 2500-2600° F., depending on the carbon content. The base metal is at room, or ambient, temperature of perhaps 65° F. when the weld is started. The result is a temperature gradient ranging from about 2600° F. to 65° F. or roughly 2535° F.

The heated metal in the weld area cannot expand laterally (sideways) because of the constricting influence of the cold base metal. Any expansion has to be in a plane essentially vertical to the surface of the base metal.

When the weld metal cools, the heated and expanded metal cools and wants to return to its original shape and dimensions but it cannot. The addition of weld metal and the rigid structure of the base metal that remained cold and unaffected by the welding heat won't permit the metal to resume its original shape. The result is distortion of the base metal due to locked in stresses, both compressive and in tension.

Thermal Conductivity

The whole idea of fusion welding with any process, including oxy-acetylene welding, is the control of heat energy input to the base metal weld area. The **thermal conductivity** of the base metal is an important consideration when selecting the welding process. For example, some metals have poor thermal conductivity while other metals have excellent thermal conductivity. To define:

Thermal conductivity = "heat transfer; the ability to transfer heat through a metal".

The term "poor thermal conductivity" means that some metals will not permit heat to spread, or transfer, except at a very slow rate. Some of the commonly welded metals that have relatively poor thermal conductivity are carbon steel, cast steel, cast iron, stainless steels, and nickel alloys. With these metals, the heat remains in the immediate area where it is introduced. It would be logical to expect more heat distortion in the metals with relatively poor thermal conductivity because of their restraint to thermal expansion in the weld area.

Commonly welded metals with good thermal conductivity include aluminum, magnesium and copper and their respective alloys. It makes sense that it would take less total heat energy input to gas weld carbon steel or stainless steel than it takes to weld copper or aluminum. The reason is that copper and aluminum will conduct the heat away from the weld joint area faster than carbon steel or stainless steel. More heat energy input is therefore necessary for copper and aluminum to overcome the heat losses; the heat that has spread rapidly through the mass because it has good *thermal conductivity.*

The metals listed in the table below show the relative heat transfer rates of some typical weldable materials. They are not listed in any particular order of ability to transfer heat.

Fast Heat Transfer	Slow Heat Transfer
Aluminum	Carbon Steels
Aluminum Alloys	Low Alloy Steels
Magnesium	Nickel
Magnesium Alloys	Nickel Alloys
Copper	Stainless Steels
Copper Alloys	Titanium

Figure 60. Relative Heat Transfer Rates of Some Metals.

The oxy-acetylene welding process is not suitable for all metals. For example, gas welding should not be attempted on refractory metals such as tungsten, columbium, molybdenum, and tantalum. Gas welding techniques are not applicable to reactive metals such as zirconium and titanium. These metals are normally welded with one of the electric welding techniques such as gas tungsten arc welding or electron beam welding.

Torch Manipulation

Probably the most important welding skill necessary for oxy-acetylene welding is the welder's ability to manipulate the torch for control of heat input to the weld joint. A steady hand, good eyesight, and good judgement in "reading the weld puddle" are all necessary factors for performing top quality oxy-acetylene welding.

Through proper torch manipulation heat energy input to the weld area can be closely controlled. Slight movement of the torch tip, by movement of the welders wrist and hand, can cause the molten weld metal in the puddle to "wet out" and flow smoothly into the weld joint. Correct torch handling techniques are most important to successful oxy-acetylene welding. Good welder techniques require patience to learn and practice, Practice, PRACTICE!

Basic Torch Handling Techniques

There are two basic torch handling techniques used with oxy-acetylene welding applications. They are called the "forehand method" and the "backhand method". Some of the details of these welding techniques are presented in the following paragraphs.

Forehand Method

The welding torch tip is pointed in the direction of travel when using the forehand welding technique. This puts the outer envelope flame ahead of the welding puddle to protect the molten metal from oxidation. The base metal is also preheated by this technique. The result is faster welding speeds with less penetration of the weld into the base metal. The forehand technique is often used for gas welding aluminum because of the high thermal conductivity of the base metal.

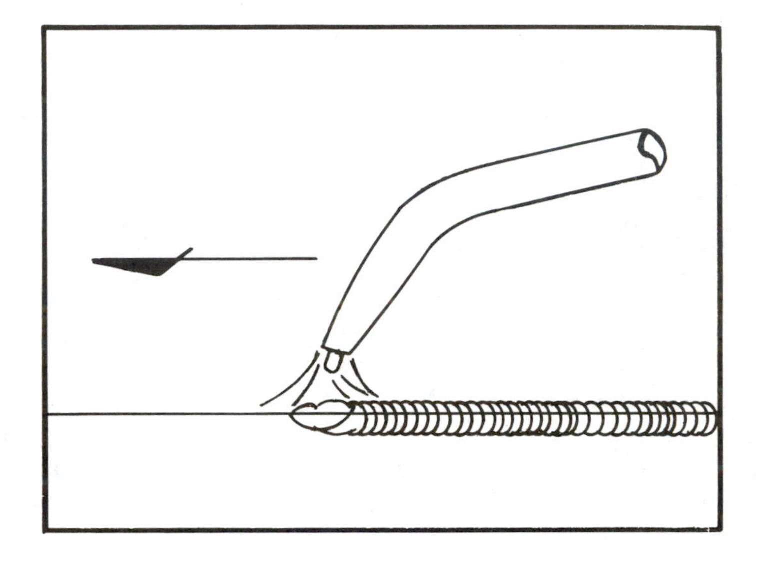

Figure 61. Forehand Welding Technique.

Backhand Method

The welding torch tip is pointed in the direction of the weld puddle, and away from the direction of weld travel, when gas welding with the backhand technique. This provides deeper penetration of the weld into the base metal. It will also tend to increase the height of the weld crown. The backhand technique is a slower method of gas welding than the forehand technique.

Summary

Oxy-acetylene welding is considered a repair and maintenance process in most modern welding shops. It is not normally used for production welding applications. There is, however, some application of oxy-acetylene welding for small diameter (2'' diameter or less) iron pipe where pressure piping codes require the process.

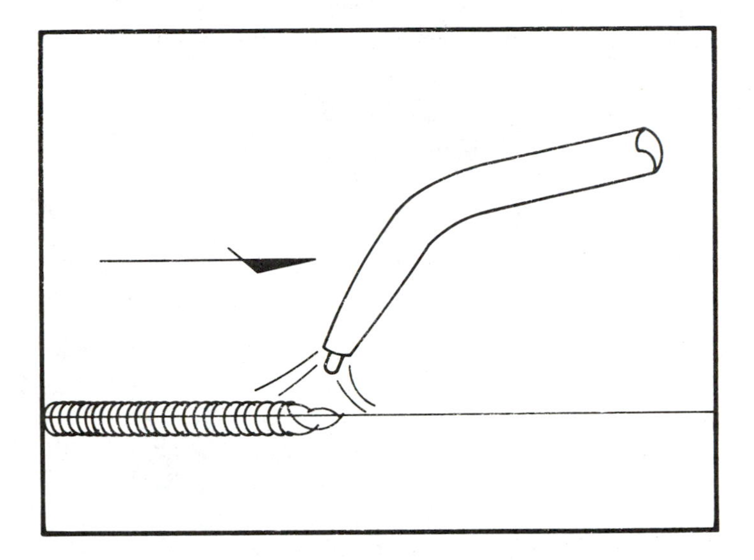

Figure 62. Backhand Welding Technique.

Flame Cutting

One of the first tools the beginning welder learns to use is the flame cutting torch. Unfortunately, the flame cutting torch is often considered so simple to operate that training of the beginning welder is neglected. This is a mistake. Flame cutting takes skill, a steady hand, and good eyes. It takes a trained cutting operator to perform a professional flame cutting job.

Fuel Gases

There are a variety of fuel gases that may be used for flame cutting purposes. Each differs from the others in flame temperature when mixed with oxygen for cutting. The various flame temperatures listed here are for neutral oxygen-fuel gas cutting flames.

Fuel Gas	Flame Temperature with O_2
Acetylene	5,800° F.
Methylacetylene Propadiene (MAPS)	5,320° F.
Propylene	5,200° F.
Natural Gas	4,600° F.
Propane	4,585° F.

Figure 63. Oxygen-Fuel Gas Cutting Flame Temperatures.

Flame Cutting Torches

Flame cutting torches are available in two specific manual styles. The true flame cutting torch is designed for cutting only and is usually a fairly heavy duty industrial tool. The second type is the flame cutting attachment that fits on a regular welding torch body. This is done by removing the welding tip and attaching the cutting attachment. We will identify the torches as follows:

Cutting torch = for cutting only.
Cutting attachment = for attaching to a welding torch.

Basic Design

There are normally two needle valves on the cutting torch. One controls the fuel gas and the other controls oxygen for the preheat flames. The cutting oxygen is controlled by a trigger or lever on the handle of the torch.

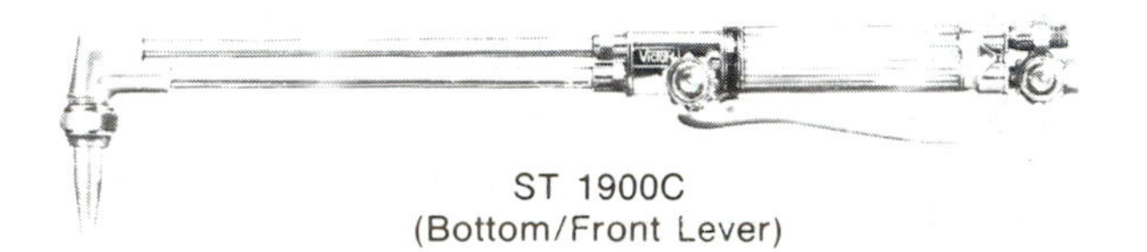

Figure 64. Industrial Cutting Torch.

The cutting attachment and welding torch body actually have three needle valves for control of preheat flames. In this design the oxygen needle valve on the torch body is opened full on at all times. The actual flow of oxygen is controlled by the needle valve on the cutting attachment. The torch body fuel gas needle valve controls fuel gas flow. There is a trigger or lever on the cutting attachment for control of the cutting oxygen.

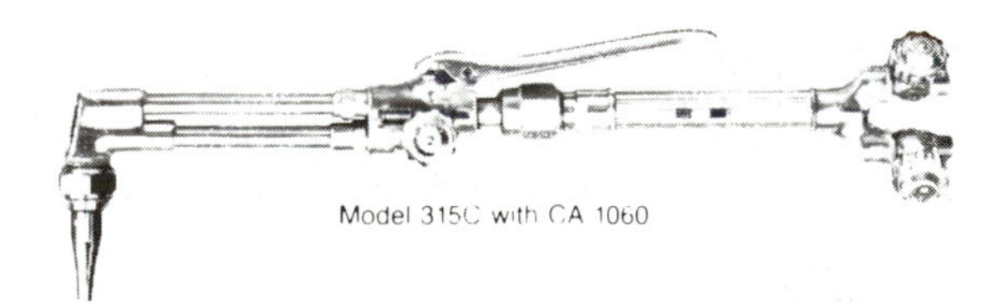

Figure 65. Welding Torch Body and Cutting Attachment.

Cutting Tips

Cutting tips are shaped considerably different from welding tips. There are small preheat flame holes in a circular pattern around the larger central orifice which is for pure cutting oxygen. The heavier and shorter body of cutting tips are designed to dissipate heat from the cutting operation.

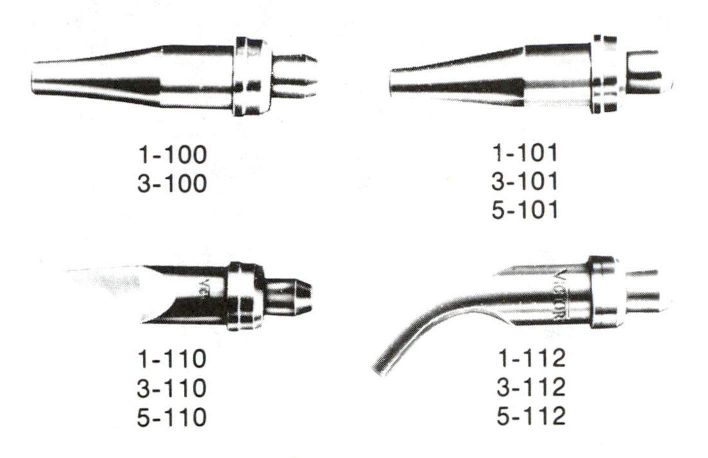

Figure 66. Flame Cutting Torch Tips.

Cutting tip sizes range from size number 000 to number 8. For most shop operations the cutting tip size should range from number 00 to 4. These tips will be adequate for metals from about 1/8'' through 3'' thickness. Data for proper cutting tip selection is furnished by the various manufacturers of flame cutting equipment. Charts and booklets may be obtained from your authorized Welding Supply Distributor.

Oxygen-Fuel Gas Pressures

Regardless of the type of cutting torch mechanism used, the oxygen and fuel gas pressures are set at the regulators. In most instances, the gas pressures are set too high for the size tip being used. The most common settings are 8-10 psig for acetylene and 40-45 psig for oxygen.

In both cases, the settings are about twice what they need to be for even the thickest materials. For example, a Number 1 cutting tip will easily make clean flame cuts in 1'' thick mild steel plate at about 4 psig of acetylene and about 25 psig oxygen. Try it!

Another common fault is to use too large a tip size for the metal to be cut. Using a number 2 cutting tip to flame cut 1/4'' mild steel plate is ridiculous! The dross from the cut fuses and welds the pieces back together as fast as you cut it. The result is no severance of the metal part at all.

This is when you see the untrained flame cutting torch operator use the torch as a hammer to break the piece loose. This action shows that the person doesn't know how to use the flame cutting equipment in a proper manner. It also costs money to repair the torch.

Always consult the cutting equipment manufacturers tip recommendations for different thicknesses of steel. Recommended fuel gas and oxygen flow rates are also provided in various tables and charts from manufacturers. The most common fuel gas used is acetylene. Some typical acetylene and oxygen gas flow rates are correlated to torch cutting tip sizes in the chart on the following page.

Cutting Tip Size	Acetylene psi Min./Max.	Oxygen psi Min./Max.
000	3/5	18/20
00	3/5	20/25
0	3/5	20/25
1	3/5	20/25
2	3/6	25/30

Figure 67. Gas Flow Rates for Flame Cutting.

Flame Cutting Techniques

Although flame cutting may be done with either manual or automatic equipment, these tips are basically for the manual cutting equipment operator. The operator of automatic flame cutting equipment may read and heed them, of course!

Flame cutting is done on carbon steels and some low alloy steels. The kindling temperature of carbon steel is about 1750° F. which shows as a bright yellow color. At this temperature the application of a pure jet of oxygen will actually burn the carbon steel.

Cutting torch flame ignition is accomplished by turning on the fuel gas and igniting it with an approved spark lighter. *Never use a cigarette lighter or matches!* Add oxygen with the torch needle valve until the preheat flames are at a neutral flame. Gently add cutting oxygen and you will probably see a slight white "feather" come from the preheat flames. Re-set the preheat flames to a neutral flame. You are ready to flame cut.

When flame cutting from the edge of a steel plate, preheat the plate edge until it is orange-yellow. Depress the cutting oxygen lever or trigger and begin the cut. Move the torch head along the cut line at a steady speed. The torch head should be tilted toward the direction of travel not more than 5-10°.

Piercing carbon steel plate is accomplished by heating a small area on the plate surface. The torch head is then tilted about 30° and the cutting oxygen is slowly added. The torch head is brought to a full 90° angle with the torch flame raised about 1/4"-3/8" above the plate surface. This prevents molten metal from blowing back into the torch tip. When the hole is pierced, make a small circle to get rid of the dross at the plate surface. Then proceed with the flame cut in a normal manner.

Summary

Flame cutting is relatively easy to accomplish with modern torches and equipment. Be sure you have the correct torch and tips for the type of fuel gas used. Always use welding goggles with correct shaded lenses when flame cutting or welding. For most flame cutting operations either a number 5 or number 6 shaded lens is adequate.

Observe the safety regulations for compressed gases and flammable gases as set forth in the ANSI/AWS Z49.1 Standard, "Safety in Welding and Cutting".

CHAPTER 4

Electrical Fundamentals

Electricity! No one knows what it is but most everyone knows that it is a magnificent servant of mankind. This chapter is designed to explain some things about electrical power that will, hopefully, take some of the mystery out of this fascinating subject.

The name "electricity" is based on the Greek and Roman words for amber. The Greek word for amber is "elektron". The Roman name for amber is "elektrum". The significance of the **fossilized resin, amber,** to this discussion is found in the section titled "Static Electricity".

The ideas discussed in this chapter include the very simple beginnings of Man's discovery of static electricity. From there we will progress to the basic theory of matter, a bit about atoms, protons and electrons, conductors and insulators, and Ohm's Law. The various methods of generating electrical power will be examined as well as the meaning of some electrical terms.

Overall, this discussion of electrical fundamentals is slanted towards electric arc welding but the physical and electrical laws presented are applicable to any use of electrical power.

CAUTION: Always observe safety rules for working with, and around, electrical power. If in doubt, consult the National Electrical Code.

Static Electricity

History tells us that Thales, a Greek, first recorded an experiment where he rubbed amber with a silk cloth. He noted that the amber would then attract small pieces of lint and paper due to some force that was imparted to the amber from the silk material. This was the first known recorded data concerning static electricity.

A simple experiment similar to that conducted by Thales will assist you to better understand static electricity. First, suspend a table tennis ball by a 12 inch minimum length of silk thread. Then rub a common glass stirring rod briskly with some silk or nylon cloth. Touch the suspended table tennis ball lightly with the glass rod, then withdraw the rod.

Now bring the glass rod slowly into the immediate area of the suspended table tennis ball—and the ball will move away from the glass rod! The static electrical charge transferred from the glass rod to the table tennis ball repels the ball from the similar static electrical charge on the glass rod.

Another simple experiment will help us to see another aspect of static electricity. With the suspended table tennis ball still in position, rub a hard rubber comb with a piece of wool cloth. The rubber comb will attract pieces of lint and paper similar to Thales' amber. If you now bring the comb slowly into the area of the suspended table tennis ball, the ball will actually move towards the comb!

Some conclusions can be drawn from these two rather simple experiments. They include:

1. There must be at least two different kinds of electrical charges in existence since in one experiment there was a **repelling** action and in the other there was an **attraction.**
2. It appears, from the action and reaction of the table tennis ball in the two experiments, that similar (like) electrical charges repel each other and dissimilar (unlike) electrical charges attract each other.
3. It is evident that some type of static electrical charges may be moved from one conductor to another by mechanical methods.

   Static electricity is electrical energy that is not in motion as generated power. Another type of electrical power is **dynamic electricity.** The term "dynamic electricity" refers to generated amperage and voltage electrical power that is apparent in electrical conductors. Dynamic electricity is working electricity in action. It is the knowledge of dynamic electricity that enables the world to have the vast electrical technology it presently enjoys.

The Basic Theory of Matter

For many years, the atom was thought to be the fundamental unit of matter. This theory was exploded with the first atomic bomb launched in World War II. Let us take a look at what some of the worlds leading physicists had suspected even prior to 1930!

The term "matter" may be defined as the physical substance of any object that occupies space. All matter is made up of particles of some substance. Molecules are one type of particle that constitutes matter. The smallest particle of a substance that contains all the elements of that substance is called a **molecule.** For example, a molecule of carbon dioxide contains the elements carbon monoxide (CO) and atomic oxygen (O). The molecule is the particle that is involved in most of the chemical changes that take place in matter.

Molecules may be divided into smaller atomic particles called **atoms.** An atom may be defined as the smallest unit particle of an element that retains all the characteristics of that element. Over 100 atomic elements have been isolated and classified in a chart known as the Periodic Table of the Elements.

It is now known that the atom, for many years thought indivisible, can be divided into smaller sub-atomic particles. The two atomic particles that are of major interest are the electron and the proton. The electron is the fundamental unit of negative (–) electricity. The proton is considered the fundamental unit of positive (+) electricity.

Every atom has a heavy central core called the **nucleus** of the atom. Most of the physical mass of the atom, and all of the positive electrical charge protons, are located in the nucleus. **Neutrons** are sub-atomic particles also located in the nucleus of an atom. They carry no electrical charge and do not affect the balance of electrical charges.

Much smaller sub-atomic particles called **electrons** are considered to be in orbital path around the atomic nucleus, much as the planets revolve around the sun. This concept was first disclosed by Dr. Neils Bohr, an eminent Danish physicist.

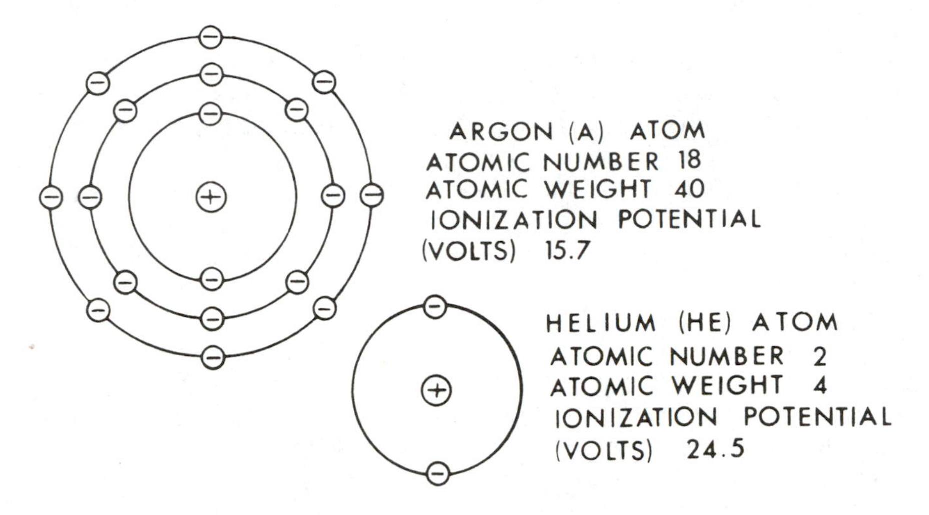

Figure 68. Diagrammatic View of Gas Atoms.

Usually there is a balance of protons and electrons in atoms. Although the proton is approximately 1800 times greater in mass and weight than the electron, the two sub-atomic particles have equal, but opposite, electrical charges. When there are an equal number of protons and electrons in an atom, the atom is considered to be electrically neutral. Remember: The proton carries a positive (+) electrical charge and the electron carries a negative (−) electrical charge.

Protons and Electrons

The number of protons in an atom usually determines the **type of element** present. As previously stated, the number of electrons normally equals the number of protons in an atom. The Periodic Table Of The Elements shows the atomic number and the atomic weight of each of the elements presently known to science. The terms "atomic number" and "atomic weight" are defined as follows:

Atomic Number = The number of electrons in planetary orbit around the atomic nucleus.

Atomic Weight = The total number of protons and neutrons within the nucleus of the atom.

Certain electrical conductor materials have what is known as a "loose electron bond". The term "loose electron bond" means that electrons can be separated from the individual atoms with relative ease. Most metals have a relatively loose electron bond as compared to other materials such as glass or rubber.

The protons are all in the nucleus of an atom. In electrical power, there is not sufficient energy to cause the integrity of the nucleus to be breached to release the protons and let them move. The electron is the atomic particle that must move, therefore, when sufficient voltage is applied. It is the movement of electrons that constitutes electrical current flow in a conductor.

An atom is in the normal condition when the internal energy is at minimum level. This is termed the **normal state.** If the energy level of the atom is raised above the normal state the atom is said to be **excited.** The excitation of an atom may occur in a number of different ways.

For example, voltage applied to an electrical conductor would excite the atoms of the conductor. Collision of the atoms with high speed sub-atomic particles may cause the energy of the particles to be imparted to the atom. If the energy, or force, is great enough it can cause an electron to leave the atom. An atom that has gained or lost an electron is electrically charged and is called an **ion.**

If the atom has lost one or more electrons it would be classed as a positive ion since it has given up some of its negative electrical charge. If the atom has gained one or more electrons it would have a negative electrical charge and would be classed as a negative ion.

In welding applications, ionization is normally related to various shielding gases used with welding processes. Shielding gases give up a negative electron when they are ionized and become positive ions. This provides a better path for welding current to flow across the welding arc since current flow is from negative to positive. Ionization that occurs in the welding arc is called thermal, or heat, ionization.

Remember: An ion is an electrically charged atom.

When electrons are removed from an atom they become free electrons. It is the movement of free electrons in any electrical conductor (solid, liquid, or gaseous) that constitutes the flow of electric current.

Conductors and Insulators

Materials that allow the movement of large numbers of free electrons are called electrical conductors. For example, **silver** is an excellent electrical conductor because it has substantial amounts of free electrons available when voltage, or electrical pressure, is applied. **Copper** is the next best electrical conductor and, because it is much less expensive than silver, it is used as the most common electrical conductor. **Aluminum,** with approximately 62% the electrical conductivity of copper, is third best as an electrical conductor. Aluminum is less expensive than copper and has been used for increasing numbers of electrical applications in industry.

Free electrons transfer from one atom to another in electrical conductors. As the electrons move the relatively short distance from atom to atom they displace other electrons in orbit around the second atomic nucleus. The displaced electrons move on to other atoms where the action is repeated.

The transfer of electrons continues until the electron flow (as electric current) is apparent all along the length of the electrical conductor. The greater the number of electrons that can be caused to move in a material, at a given level of electrical pressure (voltage), the better the electrical conductivity of the material. Good electrical conductors are those materials that have low electrical resistance to current flow and the ability to free large quantities of electrons.

Some other materials have what is called a "tight electron bond" and, therefore, few free electrons. Such materials are considered to be poor electrical conductors. In many cases, these "poor electrical conductors" are actually used in circuits as electrical resistors and insulators. Of course, all materials have some electrical resistance. There is no known perfect electrical conductor.

Even resistor and insulator materials have some electron flow, or movement, so it is equally true that there is no known perfect electrical resistor material. It is logical that the best electrical conductors, those having the least resistance to current flow, are used to carry electrical current. Those materials classed as least effective electrical conductors are often used as electrical insulators and resistors.

Summary of Section

Some of the terms used in this section of the Electrical Fundamentals chapter are defined here for ready reference.

Matter = The physical substance of anything that occupies space.
Molecule = The smallest particle of a substance that contains all the elements of that substance.
Atom = The smallest unit particle of an element that retains all the characteristics of the element.
Nucleus = The heavy central core of an atom that is electrically charged positive.
Proton = The fundamental unit, or particle, of positive electricity.
Electron = The fundamental unit of negative electricity.
Ion = An ion is an electrically charged atom of matter.
Unit Charge = The smallest unit quantity of an electrical charge. For example, a proton is a positive unit charge. An electron is a negative unit charge.

It is important to remember that protons are approximately 1800 times greater in mass and weight than electrons. That is the reason that almost all of the mass of an atom, and all the positive electrical charge, is contained in the nucleus of the atom. The protons are all contained in the nucleus of the atom.

Electrons can be transferred from one place to another within an electrical conductor. It is the movement of free electrons that constitutes the flow, or movement, of electrical current.

The negative electrical charge of the electron is exactly equal to the positive electrical charge of the proton. Equal numbers of protons and electrons in an atom result in an **electrically neutral atom.** The neutrons in the nucleus carry no electrical charge.

Conductors are those materials which easily carry electrical current with a minimum of internal electrical resistance. Such materials have loose electron bonds and the ability to free large quantities of electrons. Silver, copper, and aluminum are metals which are considered good electrical conductors.

Resistor materials are those which have high electrical resistance to current flow. Wood, rubber, and certain ceramic materials are in this material category. Such materials have a tight electron bond. It is apparent that good conductors have the ability to release large quantities of free electrons; resistors do not.

The Electron Theory of Current Flow

At this point in our discussion of electrical fundamentals you know that the electron, a sub-atomic particle, is capable of movement within an electrical conductor. As a matter of fact, electrons are capable of movement within all fundamental matter. The electron, of course, has been defined as the fundamental unit of negative (−) electricity.

It is a fact of electrical law that similar electrical charges will repel each other and dissimilar electrical charges will attract each other. This means that two positive (+) electrical charges would repel each other. The same is true of two negative (−) electrical charges. The other side of the coin is that a negative (−) electrical charge will be attracted to a positive (+) electrical charge because they are not the same type of electrical charges.

There is not sufficient energy in electrical circuits to cause the sub-atomic proton particles to leave the nucleus of conductor material atoms. It has to be the electrons, therefore, that move when electrical current is caused to flow in a circuit conductor.

The movement of electrons is caused by the imposition of an electrical force, or pressure, called **voltage** on the electrical circuit conductors. The voltage must be of sufficient numerical value to dislodge electrons from the atoms of the conductors. The voltage force is created in the generation of electrical power.

In summary, it is logical that the application of voltage to an electrical conductor will cause electrons to be dislodged, or separated, from conductor atoms. The negative electrons are attracted to the positive charge of protons in other atoms. Electrons would be repelled by a concentration of negative electrons and would try to move away from such a concentration. The electrons would be attracted to an area of high positive electrical charge; that is, an area with an excess of protons and a deficiency of electrons. It is evident that electron flow is from negative to positive in electrical circuits. It is the movement of electrons that constitutes current flow in electrical circuits. Therefore, the electron theory proves that current flow must be from negative to positive.

Direction of Current Flow

In many electrical textbooks, the mathematical formulas and theories are based on the concept that electrical current flows from positive to negative. Most of the modern textbooks for the study of electronics, however, are based on the concept that current flows from negative to positive. Which philosophy is correct?

Actually, it is only of academic interest to know which direction current flows in a circuit. Until you get involved with electronic circuits, that is, where current **must** flow from negative to positive. It was B. Franklin, Statesman, Inventor and Editor who first determined in which direction electrical current flows. This was soon after his famous kite flying experiment in which he literally pulled lightning from the skies. From his observations of electrical power, Mr. Franklin concluded that electric current must move from positive to negative.

The electron theory of current flow was first recorded by T. Edison during his development of the incandescent light bulb. Please understand that much of what we now know about electricity has been developed in the years from about 1900 to the present time.

Mr. Edison worked with direct current; that is, current which flows in only one direction in an electrical conductor.

During one experiment, he had the problem of removing a sooty, black smoke which appeared on the inside of the glass bulb of the incandescent light. Among the things he tried to overcome the problem was the placement of a small metal plate within the sealed glass bulb. The metal plate had two wires leading through the glass bulb wall, to a galvanometer (a type of electric metering device). Mr. Edison's theory was that the sooty smoke would collect preferentially on the metal plate rather than on the inner walls of his glass bulb.

For the first part of his experiment he made the bulb filament positive (+) and the metal plate negative (−). He then impressed direct current on the filament of the bulb. To his recorded disappointment the inner surface of the bulb still became sooty and blackened. He then reversed the electrical polarity of the circuit by making the bulb filament negative and the metal plate positive. Again he impressed direct current on the electrical circuit.

He recorded the fact that the bulb glass walls still became sooty and smoky but, lo and behold, there was some movement of the galvanometer needle where there had been none in the first phase of the experiment. Since Mr. Edison subscribed to the popular Franklin theory that electric current flowed from positive to negative, and it wasn't part of his experiment anyway, he made note of the galvanometer needle movement but paid no further attention to the incident. It has since been termed the Edison Effect, a notation which is found in many electrical textbooks.

The electron theory of current flow from negative to positive was again discovered by engineers working with vacuum tubes in the late 1920's. They found that electrical current would not flow through vacuum tubes from positive to negative but that it would flow through the tubes from negative to positive. Further experiments in electronics proved the electron theory of current flow. The important thing to remember is that electric current will flow in a circuit under proper electrical and mechanical conditions.

Measurement of Electrical Current

The **ampere** is the unit of electrical rate measurement. If we determine what the exact value of one ampere is we then have the basis for measuring any amperage value in an electrical circuit.

When there is one ampere in the electrical circuit it is the same as saying "one **coulomb** per second" in the circuit. A **coulomb** is a unit of electrical quantity. Since we know that electrons move in a circuit when electrical current (amperage) flows, then there must be a certain number of electrons in one coulomb. Mathematical calculations show the number of electrons in one coulomb to be 6,300,000,000,000,000,000. The number is read as 6.3 quintillion! One ampere equals 6.3 quintillion electrons moving past a given point in an electrical conductor every second.

When electric current is flowing in a circuit there could be a few electrons moving great distances or great quantities of electrons moving relatively short distances. By the calculations indicated previously, it is evident that there are great numbers of electrons moving short distances in electrical conductors.

When electrical power is used for productive purposes it is not "used up". It is actually the energy of the electrons moving in the conductor that we use. The electrical energy is derived from the process of creating the difference in the positive and negative electrical charges at specific points in the electrical circuit.

Atomic Movement in Matter

All atoms of matter are made of sub-atomic particles such as electrons, protons, and neutrons. Atoms are the basis for all matter which may be in solid, liquid or gaseous form. Atoms are in constant motion at all times.

In solid materials, atomic movement is limited to a relatively confined space. Liquid materials provide more freedom of movement for atoms than solid materials. Gases, being a free form of matter, permit very easy movement of atoms since the gases are limited in space only by the confining walls of their container.

When any atom, or atomic particle, is in motion it has some amount of energy. Energy is the ability to do work.

The **energy** of an atomic particle is dependent on the mass of the particle and the speed at which the particle is moving through space. The **relative motion,** compared to other atomic particles, determines the energy available in a specific atomic particle. It is important to know and understand that, whenever atomic particles are in motion, work is being done and energy is being expended. The result is apparent as heat in the particle in motion.

The basic theory of heat is called Kinetic Theory. Kinetic energy is available because of the motion of atomic particles. Heat energy may properly be called kinetic energy. The greater the average energy of atomic particles the higher the temperature of the material of which they are a part.

A good example of kinetic energy is when cold metal is being hammered. The more it is hit with the hammer, the hotter the metal becomes. The reason is that the energy of the hammer blow is transferred to the metal. This causes the atoms to become ''excited'' and move more rapidly. The rapid movement of the atoms is ''work being done with energy being expended.'' The faster the movement of the atoms within the metal structure, the greater the amount of energy being used. In this case, the energy is apparent as heat in the metal. Hammering cold metal can cause the metal to increase in temperature.

Electrical Conductor Concepts

Materials that are classified as electrical conductors are materials that have minimum electrical resistance to current flow in an electrical circuit. Copper and aluminum are two of the electrical conductor materials in common use. When electric current (amperage) is caused to flow in an electrical circuit, the passage of the tremendous number of electrons causes the atoms of the conductor material to be disturbed. In effect, the atoms of the conductor material are speeded up in their random motion. They are said to be excited atoms; that is, atoms which have increased motion and increased energy.

The greater energy of the atoms causes the temperature of the conductor material to rise. As has been stated before, atoms that are excited move faster, and therefore have greater energy, than when they are in the normal state. It is a fact that, the better the electrical conductivity of a material, the lower the electrical resistance of that material to electrical current flow. Of course, if the conductor is large in cross sectional area, measured in circular mils (thousandths), there is less possibility of causing excessive movement of the atoms in the conductor material and so less heat is generated.

We have mentioned electrical resistance in conductors and the fact that every conductor has some measure of electrical resistance. What has not been determined is how electrical resistance in a conductor is measured.

The unit of measure for electrical resistance is the ohm.

Electrical resistance of conductors is a constant factor based on cross sectional area, constant temperature, and material type.

Volts, Amperes, and Ohms

In working with electrical power it is necessary to know some basic electrical laws. One of the most important to know and understand is Ohm's Law.

Ohm's Law may be defined as follows:

Ohm's Law = "In any electrical circuit the current flow, in amperes, is directly proportional to the circuit voltage applied, and inversely proportional to the circuit resistance."

Using electrical symbols, Ohm's Law may be stated as follows:

where E = Voltage
I = Amperage
R = Resistance (ohms)

$$E = IR \text{ or, } I = \frac{E}{R} \text{ or, } R = \frac{E}{I}$$

Some basic applications of Ohm's Law will be apparent as this discussion progresses.

Electrical charges are termed positive and negative. Keep in mind that "like charges repel" and "unlike charges attract." It is logical that work must be done and energy expended to create a concentration of negative electrons in one place because the "like", or similar, negative charges of the electrons try to repel other electrons. The value that forces the electrons to move is called **voltage.**

It is a fact that voltage is always measured between two electrical conductors. The measurement taken is the potential difference in value between the two conductors. The value that is measured is called voltage. Other names used to describe voltage are "potential", "EMF", "electro-motive force", and "electrical pressure".

It is important to know that voltage is an electrical force which does not flow in a conductor but which causes current to flow through the movement of electrons in the conductor. As a matter of fact, voltage is the electrical pressure that forces the electrons in the conductor to move thus creating electrical current flow. The terms "current" and "ampere" are synonymous and are used interchangeably.

The ampere is the unit of electrical rate measurement.

The ampere value indicates the number of electrons flowing past a given point in an electrical conductor in a second of time. Another name for the ampere is current. Since electrons move in an electrical conductor, and electrons are electrically charged negative, they will be attracted to a positive electrical charge.

Electrical current may be defined as the time rate of charge flow. To properly explain this statement, consider this: **time** is what it says: the time the current is actually flowing in the circuit in seconds, minutes or hours. The **rate** is how many, and how fast, electrons are moving in the conductor in a given period of time. The **electric charge** is negative because it is the electrons that are actually moving in the electrical circuit.

The movement of electrons; that is, the current flow; is directly proportional to the voltage, or electrical pressure, in the circuit. For a specific diameter of electrical conductor the higher the voltage applied, the higher will be the amperage value. This statement is, of course, the first part of Ohm's Law. (Current flow is directly proportional to the voltage applied.)

When current flows in an electrical circuit there is always some measure of electrical resistance to the current flow. The electrical resistance is mostly caused by the reluctance of the electrical conductor atoms to give up electrons. The electrical resistance of a specific type and size of conductor material will remain a constant value if the conductor remains at a constant temperature. The electrical resistance of a conductor is measured in ohms, the practical unit of electrical resistance.

The meaning of the term "ohm" is defined as follows:

ohm = "The electrical resistance in an electrical circuit that allows the passage of one ampere in the circuit when the impressed voltage is one volt."

In an electrical circuit which has steady, constant voltage applied, the current flow is inversely proportional to the electrical resistance in the circuit. (This is the second part of Ohm's Law). Inversely proportional means that if the circuit resistance is doubled, for example, the current flow in the circuit would be reduced to one-half its original value.

Some Electrical Energy Concepts

There are many different forms of energy available for use in industry. Some of the forms of energy include mechanical, as exemplified by trucks, engines, automobiles, and power boats; thermal, as exemplified by fire and the welding arc; and electrical, as exemplified by lighting, radio, and welding power.

Some of the terms used in discussing electrical energy may not be as familiar to you as they should be so the following brief definitions are provided. These definitions include some metric terms. The definitions may not be totally complete in all possible applications.

work = The transfer of energy from one body to another; the product of force times distance, measured in foot-pounds.

foot-pound = The energy required to move one pound of weight through one foot of distance; a unit of work or energy.

energy = The ability to do work.

watt = The fundamental unit of power; the amount of power required to maintain a current flow of one ampere at an electrical pressure of one volt.

W(watts) = E (volts) × I (amperes). **W = EI**

power = The time rate of doing work; work done in a unit of time; the energy or force available for work; for example, one horsepower equals 33,000 foot-pounds per minute.

joule = The joule is based on the metric system of measurement; a joule is equal to 10^7 ergs as a unit of work or energy.

erg = The actual work done by one dyne moving through a distance of one centimeter; the unit of work in the cgs system (centimeter-gram-second); a metric measurement.

dyne = The unit of measure of force in the metric cgs system; the force of one dyne acting on a mass of one gram of substance for one second gives the gram of mass a velocity of one centimeter per second.

The terms "watt" and "joule" are further explained and defined in Chapter 2 of this text.

What's Watt?

Electrical energy is purchased from the utility company to operate electrical devices in both the home and factory. Since the costs of energy are rising rapidly, it is well to understand what it is you are purchasing.

The **watt** is the fundamental unit of power. It is the product of volts times amperes in DC circuits as shown in the following equation and symbols:

W = power in watts
E = circuit voltage
I = circuit amperage

W = EI (power equation)

In electrical and mathematical equations, one watt equals one joule per second. (W = joule/second). In addition, one volt equals one joule per coulomb. (E = joule/coulomb). You may recall that the coulomb is a unit of electrical quantity measuring 6.3 quintillion electrons per second. While the joule is a metric system unit of work or energy, it is similar but not equal to the foot-pound.

The **mechanical** measurement of horsepower equals 33,000 foot-pounds. The **electrical** measurement of horsepower equals 746 watts. The difference is in how the measurements are applied. The mechanical measurements would be applicable to engines, trucks, and similar devices. The electrical measurements would be used for electric motors, generators, and other electrical devices.

One of the metric terms that is in common use in the electric welding industry is "kilo". A **kilo** is 1,000 units of something. For example, a kilowatt-hour would equal 1,000 watt-hours. But what is a watt-hour? To define the term:

Watt-hour = The product of the average power, in watts × the time, in hours, during which the power is maintained.

Average power = The circuit primary voltage × the circuit amperage × the power factor (for alternating current [AC] circuits). For direct current [DC] circuits, the average power equals primary voltage × circuit amperage.

It is noted that circuit primary voltage is considered to be a constant factor. For example, most modern homes have 230 volt primary AC power, single phase, for operating ranges, air conditioners, water heaters and possibly certain power tools. Other circuitry in the home is 115 volt power for refrigerators, lamps, radios, and other household appliances. The amount of amperage drawn from the utility company primary electrical system depends on the watts the appliances may require to operate at any given time. To determine the amperage drawn by any specific appliance, simply divide the total watts shown on the nameplate of the device by the primary voltage required for operation. The answer will tell you how many amperes are required for proper operation.

For example, an electric clock is rated to operate on 115 volts at 2.5 watts. To obtain the amperage drawn and used by the clock, divide 2.5 watts by 115 volts. (Remember: volts times amperes equal watts). The correct answer is 0.02 amperes. Consider the number of electric clocks running in your home, add the refrigerator, the radios, the television, the vacuum cleaner, freezer, furnace/air conditioner fan, and all the other things that take electrical power and you can see why the demand for electrical power is increasing.

Using the data which has been previously developed, we will break down the watt-hour into energy units. Here is how it is done.

One hour = 3,600 seconds or (60 seconds × 60 minutes).

One watt = $\frac{\text{joule}}{\text{second}}$ or (one joule per second).

Watt-Hour Calculation

one hour = $\frac{\text{3,600 seconds}}{1}$ (change hour to seconds).

one watt = $\frac{\text{joule}}{\text{second}}$ (change watt to joules per second).

one watt-hour = $\frac{\text{joule}}{\text{seconds}} \times \frac{\text{3,600 seconds}}{1}$ (cancel the seconds).

one watt-hour = $\frac{\text{joule}}{1} \times \frac{3{,}600}{1}$ (multiply).

one watt-hour = 3,600 joules (result).

The joule is a unit of work or energy based on the metric system as previously stated. This is the breakdown of the watt-hour into its energy units—the units which are supplied to you by the utility company and for which you are charged each month.

The kilowatt-hour is simply a convenient method of writing the energy amount (1,000 watt-hours) without all the zeros. In the example shown above, the answer would be multiplied by 1,000 to obtain the energy in a kilowatt-hour. The answer would be 3,600,000 joules.

In welding it is common to discuss the "joules per linear inch of weld deposit." The calculation concerns the heat energy input into the weld joint. It is possible to associate joint penetration, heat-affected zone width, joint fusion characteristics, and even weld deposit microstructures to the energy input to the weld deposit, the "joules per linear inch of weld."

Electric Energy for Welding

Electrical energy is converted into heat and light when electric arc welding. This is accomplished by creating an electric arc between the welding electrode tip and the surface of the base metal to be welded. The electrode may be consumable as filler metal in the weld deposit. This is typical in the shielded metal arc welding (SMAW) and gas metal arc welding (GMAW) processes. The welding electrode may also be non-consumable (not designed for use as filler metal) as in the gas tungsten arc welding (GTAW) process. Welding filler metals are discussed in detail in those sections of the text that deal with specific welding processes.

There are just three basic values to consider in an electric welding circuit. They are **voltage** (electrical pressure), **amperage** (current), and **resistance** (ohms). Voltage is the force, the electrical potential, that causes current to flow in an electric circuit. Voltage is a force that does not flow in an electric circuit.

In the welding arc it is voltage that literally "pushes" the welding amperage across the air space between the electrode tip and the base metal surface. Air has poor electrical conductivity and, therefore, is a good electrical resistor to current flow. The electrical resistance of the air space between the electrode tip and the surface of the base metal is of major concern to the characteristics of the welding arc.

Arc Length and Arc Voltage Correlation

The length of a welding arc is considered to be the physical distance from the tip of the electrode to the surface of the base metal where the arc impinges. Arc voltage is the electrical pressure necessary to sustain current flow across the physical arc length. It is evident that arc voltage and arc length have a close correlation. This applies to any of the arc welding processes.

There are three different arc lengths illustrated in Figure 69. Consider that a welding arc can be maintained in all three conditions. For this example there are just two factors involved. They are:

1. The electrical resistance, in ohms, of the air space between the electrode tip and the surface of the base metal.
2. The arc voltage value, or electrical pressure, required to overcome the electrical resistance of the air space so welding current can flow.

The three arc lengths shown in Figure 69 are correlated to the volt-ampere curve shown in the same illustration. The "A", "B", and "C" positions on the volt-ampere curve are explained in the next few paragraphs.

Volt-Ampere Curve = "The electrical output from a welding power source plotted in volts and amperes".

At point "A" in the electrode portion of the illustration, the arc length requires some value of voltage to overcome the electrical resistance of the air space between the electrode tip and the base metal surface. This will allow welding current to flow across the arc length. The specific voltage is of no particular interest in this example. The required voltage, and the amperage, are shown at point "A" of the volt-ampere curve, Figure 69.

The arc length at electrode "B" is greater than at electrode "A". There is more physical air space, and therefore more electrical resistance in ohms, between the electrode tip and the base metal surface. The arc voltage value must be higher to overcome the higher resistance values if the arc is to be maintained.

At electrode "C", the arc length is less than at either "A" or "B". There is much less physical air space between the electrode tip and the base metal surface so the electrical

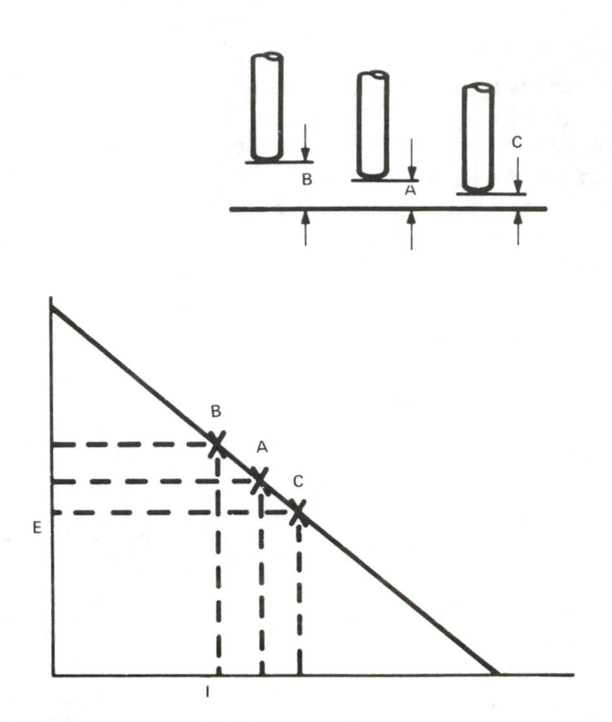

Figure 69. Correlating Arc Length and Arc Voltage.

resistance is less. (This type of short arc length is typical of E7018 electrodes with SMAW process). The arc voltage value is less since only enough voltage is developed to overcome the actual resistance of the air space in the arc length. This is shown at point "C" on the volt-ampere curve.

It is evident that the welder has the opportunity to adjust the welding voltage and amperage slightly by adjusting arc length. This is illustrated very well in Figure 69.

Resistance Heating in Welding Circuits

When using a welding power source of any kind there is some amount of amperage produced. The amperage is carried to the welding arc through various metal electrical conductors. You have already learned that all electrical conductors have some resistance to electrical current flow. The amount of electrical resistance, and therefore the power losses resulting from circuit losses, can easily be calculated. If resistance losses are excessive in a welding circuit, the results can be weld defects such as lack of penetration, lack of fusion and cold lapping.

When heat is generated in an electrical circuit, including a welding arc, the rate of heat production is measured in watts. Remember, the watt is the fundamental unit of electrical power. The watt is the product of volts times amperes. With this information, let us develop the equation for calculation of electrical circuit power losses.

Ohm's Law states that volts (E) equals amperes (I) times the circuit resistance (R). In electrical symbols it reads as follows:

E = circuit voltage
I = circuit amperage
R = circuit resistance in ohms

$$E = IR$$

The electrical power equation is somewhat similar to Ohm's Law with some small variation. Here we have watts (W) equals voltage (E) times amperes (I). In electrical symbols the equation would read as follows:

W = total power in watts
E = voltage
I = amperage

$$W = EI$$

If we consider both the equation of Ohm's Law and the electrical power equation for watts it is possible to create a formula for finding the power losses in an electrical circuit. Here is how it is done:

1. $W = EI$ [Watts = voltage × amperes]
2. $E = IR$ [substitute IR for E in equation 1]
3. $W = IRI$ [Watts = amperes × resistance × amperes]
4. $W = I^2R$ [Watts = amperes^2 × resistance]

Anytime there is current flowing in an electrical circuit there is some measure of electrical resistance which may be measured in ohms. When current flows through an electrical resistance there is energy released in the form of heat. The rate of heat production is measured in watts.

The actual amount of current used in the welding circuit is determined by the specific welding procedure. It is very important, however, to keep the circuit resistance as low as possible. This will prevent excessive heat losses in the conductor with possible damage to the conductor insulation. Of course, the larger the conductor cross sectional area for a given current value, the lower the electrical resistance of the conductor and the less the amount of heat generated in the conductor.

The power loss formula ($W = I^2R$) shows amperage as a mathematically squared term. It is a simple matter to plug in the known values in a given situation and determine the power losses in a circuit.

I^2 = circuit amperage × circuit amperage
R = circuit resistance in ohms
W = circuit power loss in watts

Using the data above, if the circuit amperage is doubled (1 × 2) in a specific size conductor, the heat generated will be four times as great ($I^2 \times 2$).

Electric Power Generation

It has been established that electrons, the fundamental units of negative electricity, move in an electrical conductor when current flows. The logical question is, "Where does this movement of electrons begin?"

The answer requires that we briefly examine the concept of power generating plants that are in production at this time. These include fossil fuel plants, nuclear plants and hydro-electric plants.

Hydro-Electric Plants

The concept of the hydro-electric power generating plant has been known for many years. Both great and small dams have been built in many parts of the world to harness the water power of major rivers for electrical power generation. One of the first and most famous of the larger dams is Grand Coulee Dam in Washington State. Hoover Dam in Nevada, the Aswan Dam in Egypt, and others too numerous to mention provide the same type of capability and perform the same basic function. All hydro-electric power plants operate on the same fundamental principles.

In essence, the river is dammed and a huge lake is created. The stored water has **latent** energy simply due to its mass. The energy is released under controlled conditions as the water is permitted to flow through the dam penstocks. The force of the moving water turns giant water wheels which, in turn, cause the turbines to rotate and generate the electrical power.

Smaller dams are sometimes used where there are no major rivers. Water is usually channeled from a lake behind the small dam through aquaducts to large tapered conduits called penstocks. Penstocks are fairly large diameter at the input end. They taper to about half their original diameter at the output end. This type of penstock is usually inclined from top to bottom down the side of a hill or mountain. The pressure created by the column of water inside the penstock is very high.

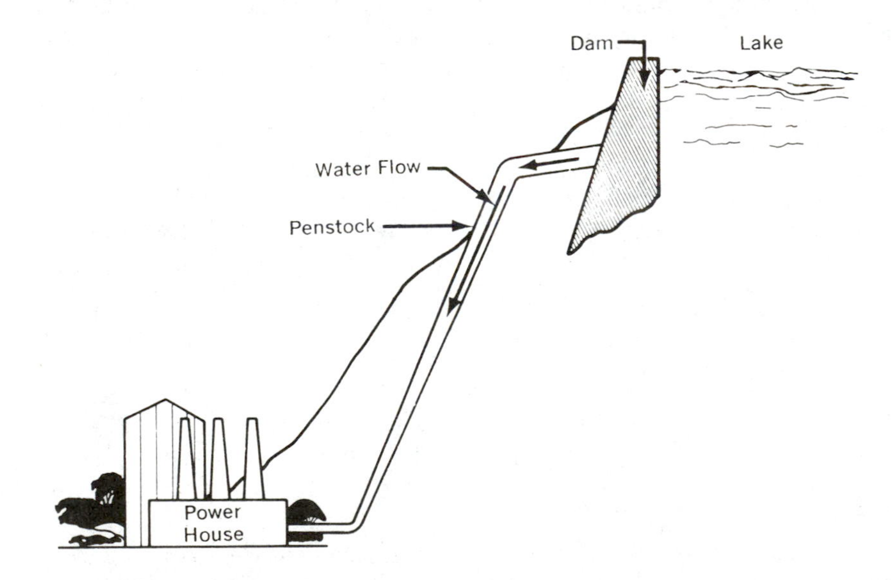

Figure 70. Hydro-Electric Plant Concept.

From the bottom of the penstock the water is directed into the generating plant where it is piped to a water wheel. The water wheel is designed with fixed position "stirrup cups" to catch the water flow. The tremendous pressure of the water column in the penstock causes the water wheel to turn. The water wheel is often coupled directly to the turbine generators where the electrical power is generated.

Fossil Fuel Plants

Fossil fueled steam generating plants have been used in many parts of the world where water power is not readily available or is not cost-effective. Some of the fossil fuels used include natural gas, oil, coal, and other energy-releasing fuels.

The energy of the fossil fuels is normally released through combustion, or burning, to heat water to steam. The heat of the released energy brings the water in steam boilers to the temperatures of super-heated steam. The steam pressure is then applied to the turbines in a manner similar to the water used in hydro-electric plants. The actual electric power is generated by the turbines and generators.

Nuclear Energy Power Plants

The use of nuclear energy for generating electrical power is much the same as in fossil fueled power plants. The main difference is the type of fuel energy used.

Controlled nuclear reaction heats water to the temperatures of super-heated steam. The steam pressure is applied to turbines and generators to generate electrical power.

Nuclear power plants have been built in many parts of the world. At first the cost of nuclear generated electrical power was considered prohibitive but in recent years the costs have dropped sharply per generated kilowatt hour. Unfortunately, excessive construction costs due to several causes have now priced this valuable source of energy almost out of the marketplace. It is to be hoped that safe, cost-effective methods can be developed for using nuclear energy for productive purposes. In this effort, governments and industries must cooperate so that mankind benefits, not perishes, from nuclear energy.

It is obvious that electrical power generation, as it is presently done, requires some form of energy to power the turbines and generators. New supplies of fossil fuels have been tapped recently to help fill this need. In addition, other alternate sources of energy are being evaluated and tested as sources of electrical power. Solar and wind energy are just two of the possible sources of electricity in the future. Experiments in both types of electrical power generation are presently in operation in the United States.

Total Electrical Power (KVA)

Electrical power is generated at pre-determined voltage and amperage levels that are not suitable for transmission over long distances. The generated voltage is therefore stepped up by means of transformers to very high numerical values for transmission purposes.

Stepping up the voltage causes the amperage to decrease proportionately to a numerical value that can be transmitted over great distances with minimum power losses in the system. **Remember:** it is amperage that flows in an electrical circuit and which must overcome the electrical resistance of the circuit. Voltage is the necessary force that causes amperage to flow in a circuit.

Multiplying the generated amperes by the generated voltage results in a value called **volt-amperes.** Dividing the number of volt-amperes by 1,000 (a kilo) provides a quotient called **kilovolt-amperes.** Kilovolt-amperes (KVA) is considered to mean total power.

For a specific KVA value the numerical product of volts times amperes must always equal the same calculated result. This is illustrated in the following discussion.

Consider the amount of 10 KVA. This value may be expressed or calculated in a number of different ways. For example:

10 KVA = 10,000 volt-amperes, or
10 KVA = 10,000 volts × one ampere, or
10 KVA = 10,000 amperes X one volt, or
10 KVA = 100 amperes X 100 volts.

It is evident that any numerical combination of volts and amperes that equals 10,000 volt-amperes (10 KVA) may be used.

This information is useful when considering electrical transmissions and the electrical losses that can occur. Voltage, of course, is the electrical pressure, or force, that causes current to flow in an electrical circuit. Voltage does not flow in a circuit. Since every electrical conductor has some electrical resistance to amperage flow, the smaller the amperage value in the circuit, the lower the power losses in that circuit. The lower the power losses in the electrical circuit, the lower the heat energy value, in watts, dissipated and lost in the conductor. This will help protect the conductor insulation and will make the total electrical system more efficient for transferring electric power.

For a total amount of electrical power (KVA), as the voltage is increased, the amperage must decrease proportionately. By stepping the generated voltage up to relatively high values through transformers, the generated amperage is reduced to a proportionately low value. This makes it much easier to transmit large values of KVA power on relatively small diameter cable.

Cables used for high voltage transmission lines are normally a composite material structure. For example, a stranded steel core cable may be overwound with various sizes of copper and aluminum conductor wire. This composition of cable provides a high voltage transmission cable having good electrical conductivity with sufficient strength to support its own weight under any reasonable operating conditions. Very often the resulting transmission cable will have an outside diameter of approximately 2 1/8" to 2 1/2".

Electrical Power Transmission

Electrical power is created by generation at the power plant, transformed to high voltages, and carried by transmission cables to its market destination. Previous data has shown that it is the movement of electrons in the conductor that constitutes current flow. The electrons actually move relatively short distances from atom to atom in the conductor. It is important to know that the electron that moves at one end of the conductor is not the same electron that moves at the other end of the same conductor.

Electrical current moves at the approximate speed of light (186,000 miles per second). In some conductors there is a delay in current movement compared to its speed of movement in free space. For all practical purposes it may be assumed that electrical power applied to a conductor will be apparent instantaneously all through the conductor. The transmitted high voltage may be several hundred thousand volts at relatively low amperage.

At the receiving end of the transmission line the electrical power is transformed again. This time the voltage is decreased through transformers until a reasonable distribution voltage is achieved. This is usually about 4,160 volts for most cities. As the voltage is decreased, of course, the amperage capability is increased proportionately.

The electrical power that is brought into a shop or plant usually comes from a line transformer located near the shop. In some major manufacturing plants, where large amounts of electrical power are used, an electrical substation may be required to supply sufficient electrical power for the plant operations.

Special recording meters are used to measure the amount of electrical power used by a customer. As we know, electricity is not "used up" when current flows and does work. What is paid for is an amount of electrical energy made available to us for use. The only charge the consumer pays is for the actual electrical energy used. This is shown as kilowatt-hours on most electric bills. As long as the total electrical system from power generation to power use is maintained and functioning, electrical power will be available.

In a sense, a practical consideration of electrical power indicates it is basically nothing more than the excitation of atoms in a conductor which causes the electrons to move faster in the conductor. The greater the input of force, or voltage, the more electrons that may be excited to movement. It is apparent that voltage in the primary supply system must be maintained at a constant level if there is to be a good strong supply and regulation of primary amperage.

KVA, Primary KW, and Secondary KW

The data presented in these next few paragraphs relates especially to welding power sources. It is furnished with the idea of providing the reader with some understanding of terms used in discussing certain electrical characteristics of welding power sources.

The total power demanded from a utility company is termed kilovolt-amperes (KVA). It is calculated by multiplying volts times amperes times the RMS value for **alternating current circuits** and dividing by 1,000. KVA includes the power losses dissipated in the plant electrical system as well as the power used in the plant. KVA is sometimes referred to as "apparent power".

In **direct current circuits,** volts times amperes equals watts. Divide the answer by 1,000 and the result is kilowatts (KW). The basic difference between the calculations for KVA and KW is there is no RMS value consideration involved with KW.

Now to the use of KVA and KW with welding power sources. You will find these terms used most often in the **specifications** for welding power sources. The definitions for KVA, Primary KW, and Secondary KW are based on the power source operating at rated amperage output at rated load voltage.

KVA is the total power demanded by a welding power source from the primary power furnished by the utility company. It is a computation based on the total power demand of the unit when the primary voltage is at a specific figure such as 460 volts. Remember: KVA is volts X amperes X the RMS value of the primary alternating current circuit.

For welding power sources, **primary kilowatts** are the actual power used by the unit when it is producing, and putting out, its rated load. Rated load, (often called rated output) is the amperage and load voltage the welding power source is designed to produce for a given specific duty cycle period. For example, a typical constant current type power source might be rated at 300 amperes, 32 load volts, 60% duty cycle. If the specifications show some value for KW, it will be primary KW.

Secondary kilowatts are the actual power output of the welding power source. In the example cited above, the calculation for secondary KW would be 300 amperes × 32 load volts, divided by 1,000. The result is 9.6 KW secondary.

CHAPTER 5

Welding Power Source Transformers

This chapter is about welding power source transformers, their component parts, and how they work. The specific component parts will be explained and their functions correlated to each other. In the first section, no effort will be made to distinguish one type of welding transformer from another since they all must function electrically in similar fashion.

In subsequent sections of the chapter, specific information is provided concerning exactly what a transformer is and what it does to the circuit power. Another section explains how the transformer functions and why it does so.

A final part of the chapter discusses welding power sources electrical efficiency and how to calculate it for any transformer type unit.

General Transformer Functions

The normal function of any transformer is to change electrical power from one voltage to another, with an inverse proportional ratio of amperage, without changing the circuit frequency. Circuit frequency is shown in **cycles per second (hertz)** which cannot be changed through a transformer.

Industrial and residential uses of primary power require voltage that may range from 575 volts down to 115 volts. The amperage used with such voltages would be fairly high in numerical value. Any appreciable amount of electrical power transmitted even a few miles at such low voltages, with proportionately high amperages, would take an enormous amount of conductor material in the transmission lines. The cost of providing electrical service would be prohibitive.

The next few paragraphs relate to Chapter 4 ''Electrical Fundamentals'' because they explain how the power generated at the electrical power plant is transformed and transmitted over great distances.

Voltage is always measured between two conductors. It is the **potential** difference in value between the two conductors. The value measured is called potential, or **voltage.** The two terms are synonymous in their meaning in electrical terminology. Voltage is the electrical pressure, or force, that causes amperage to flow in electrical circuits. Electrical power in alternating current circuits depends on the voltage between the conductors, the **amperage** flowing in the conducting circuit, and the **power factor** of the overall electrical system.

The movement of electrons in conductors constitutes the flow of electric current in conducting circuits. Circuit electrical resistance must be held to as low a value as possible to prevent severe power losses in the circuits. This is where the large transformers at generating power plants are very effective.

The use of large power transformers makes it possible to generate electrical power at any convenient voltage level, step it up through the transformers to an economical transmission voltage, transmit it great distances, and then transform the power back down to a desired level of voltage at the use end of the power transmission system.

For example, power is generated at Hoover Dam in Nevada at 13,800 volts. This voltage is stepped up by power transformers to 287,000 volts for transmission approximately 275 miles to the city of Los Angeles, California.

At the receiving end of the line the voltage is stepped down through transformers to 132,000 volts for delivery to five main receiving stations. At these stations, the voltage is again stepped down to 34,000 volts for feeding distribution sub-stations located around the city.

The voltage is further reduced at the sub-stations to a level of 4,160 volts for local city distribution. Line transformers near each residence or industrial plant make the final voltage reduction to the voltage needed at the use site.

The electrical power passes through five sets of transformers after it leaves the generators at Hoover Dam and before it reaches the ultimate customer. The total losses in each of the transformer banks is approximately five percent at rated operational load. The low loss percentage would indicate the line transformer is a very efficient piece of equipment.

Welding Power Source Transformer Concepts

There is only one reason for needing transformer type welding power sources. The relatively high voltage, low amperage primary electrical power is not suitable for welding applications. The transformer type welding power source changes, or transforms, the primary electrical power into useful welding voltages and amperages.

There are three main components that make up an iron core transformer. They are: (1) the **primary coil,** (2) the **iron core,** and (3) the **secondary coil.** All of these items are necessary for the operation of the transformer.

The **primary coil** receives the primary electrical power input from the plant primary wiring system. The power is relatively high voltage, low amperage so the electrical conductor of the primary coil is comparatively small in cross sectional area.

Remember: an electrical conductor in a circuit need only be large enough to carry the maximum amperage of the circuit.

The **iron core** is made of thin gage electrical steel laminations about 0.018''-0.020'' thick. Each lamination is insulated on both sides with a film type insulation. This makes each lamination separate from all the others and helps prevent **eddy current** formation. The iron core material is usually either a high silicon content steel (about 3-3.5% Si) or a specially rolled electrical steel that has preferred grain size and grain orientation.

The steels used in welding power source transformer cores have high **magnetic permeability** characteristics. ''Magnetic permeability'' means the ability of a material, such as electrical steel, to accept magnetic lines of force. Iron cores which have high magnetic permeability provide better magnetic and inductive coupling of the primary and secondary coils of the welding power source transformer.

The **secondary coil** of the transformer is part of the total secondary AC circuit of the welding power source. The secondary coil conductor material is larger in cross sectional area than the primary coil conductor because it will carry the relatively low voltage, high amperage welding power. The insulation for secondary coils is normally of higher temperature rating than the primary coil insulation.

The primary and secondary coils are wound on "coil forms" which shape the inside dimension of the coils. After the coils are wound, they are usually dipped in a protective varnish and then baked for a specific time. This extra coating helps protect the coils against corrosive vapors such as those found in chemical plants or in salt water coastal areas. In some cases, certain manufacturers provide an optional "double dipped" varnish coating for additional weatherproofing.

The iron core may be designed for either single phase, or three phase, operation. Single phase iron cores have two legs on which the primary and secondary coils are mounted. Three phase iron cores have three legs on which the primary and secondary coils are mounted.

Some iron cores are held together by bolts. Other iron cores are held together with welds across several points in the outer peripheral edges of the core. This may sound as if it defeats the purpose of insulating between each steel lamination. It does no harm, however, since there can be no completed circuit in the iron core as long as no weld is made on the inside, or "window", of the iron core. A weld on the inner edge, or window, would complete the iron core circuit and would permit an induced current to flow. Such a current is called an **eddy current.**

There are two types of electrical fields involved in a welding transformer. The current carrying circuits are the primary coil and the secondary coil which are the **electrical fields.** The iron core is the basis for creating the **magnetic field** necessary for the transformer to function.

A welding transformer is basically an electric field (primary AC) connected to a magnetic field (iron core) which is connected to a second electric field (secondary AC). In simpler terms, the iron core is the base on which the primary and secondary coils are located to make a welding transformer.

Transformer Types

Welding power source transformers are normally "step-down" transformers. The primary voltage is much too high for safe welding operations. Welding power source transformer primary voltage may range from 208 volts to 575 volts. By stepping the voltage down to usable welding voltage, normally a maximum of 80 volts open circuit, relatively safe welding voltages are supplied. Most of U.S. welding power source manufacturers follow the minimum specifications set forth in the NEMA EW-1 Standard for welding power source manufacturers in the United States.

There are certain special applications and processes which require a "step-up" transformer. Such applications are not for normal commercial welding with the exception of certain plasma cutting operations. In addition, all high frequency transformers are step-up transformers.

Another type of transformer used in some types of welding power source circuits is the "isolation transformer". This type of transformer usually has a 1:1 voltage ratio. The normal purpose of an isolation transformer is to isolate a circuit where some voltage value is required for a special purpose. In most cases, an isolation transformer provides power for some form of control circuit.

Transformer Coil Characteristics

Transformer coils are made of conductor materials called "magnet wire". The conductors are usually copper although some lower output and lower duty cycle power sources may have aluminum as the magnet wire. The coils are wound to a specific shape or configuration and fitted on the transformer. When voltage is impressed on the primary circuit, and current is caused to flow, the transformer will operate. Keep in mind that the current any conductor can carry is limited by the cross sectional area of the conductor.

Every electrical conductor has some electrical resistance to current flow. The work that is done to overcome the resistance converts some energy to heat which is apparent in the conductors. This indicates some power loss in watts. The power losses may be calculated easily by applying the power loss formula. The formula for finding the power loss is:

$I^2R = P$ Where I = circuit amperage
R = circuit resistance
P = circuit power loss in watts

Adequate cooling of electrical conductors is necessary. All conductors increase in electrical resistance when heated. As operating temperatures decrease, electrical resistance will decrease proportionately. Efficiency in electrical circuits increases when power losses are limited to low values.

Eddy Currents

By definition, an "eddy current" is a current that runs contrary to an established flow of something such as water, air, or electrical current. It may be called a counter-current or a contrary current. In electrical devices such as welding power source transformers, eddy currents are **induced currents.** They may be apparent in either electrical conductors or the iron cores of transformers or reactors.

Electrical eddy currents follow Lenz's Law which says, "an induced voltage is always opposite in directional force to the effect that produced it". An eddy current induced in an electrical conductor or iron core would have a voltage force opposite in direction to the general flow of current in the circuit at that point.

In an alternating current (AC) magnetic field, eddy currents may be set up in the body of an electrical conductor, increasing the effective resistance of the conductor. Eddy currents, like all electrical current flow, move at an angle which is ninety degrees from the direction of force of a magnetic field. This may be shown in the following example:

Left Hand Rule of Thumb = Hold the left hand in a loose fist, the fingers curled, the thumb extended. Consider that the thumb represents electrical current flow in the direction the thumb is pointing and, at the same time, the fingers represent the magnetic lines of force moving at right (90°) angles to the current flow. This rule is also called Flemings Law.

It is apparent that an eddy current induced in an electrical conductor would have force direction opposite to the direction of normal current flow. Since voltage is a force that causes current to flow in a conductor, there is the fact of circuit voltage pushing current in one direction and induced eddy current voltage holding back the flow of that same current. In this case, the eddy current acts as an impedance, or resistance, to the flow of current in the circuit.

The presence of eddy currents in electrical conductors will cause electrical energy to be expended as heat. Of course, as the electrical conductor increases in temperature, the electrical resistance to current flow also increases. That is why conductors used for carrying large amounts of amperage have larger cross sectional areas than conductors used for carrying lower amperage values.

Copper welding cables are designed with hundreds of small diameter copper wires encased in some type of flexible insulation such as neoprene or other similar material. The individual wires are film insulated and connected in parallel to minimize the formation of eddy currents.

The small diameter wires are twisted at regular intervals so that each conductor has a changing physical position with relation to the other conductors in the cable. The twisting of the conductor wires reduces the possibility of developing circulating eddy currents. The eddy currents would be caused by different induced voltages. This will result in lower electrical resistance in the conductor wires.

The induced eddy current flow would be within the individual conductors and would not be directly reduced by the transposition of the conductor wires. Eddy current value would be reduced by eliminating the circulating currents. Reduction of eddy currents will decrease the circuit losses expended as heat. This will improve the electrical efficiency of the total welding power source circuitry.

Transformer/Reactor Iron Core Data

Any iron core that is magnetized with alternating current (AC) must have continuous primary power supplied to maintain the magnetic field strength. Of course, there will always be some electrical inefficiency in inductive electrical devices such as transformers and reactors. The input primary KVA required for a specific iron core mass depends on the alternating current hertz (cycles per second frequency), the magnetic permeability of the iron core material, and the magnetic density at which the iron core material functions.

Iron cores used in welding power source transformers and reactors have a variety of shapes or configurations. Some common iron core shapes are shown in the illustration.

Iron Core Energy Losses

Energy losses in transformer and reactor iron cores are usually caused by either of two possible detrimental conditions. They are:

(1) **eddy currents** and (2) **hysteresis losses.**

Eddy Currents

Eddy currents are the result of induced voltage and current in iron cores. They are counter, or contrary, currents with force opposite in direction to the force which caused them.

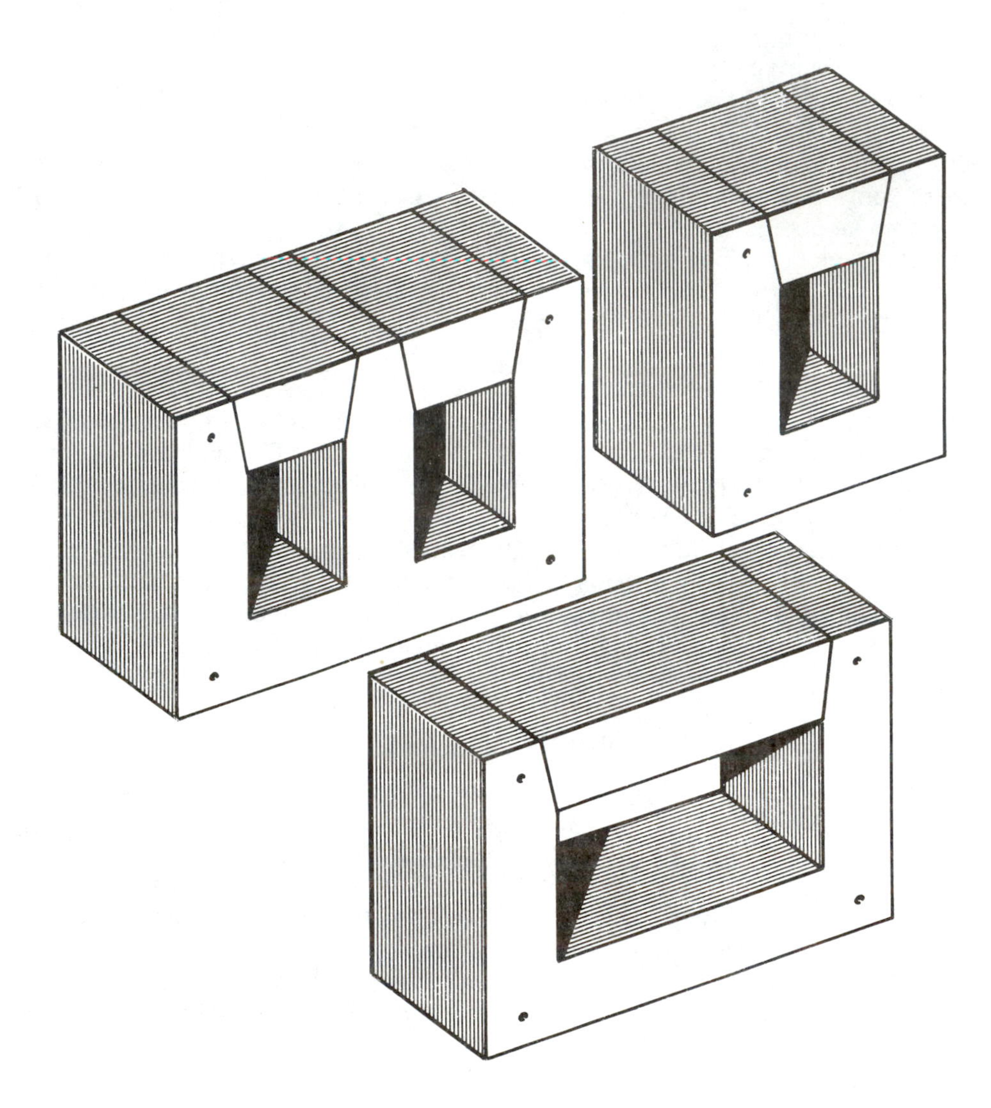

Figure 71. Some Typical Iron Core Shapes.

In iron cores, the losses from eddy currents will vary as the mathematical square of the iron core lamination thickness varies. For example, if a given thickness of iron core laminations produced an eddy current loss of 100 watts, reducing the iron core lamination thickness to one-half its original value will reduce the eddy current losses to one-fourth the original value. The equation for calculating eddy current loss is as follows:

$$\frac{E^2}{R} = \text{eddy current losses}$$

where E = circuit voltage
R = circuit resistance in ohms

Eddy current losses are almost eliminated in welding power source transformers by using thin gage laminations of sheet steel which is film insulated on both sides. The thicknesses are about 0.018''-0.020''. The possibility of forming eddy currents in material cross sections this thin are almost non-existent.

The drawing in Figure 72 shows the formation of eddy currents in an iron core that was solid rather than made of thin gage laminations. Such a solid iron core would not be usable for a welding power source transformer or reactor core.

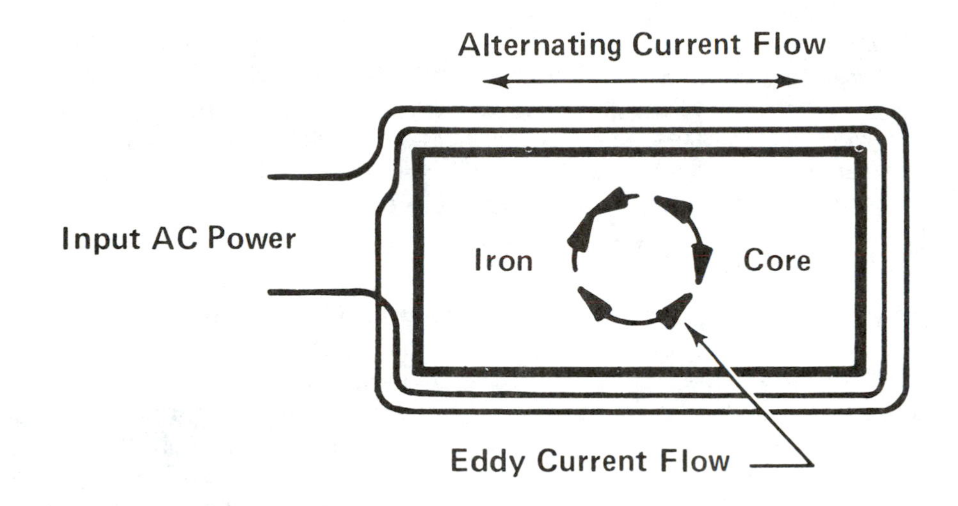

Figure 72. Eddy Current Formation in a Solid Iron Core.

The direction of flow of the primary alternating current (AC) in the primary coil is on the same geometric plane as the flow of the induced eddy current. For example, if the primary coil were wrapped around the iron core, the electrical turns of the coil conductor would be on a horizontal plane. The **magnetic lines of force** that constitute the **magnetic field** are always at right angles to the flow of current in a conductor.

The induced voltage and current (eddy current) will also be on the horizontal plane. The force direction of the eddy current will be opposite to that of the primary AC voltage.

The arrangement shown in the illustration would be excellent for an induction heater but terrible for a welding power source. There would be essentially no output from a power source designed as this illustration shows. Within a matter of minutes, as a matter of fact, there would be catastrophic deterioration of the circuit conductor insulation due to overheating of the iron core. This would be very noticeable by the smoke coming from the power source as the insulation burns.

Hysteresis Losses

In welding power source transformers and reactors, **hysteresis** means the loss of energy due to the cycling of alternating current power. Alternating current flows in the primary coil wrapped around a leg of the iron transformer core. The AC power normally has a frequency of 60 hertz (60 cycles per second). Current changes direction of flow each half-cycle (each 1/120 second).

Remember the left hand rule of thumb and apply it here. Note that, as current flow changes direction each 1/120 second (each half-cycle), so the **magnetic lines of force** must also change direction each half-cycle. The molecules of iron core are required to change polarity each time the magnetic lines of force change direction. It is the **reluctance of the iron molecules to change polarity** that produces what is called "hysteresis". Work is being done to overcome the reluctance and energy is being expended. The expended energy is dissipated as heat in the transformer or reactor iron core. In a sense, hysteresis may be defined as "magnetic friction" or "magnetic flux inertia".

Hysteresis losses may be substantially reduced by the use of special electrical steel for iron cores in welding power sources. Electrical steel with relatively high silicon content (about 3% Si) will provide increased magnetic permeability and decreased hysteresis losses. Even better is a newer electrical steel which is made by using special rolling techniques to control grain size, and orientation, in the steel. This type of steel provides low hysteresis losses without the high silicon content of other types of iron core materials.

Air Gaps in Iron Cores

It is well to keep in mind that any air gap between the iron core laminations will decrease the efficiency of the iron core significantly. It would be necessary to apply substantially more magnetizing current in order for the iron core to operate correctly.

Air acts as a very effective insulator for both electric current flow and magnetic lines of force in a magnetic field. Air has much lower magnetic permeability than iron core material used in welding power sources. This helps to explain why the sheet metal laminations of transformer and reactor iron cores are fitted extremely close together. Air gaps are used to control the output characteristics of certain iron core electrical devices in welding power sources. The air gaps are part of the design of the iron core. The air gaps may, or may not, be tapered to achieve their purpose. The use of tapered air gaps is especially important to certain types of power sources used for the GMAW process and the short circuit transfer technique of welding. They would be most useful for non-adjustable slope control reactors and inductor/stabilizer controls.

Welding Power Source Transformers

The simplest transformer type welding power source is the alternating current (AC) output unit. AC transformer power sources always operate from single phase primary power. This may be supplied as single phase primary power or it may be one phase of a three phase electrical system. The function of any transformer type welding power source is to take the relatively high voltage, low amperage primary power and change it to usable welding voltage and amperage. The maximum open circuit voltage allowed for welding power sources is specified in NEMA Standard EW-1, latest edition.

For AC output power sources, the maximum open circuit voltage is 80 volts. For welding power sources with DC output, the 80 volts maximum open circuit voltage is applicable if the unit has a ripple factor in excess of 10%. DC output power sources with less than 10% ripple factor may have open circuit voltage up to 100 volts maximum. The intent in limiting maximum open circuit voltage is safety for the welder in operating the welding power source.

To gain a proper perspective of an AC transformer type welding power source, each part of the transformer will be illustrated and its function explained. The results will show conclusively what each part does, why they do it, and how a transformer performs its function.

Basic AC Transformer Design

A basic AC transformer iron core design is shown in Figure 73. It is called the "closed U" shape. The single phase iron core has two vertical "legs" and a "key block" which is

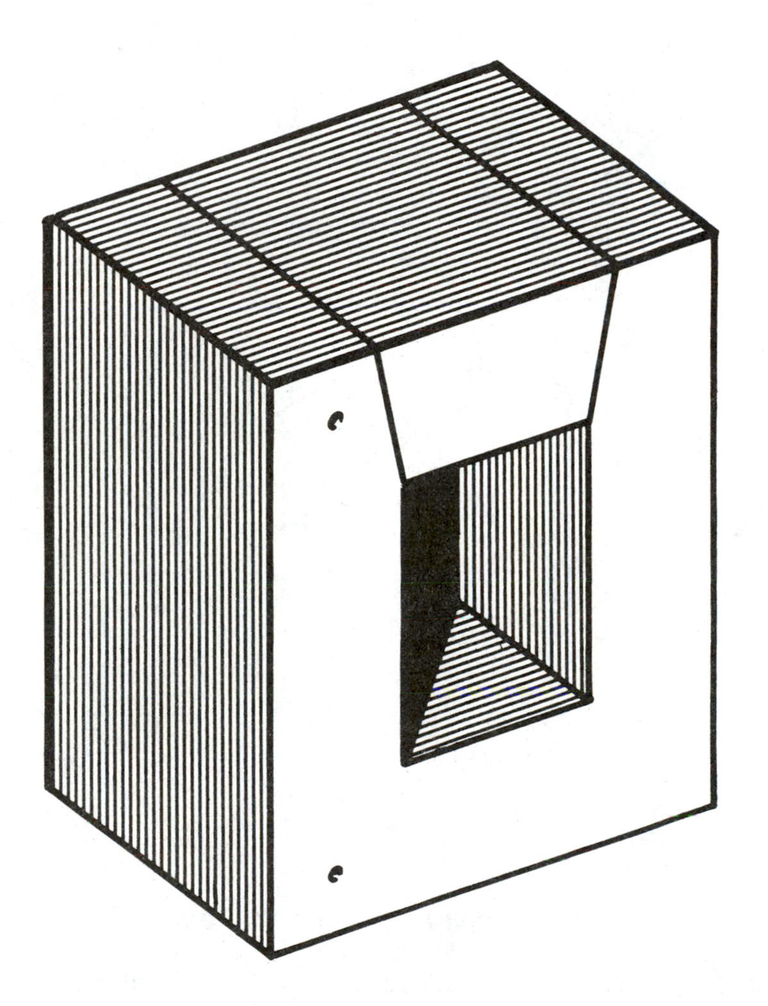

Figure 73. Closed "U" Iron Core Design.

put in place after the coils are assembled on the core. The laminated sheet steel material used in the core is very thin gage and is insulated on both sides of each lamination. The insulation is there to prevent the formation of eddy currents which would decrease the electrical efficiency of the transformer.

The iron core acts as an intermediate magnetic connecting link between the primary and secondary coils of the main transformer. The primary and secondary coils carry the actual current in the electrical circuits. The addition of a primary coil to the transformer is shown in Figure 74.

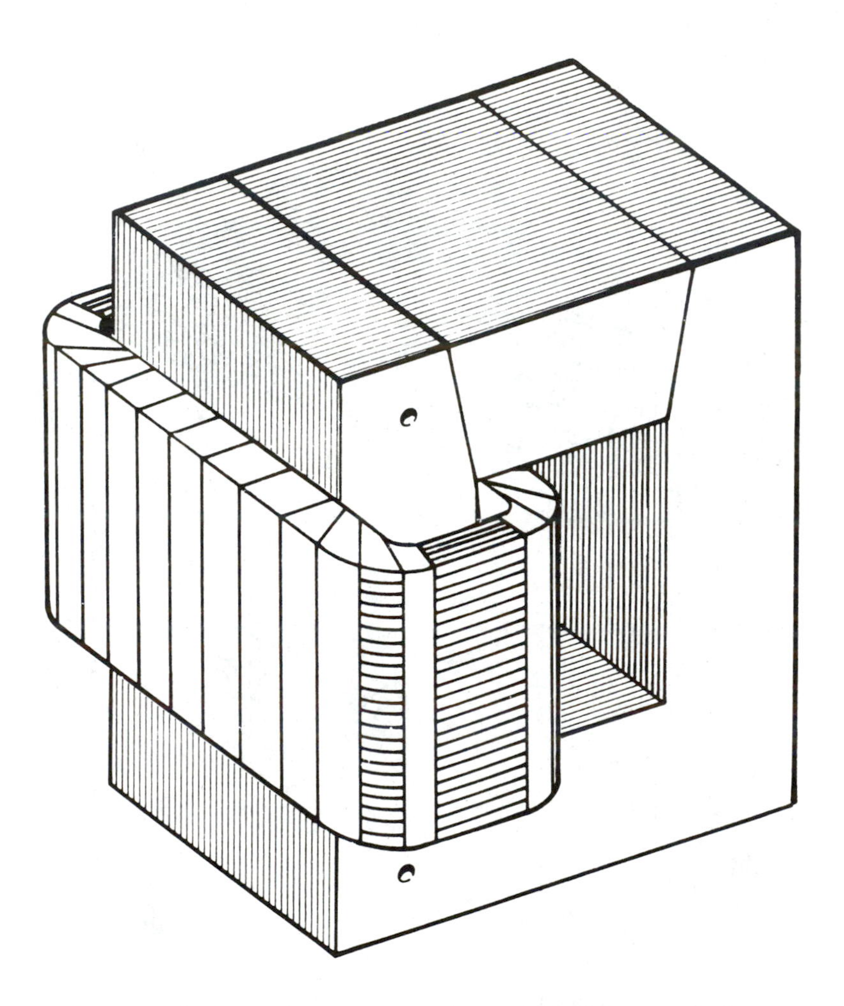

Figure 74. Transformer Iron Core With Primary Coil.

The conductor wire used in the primary coil is relatively small in cross sectional area. As you know, the conductor wire need only be large enough to carry the maximum circuit amperage. It is the primary coil which carries the relatively high voltage, low amperage primary power to the welding power source transformer.

At this point there are two of the three necessary components that make up a total welding power transformer. Figure 75 illustrates the third, and final, part of a welding transformer, the secondary coil.

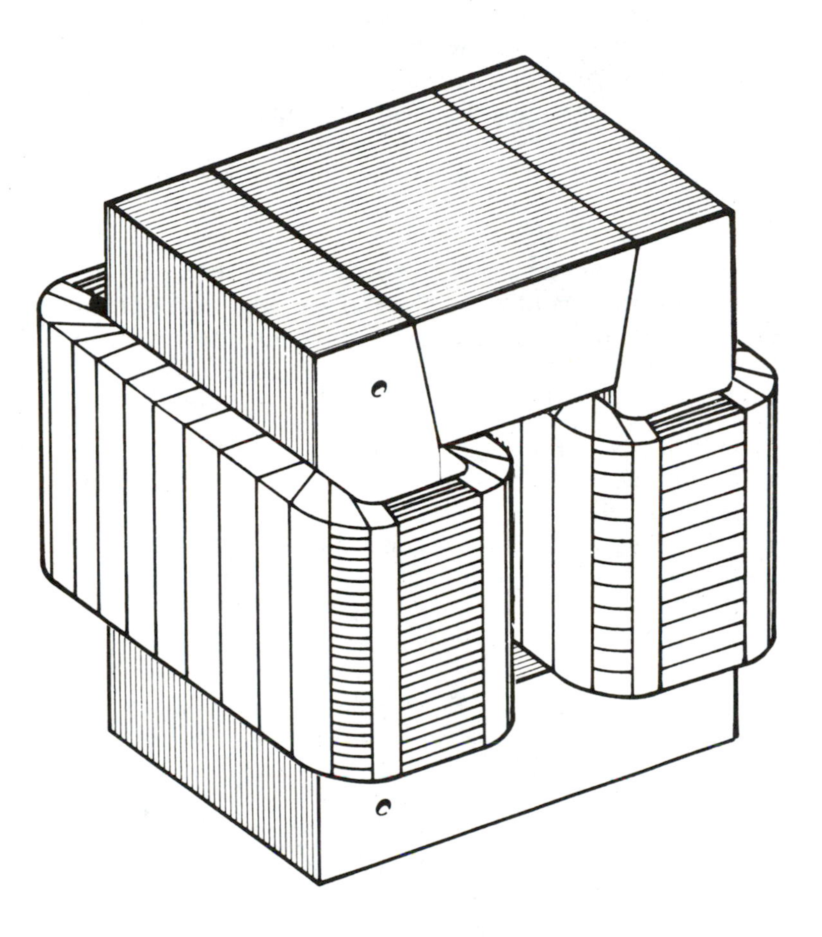

Figure 75. Simple Two Coil AC Transformer.

The secondary coil has fewer electrical turns than the primary coil. It is made of conductor material that is substantially larger in cross sectional area than the primary coil. The secondary coil carries the welding amperage, of course, which is considerably higher in value than the primary amperage.

The secondary coil is one of the two electric "fields", or circuits, that are joined by the iron core magnetic field circuit. Remember, we are referring to the **primary and secondary** coils as being the **electric fields** and the **iron core** of the transformer as being the **magnetic field.**

Electrical Symbols and Diagrams

It is not feasible to show electrical circuits in pictorial drawings because they would not show how the electrical circuits work. Technical electrical people have developed a series of electrical symbols that show what circuits do and how they are connected together. The two common types of diagrams are defined here for your information.

A **circuit diagram** will show the **electrical relationship** of welding power source parts without necessarily showing their physical relationship. The circuit diagram is also called a "schematic" diagram. It is the circuit diagram that is used in tracing electrical difficulties in circuits.

A **wiring diagram** will show the **physical relationship** of welding power source parts. This includes placement and location of terminal connections, relays, diodes, and all other electrical devices in the circuit. A wiring diagram is used for constructing electrical devices. The electrical symbols used by most welding power source manufacturers are the **American National Standards Institute (ANSI)** approved symbols. Some common electrical and electronic symbols are shown in a Data Chart in the Appendix of this text.

Welding Transformer Parts

The only reason for having a welding transformer power source is to change the primary power to usable welding power. The following illustrations show the relative parts of a welding transformer in a circuit diagram.

The first illustration (Figure 76) shows a coil symbol. The two lines attached to the coil are for the input AC power to enter and exit the circuit. The "Pri." indicates that this is primary power.

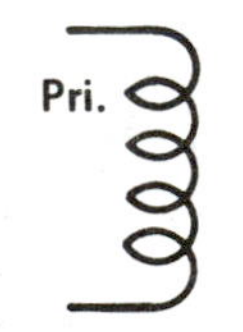

Figure 76. Transformer Primary Coil.

Figure 77 shows the addition of three straight lines to the drawing. This indicates an iron core of some type and shape. The fact that the iron core and primary coil symbols are so close together indicates that the primary coil is physically wrapped around some part of the iron core.

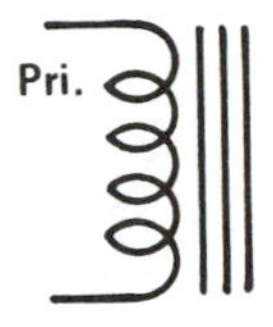

Figure 77. Transformer Pri. Coil, Iron Core

The next illustration (Figure 78) shows another coil symbol which is labeled "Sec.". This is the secondary coil of the transformer. The two lines going from the secondary coil carry the welding voltage and amperage to the power source output terminals.

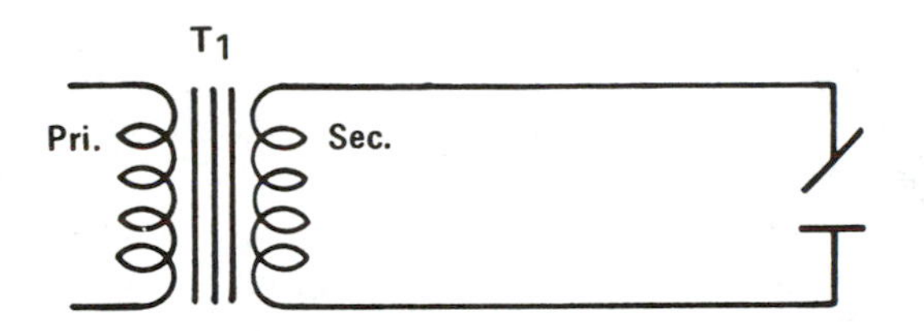

Figure 78. Complete AC Transformer.

Note that the secondary coil and iron core symbols are very close together. This indicates that the secondary coil is physically wrapped around some part of the transformer iron core. This illustration shows all the necessary parts that make up a welding transformer: the primary coil, the iron core, and the secondary coil. The electrical symbol is shown as "T_1" which is the total welding transformer.

What the Transformer Does

The next step is to add the electrical quantities of voltage, amperage, and electrical turns to the drawing. The voltage and amperage values refer to primary and secondary values. The "electrical turns" refer to the actual number of wraps of conductor wire around the individual coil. To define:

Electrical turn = "An electrical turn is one complete wrap of the conductor wire around the periphery of the specific coil". For example, a 10 turn coil would have 10 physical wraps of conductor wire around the internal dimensions of the coil.

It must be understood that the numbers used in the following examples are fictitious and do not refer to any known design of welding power source. Only the concepts of the transformer operation are to be examined.

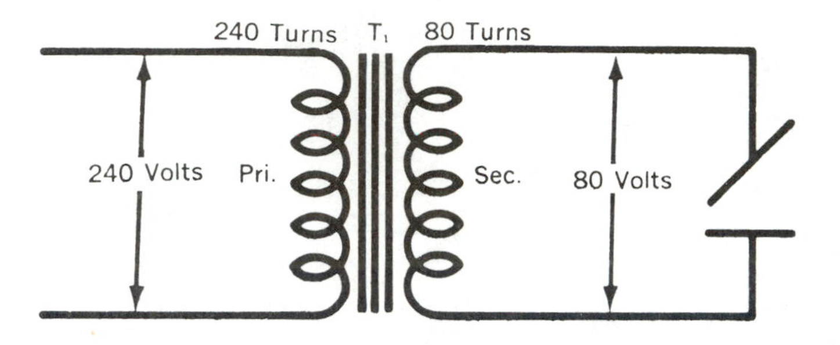

Figure 79. Voltage and Turns Ratios, Simple Transformer.

In Figure 79, the primary voltage is shown as 240 volts. The number of turns in the primary coil is shown as 240 turns (240T). Dividing the primary turns by the primary voltage gives the result of one volt per turn (1V/T).

It is a theoretical fact that, whatever the volts per turn of the primary coil of the transformer, the volts per turn of the secondary coil will be the same. Thus, the one volt per turn (1V/T) of the primary coil is the same for the secondary coil of the transformer.

The illustration in Figure 80 assumes the 1V/T of both the primary and the secondary coils. In addition, the secondary voltage is shown as 80 volts which is the maximum open circuit voltage allowed for AC output welding power sources. It is easy to figure the number of turns necessary for the secondary coil. At one volt per turn, it is necessary to have 80 turns in the secondary coil to provide the 80 volts open circuit.

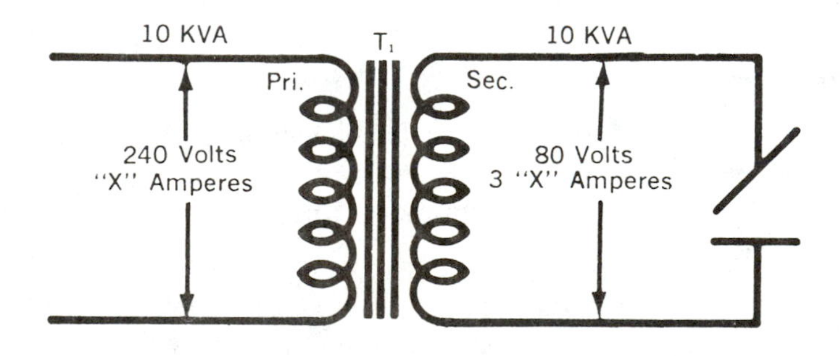

Figure 80. Amperes and Voltage Ratios, Simple Transformer.

Looking at the illustration, Figure 80, it is apparent that there is a ratio between the 240 primary, and 80 secondary, volts. Dividing the secondary voltage into the primary voltage results in a ratio of 3:1. The same is true of the primary and secondary electrical turns. It is possible therefore to develop a rule that states, "The ratio of primary to secondary volts in a welding transformer is the same as the ratio of primary to secondary electrical turns."

The amount of primary amperage demanded from the primary electrical service will depend on the amount of welding amperage drawn by the welding power source. This will vary with the output amperage rating of the power source and, because of this fact, an "x" is used to designate the primary amperage. This is shown in Figure 80. You know that voltage times amperage equals volt-amperes which, divided by 1,000, results in kilo-volt-amperes (KVA). This is the total primary power required to operate the welding power source at its rated amperage output.

If we discount the electrical losses through the transformer, we would have the same total KVA on the secondary side of the transformer as on the primary side. But there is only 1/3 the voltage (80 volts) on the secondary side. To obtain the same numerical KVA, there would have to be three (3) times the amperage on the secondary side as there is on the primary side of the transformer. This is shown in Figure 80 as "3x." The second part of the rule previously stated is, "The ratio of primary amperes to secondary amperes is inversely proportional to the voltage and turns ratio."

In summary of this portion of the discussion, the complete theoretical electrical rule for welding transformers is declared as follows: "The ratio of primary volts to secondary volts in a welding transformer will be same as the ratio of primary electrical turns to secondary electrical turns. The ratio of primary amperes to secondary amperes will be inversely proportional to the volts and turns ratio."

Welding power sources are called "step-down" transformers and it is apparent that what is stepped down is voltage. This, then, is what a transformer actually does.

In some electrical discussions of transformers, the term "winding" is often used. A winding is an electrical coil. The terms "winding" and "coil" are synonymous in their electrical usage. The term "winding" is a colloquial term. The term "coil" is technically correct.

How the Welding Transformer Functions

You have studied the theory of what a welding transformer does. By changing, or transforming, high voltage, low amperage primary power to usable welding voltage and amperage, the transformer provides relatively safe electrical welding power. Now we will examine how the electrical forces work to obtain that welding power.

The theory and practice involved here will also apply to the electrical control concept of welding power source output. These same concepts will be used later to describe some methods of variable "slope control" in constant potential/constant voltage welding power sources.

A circuit diagram similar to those previously used is shown in Figure 81. The same voltage and amperage figures are provided for the simple two coil transformer illustrated. One additional feature has been added. The induced secondary voltage is shown as being opposite in force direction from the primary voltage force direction. This is explained by the natural occurring phenomena called Lenz's Law.

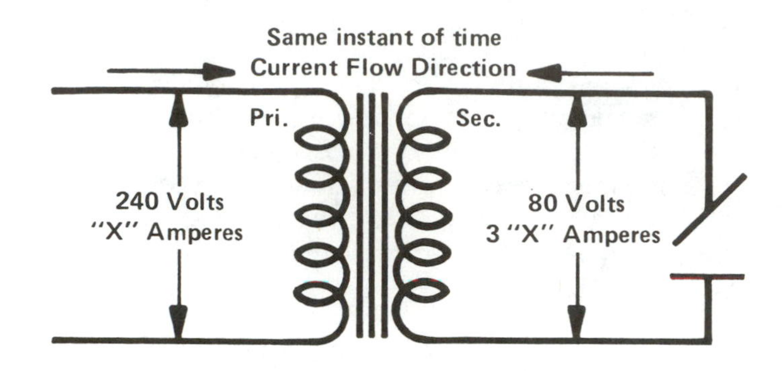

Figure 81. Welding Transformer Electrical Concepts.

Remember the basic rules already learned. Voltage is an electrical force that does not flow in a circuit. When there is sufficient voltage impressed on an electrical circuit, amperage will flow in the circuit. In the illustration, Figure 81, the primary and secondary coils are wrapped around some portion of the transformer iron core. The primary and secondary coils are considered to be the "electric fields" which are connected by the iron core, the "magnetic field".

Anytime there is amperage flowing in a coil wrapped around an iron core, as there is here, there will be a magnetic field created. The strength of the magnetic field will depend on three factors:

1. The mass and type of electrical steel in the transformer core.
2. The number of effective electrical turns in the coil.
3. The numerical value of the amperage flowing in the coil.

Changing any one of the three factors will also change the magnetic field strength.

In the welding transformer illustrated in Figure 81, consider that the mass and type of iron stay constant once the core is placed in the power source. The number of effective electrical turns in the coil will also be a constant factor once it is placed around the iron core. In this situation, the only variable factor is the alternating current amperage value flowing in the primary coil. This is shown by the **sine wave** form in Figure 82. It is evident that the strength of the magnetic field is dependent on the constantly changing amperage value in the primary coil.

The sine wave form is one complete cycle of alternating current. As indicated by the arrows, direction of current flow changes each half-cycle of alternating current. The straight line through the middle of the sine wave form provides a division point for the positive and negative half-cycles. For 60 cycle per second (60 hertz) power, each half-cycle will be 1/120th of a second.

The magnetic field of the transformer depends on the primary amperage for strength and direction. As the amperage increases each half-cycle, the magnetic field strength increases proportionately. The important thing to remember is this: **energy** is used to create

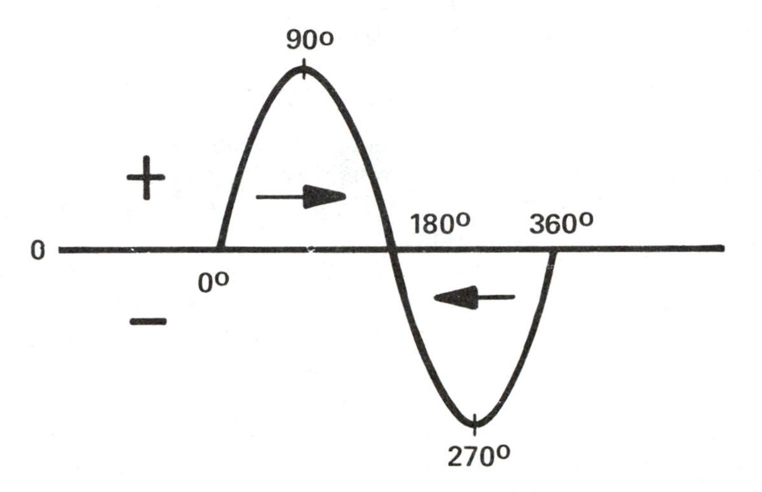

Figure 82. AC Sine Wave Form With Magnetic Field Data.

the magnetic field but it is not "used up". Instead, the energy is momentarily stored in the magnetic field until it reaches maximum strength and amperage.

At the points of maximum amperage (90 and 270 electrical degrees), the sine wave form shows that amperage immediately begins to decrease. Since the increase of amperage also increased the magnetic field strength, it makes sense that a decrease in amperage will also decrease the magnetic field strength. This is exactly what happens. It is called the "collapsing" of the magnetic field.

At the same time the amperage begins to decrease, and the magnetic field begins to collapse, the energy that is momentarily stored in the magnetic field has to go somewhere. There are three possible places where it can go. They are the primary coil, the transformer iron core, or the secondary coil. Let's examine what must happen electrically.

REMEMBER: Electrical power will always follow the path of least electrical resistance.

The **primary coil** is made of electrical conductor with relatively small cross sectional area. The coil already has primary current flowing in it and so would not be the path of least electrical resistance.

The **iron core** is the largest individual piece in the transformer and it should be the logical place for the energy to be dissipated. However, each insulated lamination of steel is only about 0.020" thick. This will not accommodate an induced current.

The last part of the transformer is, of course, the **secondary coil** which is made of electrical conductor with a relatively large cross sectional area. It has the least electrical resistance of any of the transformer parts. It is to the secondary coil that the electrical energy is induced from the magnetic field. The induced energy will be apparent as open circuit voltage at the output terminals of the welding power source. When an arc is initiated, both welding current and load voltage are available for the arc energy.

The information about what welding transformers do, and how they accomplish it electrically, is applicable to any type of welding transformer. Electrically, all welding transformer type power sources must function in the same manner. Any other devices in the power sources are some form of control apparatus. This includes main power rectifiers, stabilizer/inductors, reactors, and various solid state electronic components.

Welding Power Source Transformer Maintenance

Welding power source transformers are easy to maintain as part of the power source. Shop dust and dirt can accumulate on the transformer coils and iron core. This should be removed by either blowing the dust out of the unit with compressed air or it may require washing the transformer with an approved solvent.

If the dust and dirt is dry, an air blast will usually remove it easily. If there is substantial oil and/or other moisture in the atmosphere, the dust and dirt may not be removable with compressed air. Washing greasy dirt from the transformer should be done by trained personnel who are competent to use approved degreasing solvents.

Be watchful for any buildup of iron dust on transformers. This may occur in areas where there are heavy grinding activities. The iron dust can bridge across from the transformer coils to the transformer iron core. An induced current can create eddy currents in such a situation. This could lead to a short circuit between the coils and iron core. The result would be catastrophic failure of the welding transformer.

Welding Transformer Efficiency

The calculation of welding power source efficiency is based on the relationship of **primary kilowatts used** compared to **secondary kilowatts produced** for the welding arc. Constant current (CC) type welding power sources are intentionally designed to be electrically inefficient. The main reason is explained in the following paragraphs.

Constant current type welding power sources are specifically designed for the shielded metal arc welding (SMAW) process although they may be used with other welding processes. This type of power source transformer has "loose magnetic coupling" between the primary and secondary coils. The loose magnetic coupling is accomplished by physically separating the primary and secondary coils with an air space.

The electrical inefficiency of this design creates the negative volt-ampere output curve necessary for the SMAW process. The important thing to recognize is that the maximum short circuit current output of the welding power source is limited. This protects both the power source and the welder in case a dead short circuit condition should occur. This could be caused by the welder "sticking" a large diameter electrode to the workpiece or by a piece of metal shorting out the welding leads or terminals. Figure 83 illustrates a typical volt-ampere curve for a constant current power source. Note the limited maximum short circuit current.

An example for calculating welding power source electrical efficiency is provided in the following data. The power source selected conforms to the NEMA Class I constant current DC power source rating as follows:

300 amperes, 32 load volts, 60% duty cycle

The steps to arrive at the correct electrical efficiency are simple. First, multiply the secondary output 300 amperes by the 32 load volts. The correct answer is 9,600 watts.

Now divide the 9,600 watts by 1,000 to obtain the correct answer of 9.6 secondary kilowatts (KW).

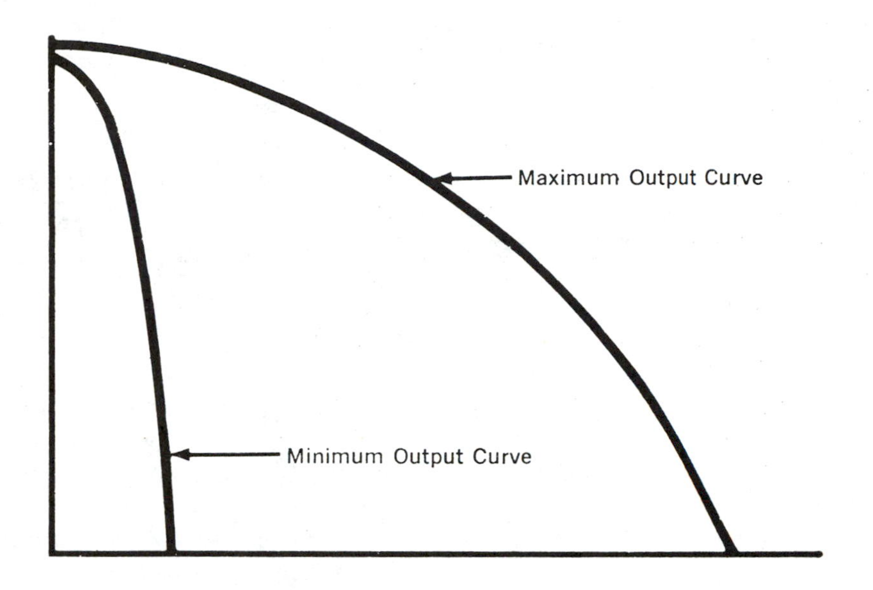

Figure 83. Typical Constant Current Volt-Ampere Curve.

The definition of primary kilowatts is as follows:

Primary kilowatts = the actual primary power used by a welding power source to produce its rated amperes at rated load voltage.

If the primary KW for this power source is 15.2 KW, we divide that number into the secondary KW 9.6 to obtain the electrical efficiency.

$$\frac{9.6 \text{ KW (secondary)}}{15.2 \text{ KW (primary)}} = 63\% \text{ electrical efficiency}$$

In summary, dividing the total output you obtain by the total input expended gives you the electrical efficiency!

CHAPTER 6

Power Factor

Any discussion of electrical power distribution invariably leads to the subject of power factor and power factor correction. Unfortunately, discussions about power factor are usually conducted in mathematical and engineering terms that are almost incomprehensible to the average person.

The term "power factor" usually is applied to alternating current (AC) primary power. It is the amount of electrical power used, expressed as a percentage, with reference to the total electrical power supplied. Power factor is especially important to single phase electrical systems.

Power factor correction is normally used only on welding power sources which operate from single phase primary power. This would include all AC and AC/DC output welding power sources and some power sources with DC output. It is not presently feasible to manufacture a single operator AC or AC/DC welding power source with a three phase transformer.

The basic function of power factor correction circuits in welding power sources is to reduce the amount of primary amperage demanded from the primary power supply. In this way, more efficient use is made of the power supplied by the utility company.

It is the intent of this short chapter to explain power factor and power factor correction in relatively simple electrical terms. To do this properly we will first provide some definitions of terms used in discussing power factor. The terms used are related only to power factor in this writing although they may have other meanings as well.

Kilo = 1,000 units. (One kilowatt = 1,000 watts).

Primary kilowatts = The actual primary power used by a welding power source to produce and deliver its rated amperage at rated load voltage (rated load). Usually measured by a wattmeter, primary kilowatts are the portion of total demand electrical power registered on a kilowatt-hour meter. Primary kilowatts are the power that can be metered and charged for by the electric utility company.

Primary kilovolt-amperes (KVA) = The total demand power that is taken from the primary supply system is termed kilovolt-amperes. For single phase power, KVA is determined by multiplying primary volts times primary amperes times the effective RMS value. The answer is divided by 1,000 to obtain kilovolt-amperes (KVA). The following illustration shows the steps and mathematics involved.

a. primary line voltage × primary amperage = **volt-amperes**
b. volt-amperes × effective RMS value = **net volt-amperes**
c. net volt-amperes divided by 1,000 = Kilovolt-Amperes (KVA).

For three phase primary electrical power, KVA is determined by multiplying primary volts times primary amperes times the effective RMS value times 1.73 (the square root of 3). The product is divided by 1,000 to obtain KVA. It looks like this as an equation:

$$\frac{\text{Primary Volts} \times \text{Primary Amperes} \times \text{RMS } (0.707) \times 1.73}{1{,}000} = \text{KVA}$$

Vector = Direction and magnitude values based on a stable reference point.

Vector sum = The sum obtained in vector addition.

Apparent power = The vector sum of the real power and the kilovars is the apparent power. (Apparent power = KVA).

Kilovars = Kilovars may be either inductive or capacitive. The abbreviation for kilovars is KVAR. **Inductive KVAR** always has current lagging voltage by up to 90 electrical degrees (Figure 84).

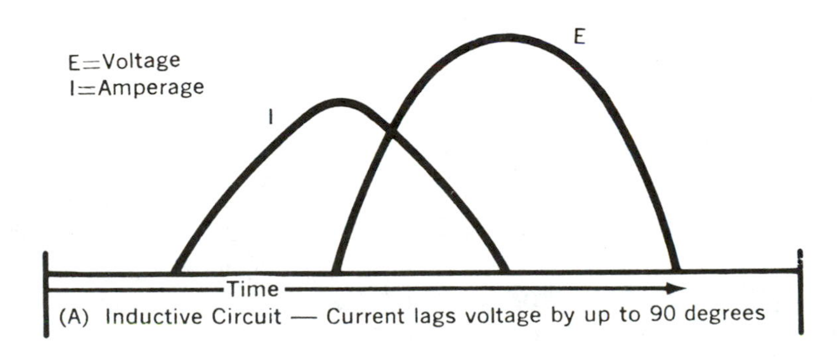

Figure 84. Inductive Circuit KVAR.

Capacitive KVAR is illustrated in Figure 85. Note that current leads voltage by 90 electrical degrees.

Kilovars are the reactive, or "wattless", component in electrical systems. The magnetizing current required to produce the magnetic field necessary for the operation of any inductive equipment is termed inductive KVAR. Inductive parts would include welding transformers and reactors. Without a small amount of inductive KVAR, energy could not transform through the iron core of the main transformer. Specific KVAR values are determined by a wattmeter which has a special phasing transformer. This type of calculation is done in development laboratories of welding power source manufacturers.

It can be stated that the unit of measure of **real,** or used, power is the primary kilowatt. The unit of measure for **total,** or apparent, power is the kilovolt-ampere.

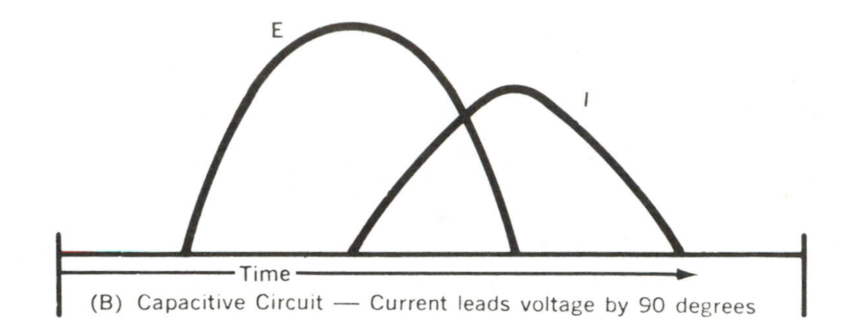

Figure 85. Capacitive Circuit KVAR.

This brief discussion of power factor relates to its determination and use in the primary circuitry of welding power sources. The power factor of electrical distribution circuits in a shop or plant will certainly be affected by the shop power loading of those circuits.

Probably the most asked question regarding power factor correction is, "Why do we need it?"

There are several pertinent and correct answers to the question. Some of the benefits of power factor correction are for the customer-user of electrical power and some benefit the electrical utility company. In other ways, power factor correction is beneficial to both the customer and the utility company.

Some of the benefits, and who they help most, are listed as follows:

BETTER PRIMARY VOLTAGE REGULATION (Utility company).
INCREASED PLANT ELECTRICAL SYSTEM CAPACITY (customer).
SMALLER LINE TRANSFORMER REQUIREMENT (Utility and customer).
REDUCED ELECTRICAL SYSTEM LOSSES (Utility and customer).
REDUCED PRIMARY AMPERAGE DRAW (Customer).

Of the various benefits of power factor correction, probably the most important to the welding power source user is **reduced primary amperage draw.**

Another question often heard is, "Just what is power factor and power factor correction?"

Power factor is the quotient, expressed as a percentage, of kilovolt-amperes divided into primary kilowatts.

$$\frac{\text{KW (primary)}}{\text{KVA}} = \text{\% Power Factor}$$

Power factor correction is a system of capacitors and conductors in the primary portion of the welding power source circuit that reduces the amperage value demanded of the primary power supply system. Shop primary wiring carries the total power drawn from the primary distribution system of the utility company. The shop wiring would naturally have to carry less primary current when power factor corrected welding power sources are used.

As a general rule, power factor correction capacitors and allied circuitry are used only in AC and AC/DC power sources. AC and AC/DC welding power sources always operate from single phase primary electrical power.

Welding power sources having DC output only normally operate from three phase primary power. DC welding power sources have inherently good power factor and do not require power factor correction.

Power Factor Correction Analysis

The step-by-step analysis following is for a typical NEMA Class 1 AC transformer welding power source. All data is based on published literature from the manufacturer.

The primary voltage is supplied by the utility company and is considered to be constant in value. Primary amperage is the amperage drawn by the transformer power source when it is producing its rated load in volts and amperes. The primary power specifications for the welding power source are shown in Figure 86. Remember, only the primary input electrical circuit is affected by power factor correction.

Primary Amperage	=	106
Primary Voltage	=	230
Primary KVA	=	24.4
Primary KW	=	15.2
Single Phase	=	60 Hertz

Figure 86. AC Power Source Primary Specifications.

Step 1.

Primary KVA is determined by multiplying primary voltage times primary amperage to obtain volt-amperes. Dividing the volt-amperes by 1,000 (a kilo) produces kilovolt-amperes (KVA). This is illustrated as follows:

$$\text{Primary voltage} \times \text{primary amperage} = \text{volt-amperes}$$

$$(230) \times (106) = (24{,}400)$$

$$\frac{24{,}400}{1{,}000} = 24.4 \text{ KVA}$$

Step 2.

Primary kilowatts are measured by a special wattmeter. The measurement is made in the manufacturers laboratory while the welding power source is under a static electrical load producing rated amperes at rated load voltage. Primary kilowatts for this power source is 15.2 KW. It is always primary kilowatts that are used in calculating power factor in a welding power source.

Step 3.

Normal welding industry practice is to power factor correct welding power sources to 75% at rated amperage and voltage output. Small variations of this value, either plus or minus, will be noted in actual practice due to the increment value of power factor correction capacitors.

The vector diagram, Figure 87, illustrates a method for determining the necessary power factor correction for this NEMA Class 1 AC power source. The vector diagram uses the

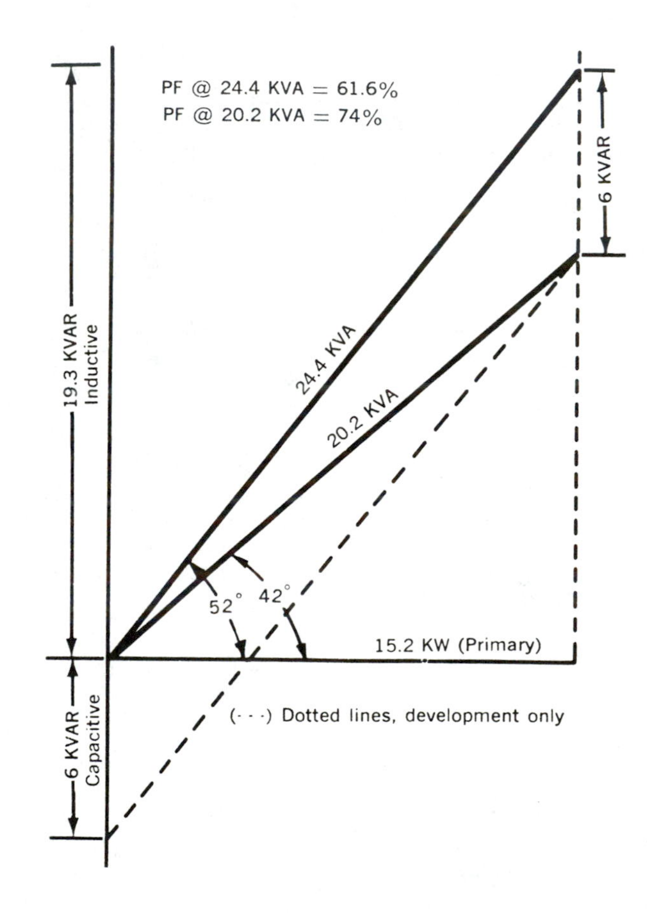

Figure 87. Vector Diagram for Power Factor Determination, NEMA Class 1 AC Welding Power Source.

same numerical measurement value to equal one KVA, one KW, or one KVAR. The following data are noted in the vector diagram:

	W/O PFC	W/PFC
Primary amperage (I)	106	88
Primary voltage (E)	230	230
Primary KVA	24.4	20.2
Primary KW	15.2	15.2

To summarize this discussion, power factor correction is a method of making an electrical distribution system more efficient. Power factor and power factor correction concerns only the power drawn from the primary supply by a welding power source. The secondary output of a welding power source is not affected by power factor correction.

Power factor correction for any electrical device, such as a transformer welding power source, reduces the primary amperage draw. It also reduces the primary KVA demand from the utility company power lines. The total current demand from the primary power supply system must be carried through the shop electrical circuits. Conductors used in an electrical system with good power factor correction can usually be smaller in cross sectional area than conductors in a similar system without power factor correction.

It is significant that in some areas the unit cost of electrical power in kilowatt-hours is less where a high average power factor is maintained.

The main advantages of power factor correction are in reduced electrical system losses, increased electrical system capacity, smaller line transformer requirements, and reduced primary amperage demand from the primary supply system.

Remember: Power factor correction is not line voltage compensation for fluctuating primary line voltage!

CHAPTER 7

Shielded Metal Arc Welding

The shielded metal arc welding (SMAW) process is probably better known to welders as the "stick electrode" process. This is because there has been very little effort to communicate correct terminology to welding people working at the trade. The process name breaks down like this:

Shielded = This word comes from the dry flux covering on the metal electrode. The flux covering which decomposes under the heat of the arc creates both a shielding gas (basically CO_2) and a slag covering for the deposited weld metal. The weld is "shielded" by these effects.

Metal Arc = The electrode is consumed under the heat of the arc and flows into the weld joint as filler metal.

Welding = The fusion that takes place between the molten filler metal from the electrode and the molten base metal is called "fusion welding".

The basic equipment required for shielded metal arc welding includes an electric arc welding power source with either AC or DC output, electrode and work welding cables, a suitable ground clamp device on the work lead, and an electrode holder with suitable amperage rating. The personal working equipment of the welder will include a welding helmet with properly shaded lens, leather gloves, a wire brush, a chipping hammer, hardened safety glasses and such other protective clothing as the job requires.

Welder

A welder is a person who knows how to join metals with one or more welding processes. A welder is the person who makes a weld. A professional welder observes all safety rules. The person who see's and controls the flow of molten metal into a weld is called a welder. The professional welder has great responsibility in creating a safe world of work.

Safety for Arc Welding

The controlling safety standard for arc welding is published by the American Welding Society (AWS). It is Standard Z49.1 latest edition, "Safety In Welding And Cutting". The Z49.1 Standard has all the necessary safety information for both arc and gas welding.

Some general, common-sense safety rules are listed here for your information. These come from experienced welders who are concerned with the safety of their fellow workers as well as themselves. Read and heed!

1. All primary electrical power to a welding power source must come through a fused line disconnect switch. There are no exceptions to this rule!
2. It is essential to work in a safe manner while welding. Wear hardened safety glasses with side shields. Eyes cannot be replaced.
3. Clothing should be closely woven wool or cotton fabric. Synthetic fabrics must never be worn when welding or cutting metals. Wear long sleeved shirts buttoned at the wrists. Leather jackets, aprons, and gloves will help protect you from arc burns and radiation.
4. A welding helmet with the proper protective filter lens must be worn for arc welding with any arc process. The helmet should protect your eyes, ears, neck and face from arc burns and radiation.
5. Always be sure there is adequate ventilation when welding or brazing. Welding fumes, and concentrations of certain shielding gases, can be hazardous to your health.
6. Welding power sources must be placed in a well ventilated area with the fan discharge not closer than 18 inches from any wall or obstruction. The cooling air must circulate over the power source components to cool them properly. **Never use any form of filter on welding power source air intake or air discharge openings.**
7. All electrical connections shall be clean and tight. Electrode holders shall be cleaned of all spatter before welding. Ground clamps will be of adequate amperage rating.
8. **Never strike an arc on a compressed gas cylinder of any kind.** The side wall of the cylinder will be weakened by the electrical contact. This could result in an explosion with possible injury or death. If an accidental arc contact is made on a cylinder, remove the cylinder from the shop immediately. Call the gas supplier and report the problem.
9. All metal in a welding shop should be considered "hot". When you have proven otherwise it is safe to handle it. Many people have been burned severely for picking up what they thought was cold metal.
10. High voltage is used to power welding power sources and other shop electrical equipment. Always be sure the fused line disconnect switch is in the off position before working on any part of welding power source circuitry. Electrical power is very unforgiving.
11. Be sure that all flammable materials are removed from the welding area before any welding or cutting operation. Pressurized spray cans shall never be kept in the arc welding area. They are potential bombs that can seriously injure or kill people.
12. Never change polarity switches or range switches when the power source is being used for welding. Serious injury could result.
13. Never wear cuffed trousers while welding or cutting. Sparks and molten metal spatter can fall in the cuffs and cause fire and injury. Always wear high top shoes or boots when welding or cutting. Keep in mind that metal-burned skin heals very slowly.

14. The welding arc supplies both heat and light to the work area. The heat is for welding, of course, and the light is only incidental. The light from a welding arc can seriously damage unprotected eyes. Never look directly at a welding arc without proper eye shielding.

There are many other safety rules that could be written here but the most important thing to remember is to think before you act. Use your good common sense and don't take part in any horseplay in the shop. Always vent enclosed areas so that adequate ventilation is present for the workmen in those areas.

Above all, don't weld on containers that have held combustible materials, such as gas tanks, without first cleaning them out. This should be done according to the American Welding Society Standard "Cutting and Welding of Containers Which Have Held Combustible Materials". Be safe in all welding operations.

Fusion Arc Welding Carbon Steel

There are four specific types of carbon steels. They are: (1) low carbon steel, (2) mild steel, (3) medium carbon steel, and (4) high carbon steel. For most welding applications it is low carbon and mild steel that is used. Mild steel and low carbon steel are not hardenable by heat treatment.

It is the amount of carbon (C) in steel that determines its ability to be hardened by heat treatment. The higher the carbon content, the easier it is to harden by heat treatment. On the other hand, the higher the carbon content the harder it is to weld the steel.

Carbon content for the carbon steels is approximately as follows:

Low carbon steel	=	up to 0.15% C approximately
Mild steel	=	0.15%-0.29% C "
Medium carbon steel	=	0.30%-0.59% C "
High carbon steel	=	0.60%-1.50% C "

Medium, and high, carbon steels are more difficult to weld because they can develop hard spots in the weld deposit and the heat-affected zone (HAZ). These hard spots can make the weld zone brittle and subject to cracking.

Fusion welding is a method of joining metals by controlled application of heat. Controlled application of heat also helps decrease distortion of the parts being welded. The means of calculating heat input to the weld is by use of the mathematical equation for joules of heat input per linear inch of weld. This equation is shown as follows:

$$\text{H (joules per linear inch)} = \frac{\text{E (volts)} \times \text{I (amperes)} \times 60}{\text{S (speed in inches per minute)}}$$

All that is required is to substitute the actual numbers for the letters. For example, consider that you are welding with a 1/8" diameter E6010 mild steel electrode. Welding amperes are 105, load voltage 25, and speed of travel is 3.5 inches per minute (IPM). Using the equation shown, the problem would look like this:

$$\text{H (joules)} = \frac{25 \times 105 \times 60}{3.5 \text{ IPM}}$$

$$\text{H} = 45{,}000 \text{ joules per linear inch.}$$

SMAW Arc Initiation Methods

There are two basic methods of initiating the SMAW electric arc. They are called the "tap" method and the "scratch", or "brush", method. Both methods of arc initiation should be practiced until the welder can easily start the arc without sticking the electrode to the base metal.

1. The Tap Method

The electrode is inserted into the electrode holder and gripped firmly. The electrode is brought to the base metal in a position vertical to the surface of the base metal. The tip end of the electrode is tapped against the surface of the base metal. As the electrode is withdrawn an electric arc is established. The electrode tip is then adjusted to the correct arc length. This is normally about the same length as the diameter of the electrode core wire. The tap arc start method is illustrated in Figure 88.

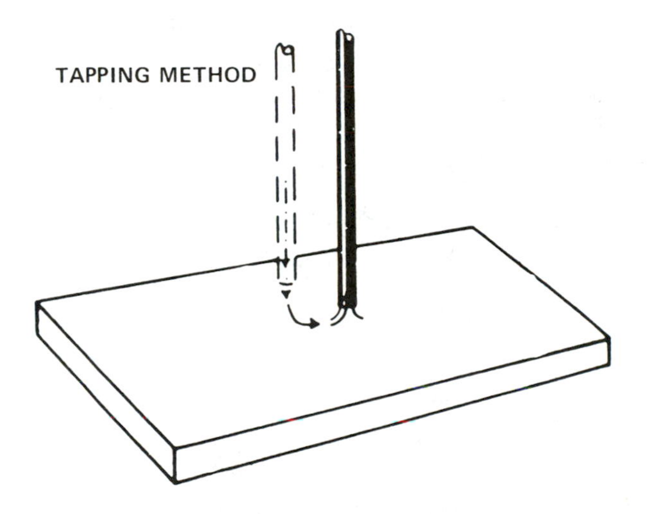

Figure 88. Tap Arc Start Method

2. The Scratch/Brush Method

The electrode is inserted into the electrode holder and gripped firmly. The electrode tip is brought to the base metal by brushing, or scratching, it along the surface of the base metal. At some point there is sufficient contact between the electrode tip and the base metal so that an electric arc is initiated. At that time the electrode tip is withdrawn from the base metal surface to the correct arc length. The scratch/brush arc start method is illustrated in Figure 89.

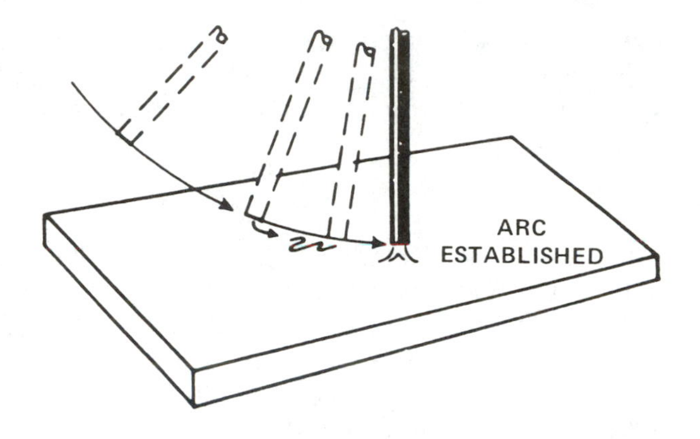

Figure 89. Brush/Scratch Arc Start Method.

To practice arc initiation it is a good idea to get the arc started and then weld for an inch or two along a straight line. It is good practice to stop forward progress and practice "crater filling" the weld crater. Crater filling is done by stopping weld travel with the arc still active, then moving the electrode tip back over the last 1/2" or so of weld. This will allow molten filler metal to flow into the weld crater and fill the cavity. The welding arc is then broken by a quick movement of the electrode tip away from the base metal weld joint.

The purpose of crater filling at the end of every weld is to eliminate cracking in the weld crater. The illustration in Figure 90 shows a typical crater filling movement for SMAW.

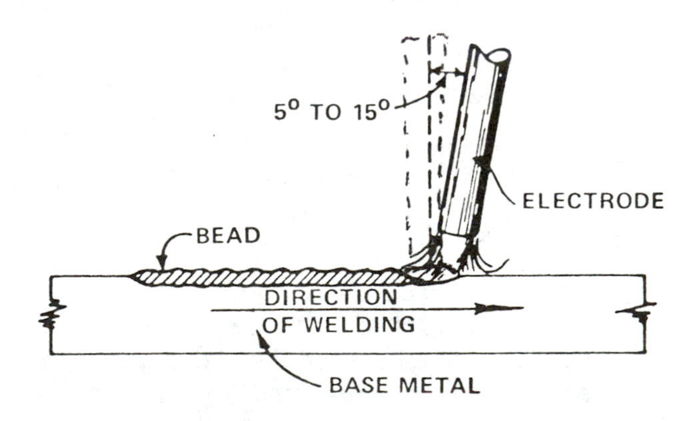

Figure 90. SMAW Crater Filling Technique.

Welding Nomenclature

There are a variety of names given to specific locations in weld joints and in deposited welds. Some of the common terms used with groove weld joints are shown in the illustration.

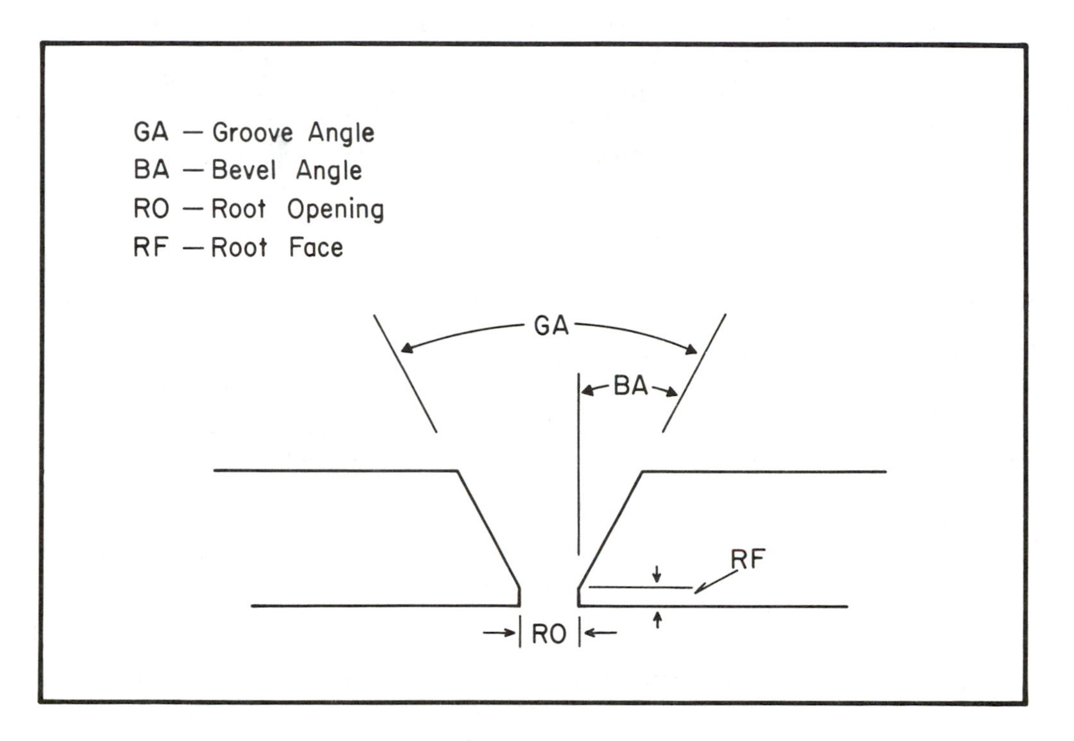

Figure 91. Weld Joint And Weld Terminology.

SMAW Electrodes

The first metal electrodes used for arc welding did not have any type of flux coating. It was found that better arc stability could be achieved with some type of covering (even rust) on the electrodes. This led to the development of very thin wash coatings for the electrodes.

Eventually electrodes were developed which had extruded flux coatings. Although the idea of coating electrodes was patented in 1907 (Sweden), it was not until the 1920's that extruded coated electrodes were commercially available. For some metals, notably aluminum, dipped coatings were still popular until the late 1950's.

With the shielded metal arc welding (SMAW) process heat is produced between the tip of the electrode and the surface of the base metal workpiece. Shielding of the electrode tip and the molten weld puddle is accomplished by the decomposition of the flux coating of the electrode under the heat of the arc. It was the analysis of the atmosphere around the welding arc that showed substantial amounts of carbon dioxide (CO_2) in the arc area. From this information has come the use of CO_2 for the GMAW and FCAW processses with carbon steels.

The American Welding Society (AWS) publishes the U.S. Specifications for all welding filler metals. The various AWS covered electrode Specifications for the SMAW process are as follows:

A5.1 Specification for Carbon Steel Covered Arc Welding Electrodes.
A5.4 Specification for Corrosion-Resisting Chromium and Chromium-Nickel Steel Covered Welding Electrodes.
A5.5 Specification for Low-Alloy Steel Covered Arc Welding Electrodes.
A5.6 Specification for Copper and Copper Alloy Covered Electrodes.
A5.11 Specification for Nickel and Nickel-Alloy Covered Welding Electrodes.
A5.15 Specification for Welding Rods and Covered Electrodes for Welding Cast Iron.

The AWS Specifications for filler metals should be available to the welders as well as the welding Supervisors. In all of the Specifications there is an Appendix. In this portion of the publication, AWS explains informally many aspects of the specification that cannot be properly discussed in the specification itself.

The Appendix is a wealth of information to the welder. It explains how the electrodes are classified. It may tell how to condition electrodes, which electrodes need to be maintained in temperature controlled ovens, the temperatures to use, how long such electrodes can be outside the oven before they are no longer usable, and many other facts. This information is particularly of value when considering the E7015, E7016, E7018, E7028 and E7048 low hydrogen type electrodes.

In the section titled "Description and Intended Use of Electrodes", you will find each of the electrode classifications listed. In most cases the type of flux is indicated and the penetration capabilities are listed. Iron powder additions and/or low hydrogen characteristics are explained as required.

In the case of high alloy electrodes which are classified by composition, the major alloys are listed by percentage. Any unique qualities of the electrodes are pointed out for each classification.

It is highly recommended that each welding shop have available those AWS filler metal specifications that are pertinent to the types of welding being done in the shop. In all purchases of electrodes or other filler metals, the purchase order should always show the AWS Specification and the electrode AWS Classification. The purchaser should always ask for "Certificates of Conformance" for each lot and type of electrodes ordered.

The Electrode and the Arc

Once the welding arc is initiated, the tip of the electrode begins to melt. The molten metal from the electrode is moved across the arc by the voltage force in the electrical circuit. The molten electrode filler metal fuses with the molten base metal in the weld joint to form the weld puddle. As the weld progresses, the deposited weld metal cools and solidifies. This is the fusion weld described earlier.

The illustration, Figure 92, is an example of what is happening in the shielded metal arc welding process area. The depth of penetration is relatively shallow. Actual penetration will depend on the electrode AWS Classification and how it is applied. For example, the E6010 electrode classification has the deepest penetration of any SMAW electrode. Yet this mild

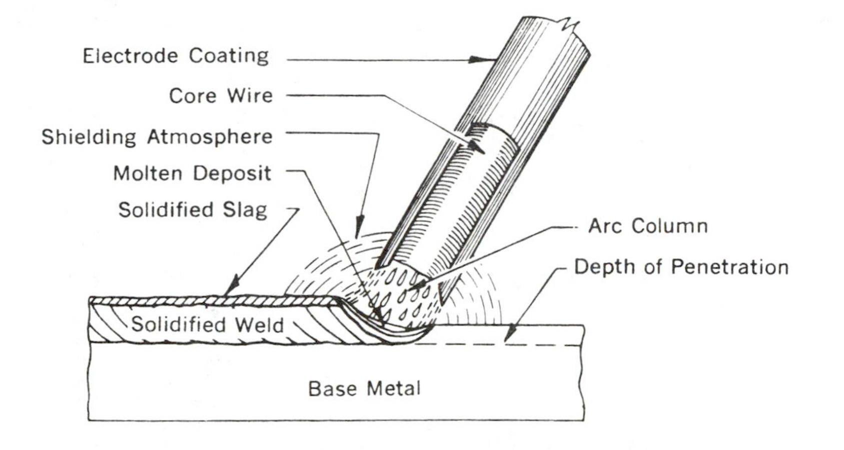

Figure 92. Shielded Metal Arc Welding.

steel electrode, operating DCRP (electrode positive), will have a maximum penetration into the base metal of about 1/8''. It makes sense that other electrodes used with the SMAW process will have substantially less penetration characteristics.

The minimum penetration characteristics of the SMAW process explain why the groove joint designs used with the process have such wide open angles. It also explains why the root face (sometimes called the ''land'') is seldom more than 1/16'' wide. The electrode could not penetrate the joint and produce a weld underbead dropthrough.

The electrode flux covering performs several functions when it is melted and thermally decomposed under the heat of the welding arc. It produces a shielding gas (basically carbon dioxide) that excludes the atmosphere from the weld area. The heat promotes electrical conductivity across the welding arc column by ionizing the developed shielding gas (CO_2). Slag-forming materials are added to the molten weld puddle to help in grain refinement and, in some cases, to provide alloy additions to the weld metal.

The slag residue that is formed over the weld metal puddle helps prevent oxidation of the weld deposit while it is cooling to the solid state. The slag is usually easy to remove once it is cooled and solidified. Various tools including slag hammers, multineedle guns, and wire brushes are used for slag removal.

The electrode core wire may, or may not, be similar in chemical content to the base metal being welded. In the case of some low alloy steels, for example, the core wire may be plain low carbon steel. The alloying and deoxidizing elements needed are contained in the flux covering of the SMAW electrode. The resulting mixture of deposited weld metal and base metal will normally produce metallurgical and physical characteristics similar to the original base metal.

The melting rate, sometimes referred to as the ''burnoff rate'', of the electrode is directly related to the amount of heat energy (electrical energy) in the welding arc. The arc energy is divided with a percentage used to heat the base metal and a percentage used to heat and melt the welding electrode. The electrical polarity (when using DC welding power) determines where the percentages of heat in the welding arc are located. Figure 93 shows the approximate thermal (heat) energy distribution in the welding arc. For AC welding power, the heat distribution is approximately 50% positive and 50% negative.

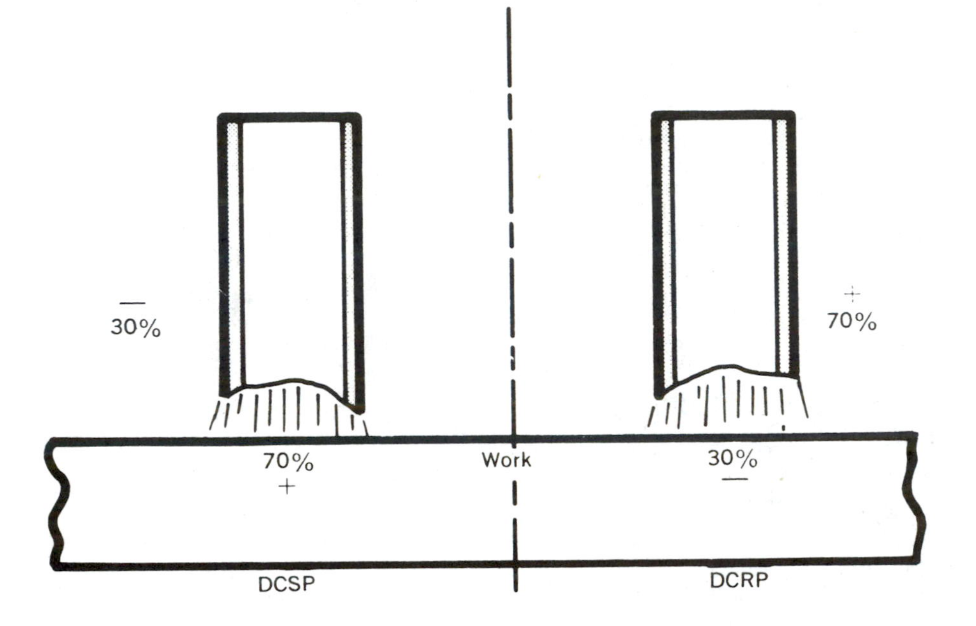

Figure 93. Heat Distribution in DC Arc Welding.

In most applications the deeper penetrating arc obtained with direct current, reverse polarity (electrode positive), is preferred where penetration and out-of-position welding capability are required. Typical of this type electrode are the E6010 and E7018 AWS classifications.

Some welding applications require fast welding speeds with relatively high deposition rates. Direct current, straight polarity (electrode negative), is normally specified for these applications. Much of this type of welding is done in the downhand (flat) or horizontal fillet position because of the high fluidity of the welding puddle. AWS electrode classifications typical of this type are E7014 and E7024. Many of the electrodes used with DCSP (DCEN) have substantial percentages of iron powder in their coatings.

Alternating current (AC) has been used widely with the shielded metal arc welding process. One of the outstanding advantages of AC arc welding is the almost total elimination of magnetic arc blow. Magnetic arc blow is caused by direct current (DC) magnetic fields set up in the workpiece because of welding current flowing in the circuit. Magnetic arc blow will cause the welding arc to move wildly about and, in some severe cases, has actually caused large amounts of molten weld metal to be expelled from the weld puddle.

The E6011 classification electrode is designed for either AC or DCRP welding current. It has essentially the same welding characteristics as the E6010 classification. The reason that the E6011 electrode can be used with AC, and the E6010 electrode cannot, is that the E6011 has some potassium in the electrode coating. Potassium is an ionizer and maintains the ionized condition of the gas around the arc area through the arc outage time each half-cycle with AC.

Electrode Angles Used With SMAW

There are a variety of names given to the electrode angles used with the arc welding processes. If you consider the direction in which the electrode is pointed while welding, the terms explained here will make some logical sense.

Lead Angle (Forehand)

The electrode tip is pointed in the direction of weld travel when using the lead angle, or forehand, technique of welding. The heat of the welding arc precedes the weld puddle. This permits faster welding speeds but with less penetration of the weld into the base metal. This technique of welding has also been referred to as the "push" angle.

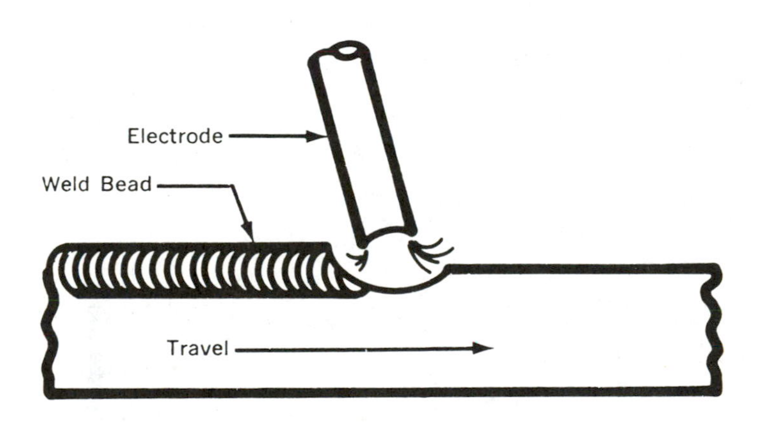

Figure 94. Lead Angle (Forehand) of Electrode.

Lag Angle (Backhand)

When using the lag angle, or backhand, technique the welding electrode tip is pointed in the direction of the weld puddle and away from the direction of weld travel. This puts more heat in the weld puddle and provides deeper penetration into the base metal. The welding speed is slower than when using the forehand technique. This technique is also referred to as the "drag" angle of welding.

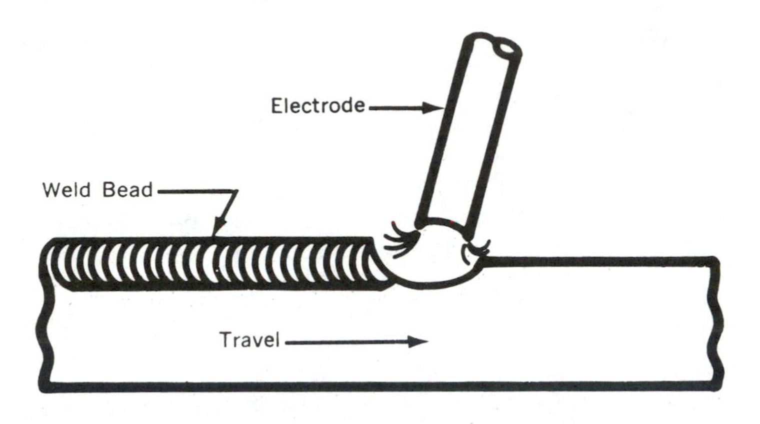

Figure 95. Lag Angle (Backhand) of Electrode.

Work Angle

The work angle is the electrode angle with relation to the surface(s) of the base metal. For example, if the weld joint is a flat groove configuration, the work angle of the electrode would be 90 degrees, or right angles, to the work surface. If the weld joint is a horizontal fillet the base metal would be in both the horizontal and the vertical position. Logically, the electrode work angle should be at 45 degrees. Both work angles are illustrated in Figure 96.

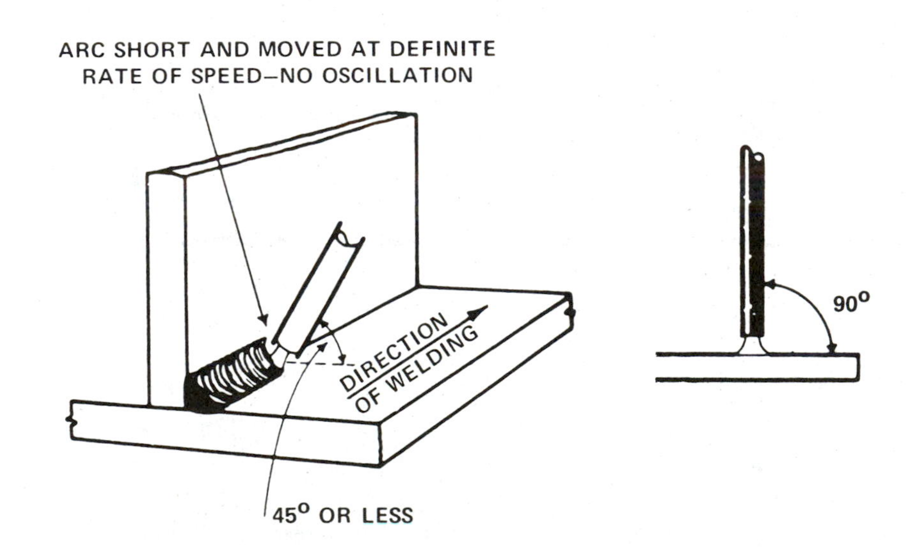

Figure 96. Typical Electrode Work Angles.

It is important to know how to correctly weld with all the electrode angles. It is equally important to know when to use the correct electrode and work angles for the application you have to weld.

The Arc Voltage and Arc Length Correlation

For many years welders have known that they could change voltage and amperage slightly while they were welding with the SMAW process. Most welders have never had the principle of arc length and arc voltage explained so they were not aware of how this phenomenon occurred. This examination of facts is to explain a principle so no actual voltages or amperages will be required.

In the illustration, Figure 97, there are three electrodes shown with different arc lengths. Please consider that a welding arc can be maintained in all three electrode-base metal relationships.

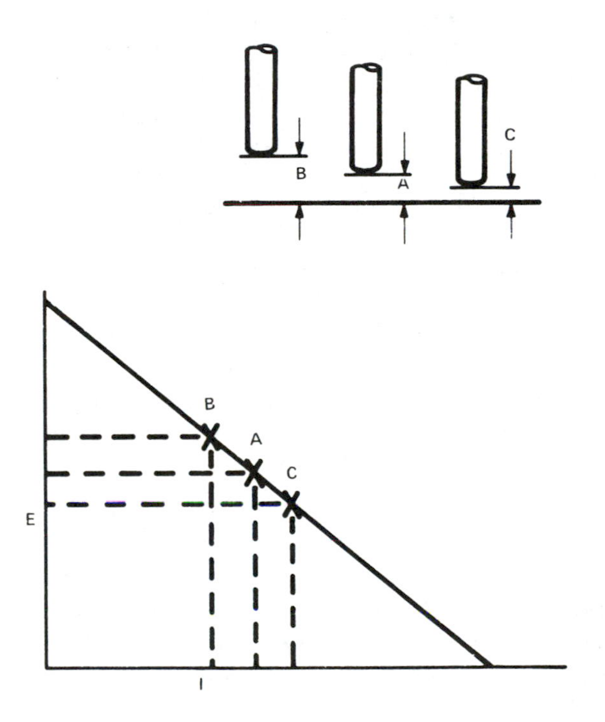

Figure 97. Arc Voltage and Arc Length Correlation.

In the lower part of the drawing, a volt-ampere curve for a constant current type power source is illustrated. We will show the correlation of arc length and arc voltage, with the accompanying amperage changes, using both the volt-ampere curve and the electrode-base metal drawings.

The center electrode has been selected as point ''A''. The arc length is normal and will be considered to provide a suitable welding condition. The arc length is an air space between the electrode tip and the base metal surface. You may remember that air is a poor conductor of electrical current or, conversely, it is a good electrical resistor.

You may also remember that voltage is an electrical force that does not flow in electrical circuits. Voltage is the electrical pressure, or force, that causes amperage to flow in electrical circuits.

In the illustration, electrode ''A'' has some arc length; that is, air space between the electrode tip and the base metal surface. Some amount of voltage is required to overcome the electrical resistance of the air space and permit current to flow. The numerical value of the voltage is not pertinent to this discussion. The voltage value of ''A'' is shown on the volt-ampere curve in the illustration. (Voltage is represented on the vertical ordinate [E] and amperage is shown on the horizontal ordinate [I]). Point ''A'' on the volt-ampere curve shows some value of voltage and amperage based on the actual arc length of electrode ''A''.

Moving now to electrode ''B'' in the illustration, it is apparent that there is an increase in the arc length as compared to electrode ''A''. This means there is more physical air space between the electrode tip and the base metal surface. It is logical that there would be greater electrical resistance due to the greater air space. More electrical pressure (voltage) will be required to overcome the increased electrical resistance of the air space. This is shown at point ''B'' on the volt-ampere curve with higher voltage but, in this case, lower amperage. This is typical of standard constant current type welding power sources.

Electrode ''C'' in the illustration shows a decrease in the arc length as compared to both electrodes ''A'' and ''B''. The physical air space between the electrode tip and the base metal surface is less. This results in less electrical resistance across the welding arc. Less voltage is required to overcome the electrical resistance in the arc. This is illustrated by point ''C'' on the volt-ampere curve which shows the lower voltage but with increased amperage. (This very tight arc would be typical of SMAW with E7018 electrodes).

It should be very apparent that it is the physical air space between the electrode tip and the base metal surface that actually determines arc voltage. The electrical resistance of the electrode, by the way, is a minor value which has no real effect on the arc voltage.

Current Density

The terms ''amperage'' and ''current'' are used interchangeably in electrical terminology. In considering electrodes, it is important to understand what is meant by ''current density'' and how the term is used. As the current value is increased when using a given diameter of electrode, the current density may also increase. Current density is calculated by dividing the electrode cross sectional area (in either square inches or square millimeters) into the welding amperage being used. The result will be *amperes per square inch or amperes per square millimeter.* A chart in the appendix of this book (Data Chart 3) provides the formula and other information for determining current density, in square inches, for electrodes from 0.020'' diameter to 3/8'' diameter.

It is current density that determines the melt-rate of an electrode. If current density is too high for a given diameter of electrode, there will be a rapid buildup of heat in the electrode. This can destroy the flux coating of the electrode.

Welding Positions Used With the SMAW Process

There are four basic positions for arc welding. They are: Flat, Vertical, Horizontal, and Overhead. In many instances the welding position is called out on work orders or engineering drawings by abbreviations such as "F", "V", "H", and "O".

Each of the welding positions will be briefly explained and the actual weld joint location with reference to the electrode illustrated.

Flat (F)

The weld joint root and base metal surface is normally parallel to the floor. The electrode is pointed down at a 90 degree work angle to the surface of the base metal. The electrode lead or lag angle would usually be 10-15 degrees based on the direction of weld travel.

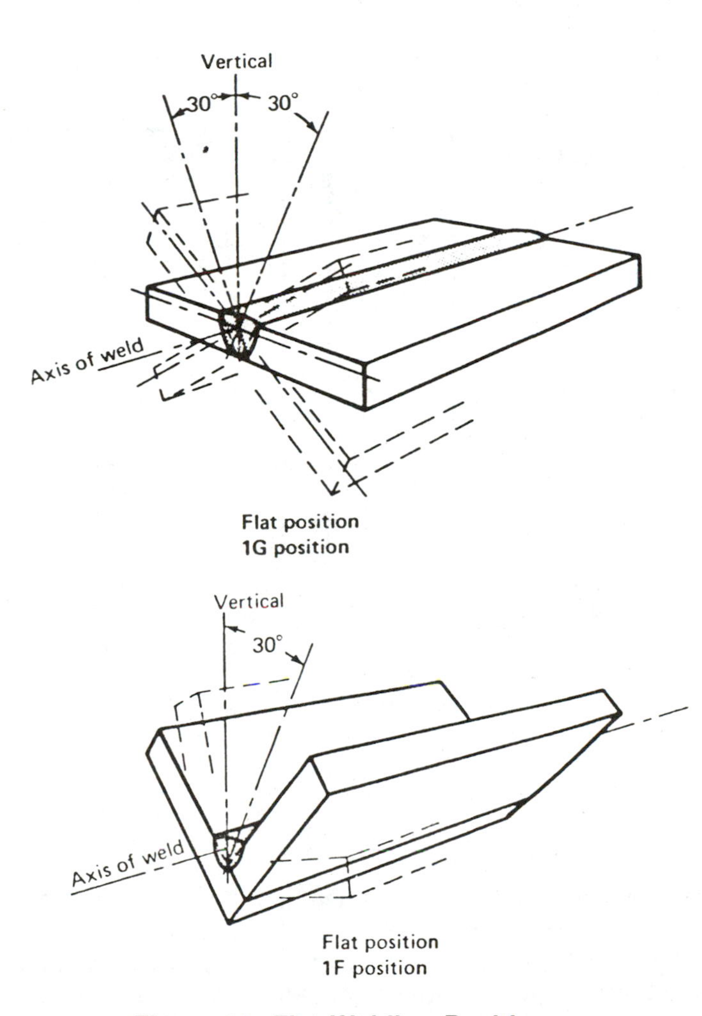

Figure 98. Flat Welding Positions.

Vertical (V)

The weld joint in the base metal is in a vertical plane at a 90 degree angle to the floor. This angle may vary considerably and the weld joint still be within the vertical position category. In sheet metal up to 1/8" thickness, it is common to make the vertical weld from top to bottom of the joint. This technique is called "vertical down" welding.

For heavier thicknesses of metal it is common to weld from the bottom to the top of the weld joint. This technique is called "vertical up" welding.

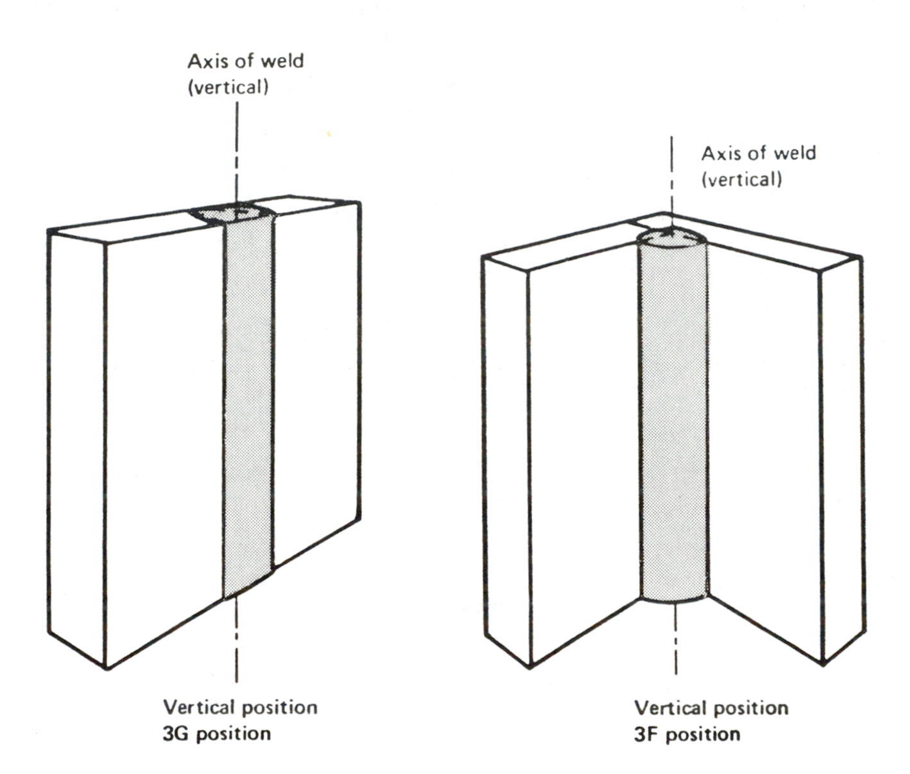

Figure 99. Vertical Welding Positions.

Horizontal (H)

For horizontal welding the base metal is normally at 90 degree angles to the floor. The weld joint is parallel to the floor and is located at the seam where two parts come together. The weld joint design is critical to assure making a satisfactory horizontal groove weld. The horizontal fillet weld joint design is often mistakenly called a flat fillet.

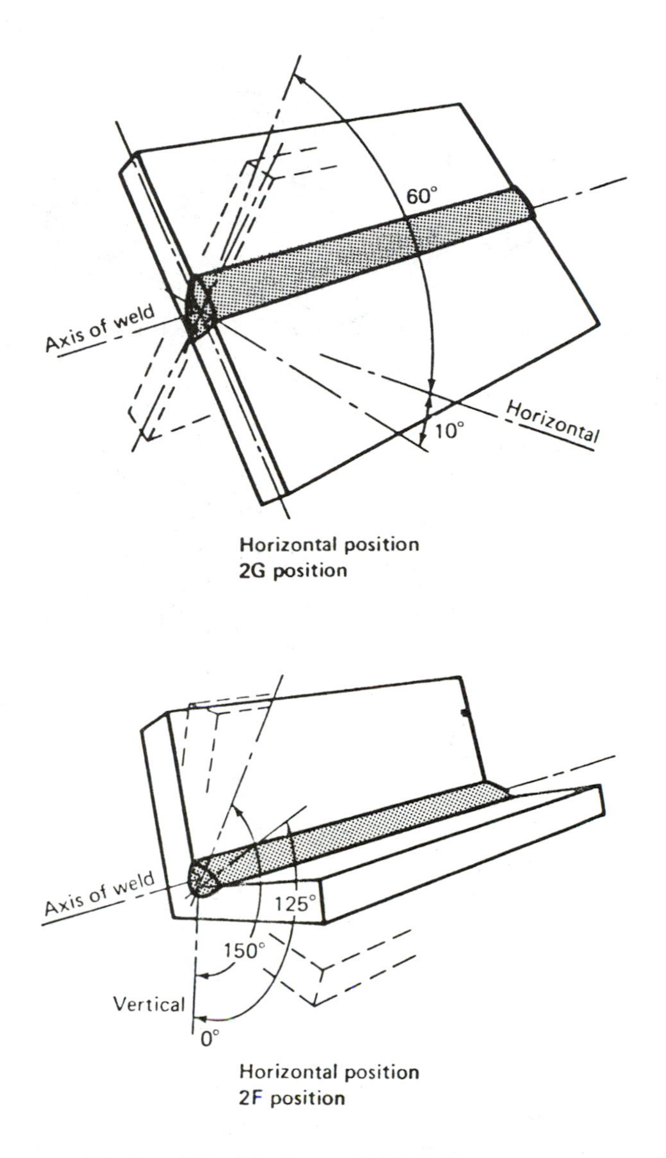

Figure 100. Horizontal Welding Positions.

Overhead (O)

As the name implies, overhead welding is accomplished with the weld joint inverted as compared to the flat position. The weld joint is parallel to the floor. Overhead welding requires using a smaller diameter electrode and making smaller size weld beads than a similar

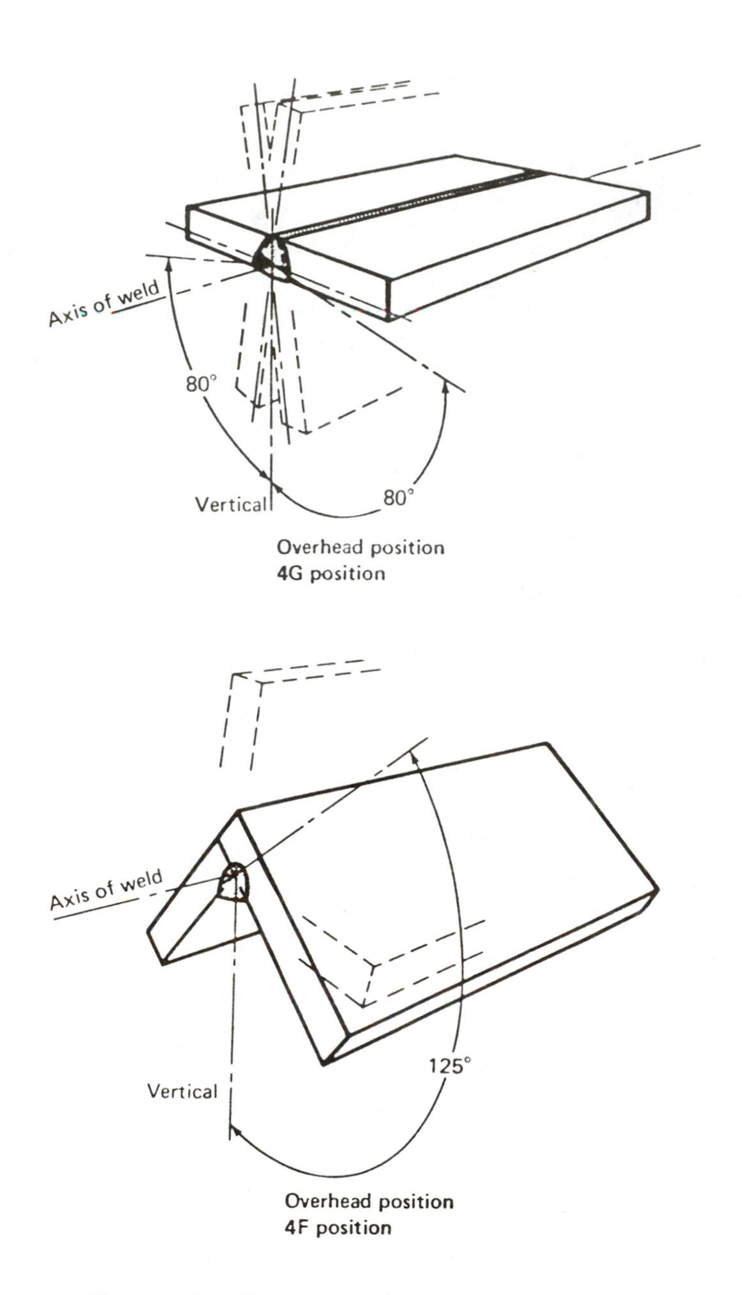

Figure 101. Overhead Welding Positions.

weld in the flat position. A leather jacket, buttoned to the collar, should always be worn when welding overhead. This will help protect the welder from falling molten metal sparks.

Some Techniques of Welding With SMAW

In all the welding techniques that could be discussed, the most important for the welder is this: find the most comfortable position to weld that is possible. This takes physical strain off the muscles of the body and permits the welder to hold a more steady arc. Attempting to weld when the body is in a strained position is self-defeating.

Be certain the welding cables are in good repair before you begin welding. Tight, clean connections are necessary to both the power source terminals, the electrode holder, and the ground clamp to work lead. The welding cable must be of the correct cross sectional area to carry the maximum welding current you will use. The use of smaller diameter "whip" leads for the electrode holder is to be discouraged.

It is the welder's responsibility to properly ground the work lead connection to the base metal workpiece. This often means grinding a small place on the base metal to provide the best ground clamp-to-clean base metal contact. The ground completion of the weld power circuit is very important to weld quality. Poor welding circuit grounding to the workpiece probably causes many preventable welding defects.

Flat Position Welding

It is common for welders in the craft to believe that flat position welding is the "easiest" position to weld. That is not necessarily true. The flat position is the easiest to get to and weld comfortably. This often leads to sloppy welding practices. The result is that many defects are welded into flat position welds because of carelessness.

Some production supervisors insist on large diameter welding puddles for flat position work. The maximum diameter of the welding puddle should not be greater than approximately 2 1/2 times the electrode core wire diameter when using the SMAW process.

The tendency to carry large diameter weld puddles when welding in the flat position increases the chances of lack of fusion at the sidewalls of the weld joint. Interpass lack of fusion is another severe problem since the molten weld puddle cools and solidifies before fusion can take place with previously placed weld metal. Wide weave beads made in the flat position seldom achieve the quality desired due to lack of fusion, excessive grain growth in the weld deposit and heat-affected zone, and non-metallic slag inclusions in the weld.

It is always tempting to use more amperage than the electrode diameter and base metal thickness require when flat welding. After all, the metal can't fall out of the weld puddle! Unfortunately, excessive amperage will create defects in the weld deposit. These may include gross porosity, non-metallic slag inclusions, erratic penetration, overlap, and possibly undercut in some areas.

It is not uncommon for welders to confuse the horizontal fillet weld with the flat fillet weld. The flat fillet weld joint is in a trough with the base metal at 45 degree angles to the flat plane. The electrode is held at a 90 degree work angle to the flat plane and 45 degree angles to each side of the joint. The horizontal fillet weld joint has one surface in the horizontal position and the other surface in the vertical position. The electrode work angle should be at 45 degrees to each side of the joint.

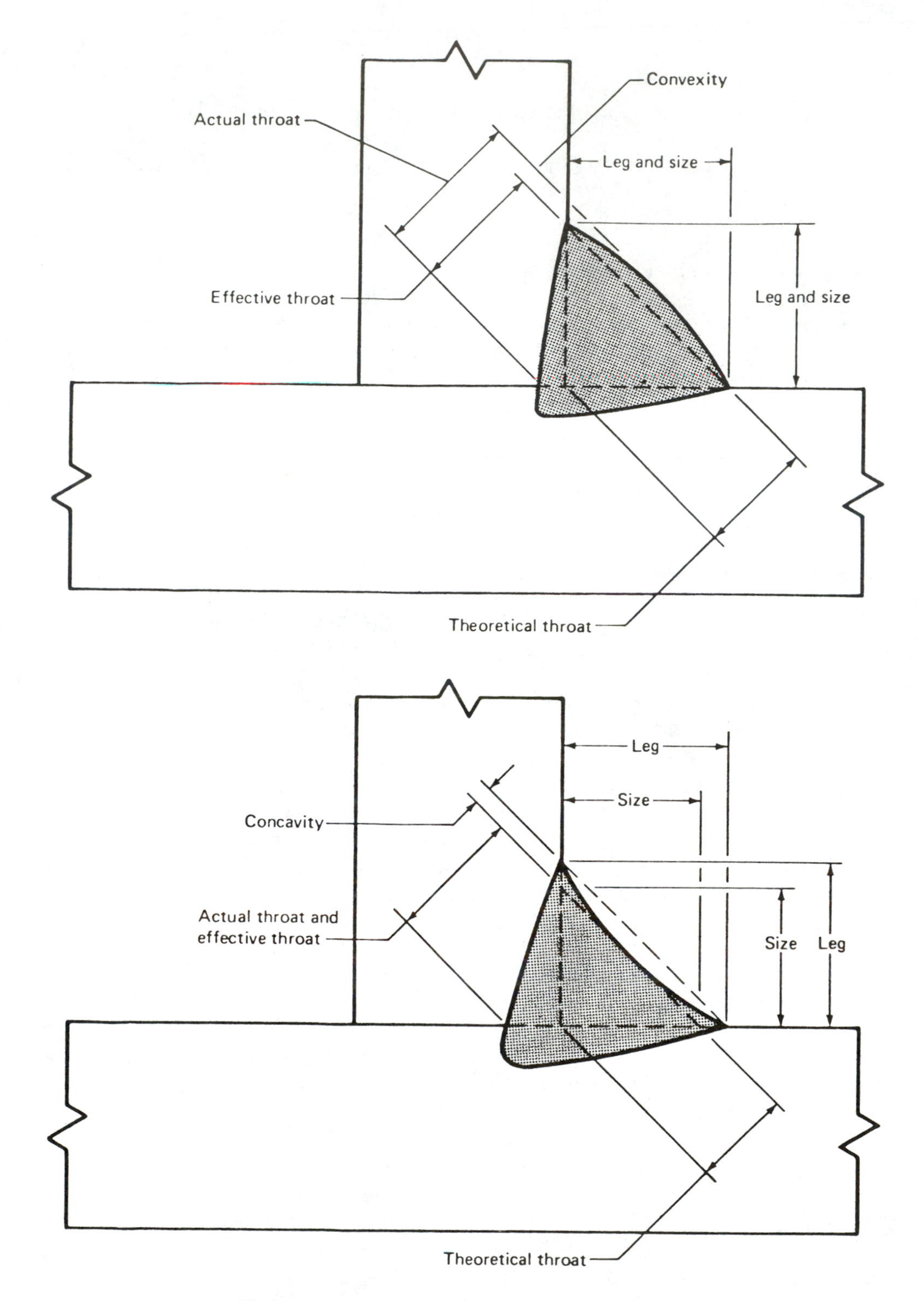

Figure 102. Typical Fillet Weld Joints.

Horizontal Position Welding

The horizontal position groove weld joint is parallel with the floor and located on the vertical surface of a workpiece. One of the most common horizontal joint designs is the 45 degree single bevel illustrated in Figure 103. Note that one joint has a backing bar while the other joint has an open root. The bevel is always located on the top plate of the joint as indicated.

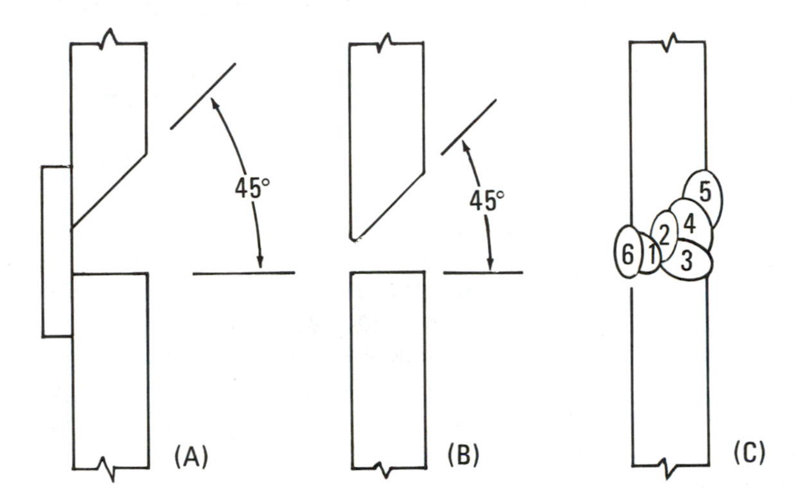

Figure 103. Horizontal 45 Degree Single Bevel Groove Joints. (A) with backing bar, (B) open root, and (C) open root, welded, backgouged.

The critical weld pass is the root pass. This is true of all multipass welds. It is important that the electrode used for the SMAW root pass be not greater than 1/8'' in diameter. The root opening of the joint must be adequate to allow the 1/8'' dia. electrode to be easily brought to the root of the joint.

For full penetration welds of the highest quality, a backing bar is recommended behind the weld joint. The root pass, or passes, fusion welds the backing bar to the root of the weld joint. When the joint is completely welded out and capped, there is a full penetration weld.

The 45 degree single bevel joint design with open root requires a restricted root opening. In this design, a single root pass will close the joint with full fusion of both sides of the joint in the root. The weld is then made as illustrated in Figure 103 to fill the joint. At that point the joint underside is backgouged with air carbon-arc (AAC), or mechanically ground out, until clean weld metal is visible. A backing weld pass would then be made. This would be a full penetration weld.

The technique of horizontal welding with the SMAW process involves relatively small weld beads well fused into the preceding weld passes. It is to the welder's advantage to make each pass with as little convexity as possible. Each pass must be thoroughly cleaned and brushed before the next pass is made. The interpass temperature should be controlled and should seldom exceed 300° F. at any time. One way to control heat input and interpass temperatures is to sequence the weld deposits in some manner such as backstepping.

The top, or capping, passes of a multipass horizontal weld must be small in size and relatively flat in shape. The objective is to have complete fill of the weld joint with no more than about 1/16" minimum to 1/8" maximum weld reinforcement.

Vertical Position Welding

Many welders dread vertical position welding because it requires very good welding skills. It is no accident that AWS and ASME welding Codes permit a welder to qualify for flat, horizontal, and vertical position welding by successfully passing a vertical weld test. Yet vertical welding is one of the easiest positions to weld if the welder is alert to what is being done at all times. The key to vertical welding is heat control.

Vertical welds are normally made from bottom to top of a weld joint with the exception of sheet metal of 1/8" thickness or less. The first step in welding each "vertical up" pass is to establish a small shelf of metal at the bottom of the joint. This becomes the base for the weld metal deposited in that pass.

I suggest that vertical weld passes be made with minimum oscillation side-to-side and minimum, if any, "whipping" of the electrode tip out of the weld puddle. A steady, continuous movement of the electrode tip along the weld path will provide strong, smooth weld stringer beads. The beginner at this vertical up technique requires plenty of practice to produce smooth, sound welds. The final result is a multipass weld with excellent fusion and strength.

In welding long seams, such as would be involved in replacing a section of steel plate with several feet of dimension in both width and length, it is often necessary to sequence the welding to minimize distortion. One excellent method of sequencing for heat control is to backstep the weld.

In a vertical weld, the first root pass would be made near the top of the joint. Clean all slag residue from the weld and wire brush carefully. Grind out the starting point of the first root pass for about 1/2"-3/4" and slope it to the top of the weld bead. This makes a place for the second root pass to tie in with the first pass without creating a hump in the weld bead.

When making multipass welds in any position, always stagger the weld starting and stopping points in each layer of weld. It is important that no two layers of weld deposit ever begin and end at the same place. To do so could create a notch effect through the weld and weaken it considerably.

Overhead Position Welding

The overhead position groove joint designs are normally welded with SMAW electrodes of 5/32" diameter or less. Control of the molten weld puddle is a bit more difficult so smaller weld bead sizes are necessary. The molten weld metal is transferred by voltage force to the base metal. Gravity is working against the welder in this position.

As illustrated in Figure 104, the welding technique is to hold a close arc length with the electrode held at about a 10-15 degree lag angle. The welding puddle is small and the speed of travel is normally slightly faster than with other welding positions.

It is necessary to wear protective leather clothing when welding in the overhead position. There is unavoidable hot metal spatter and sparks falling from the arc area. If not properly protected, the welder could be seriously burned.

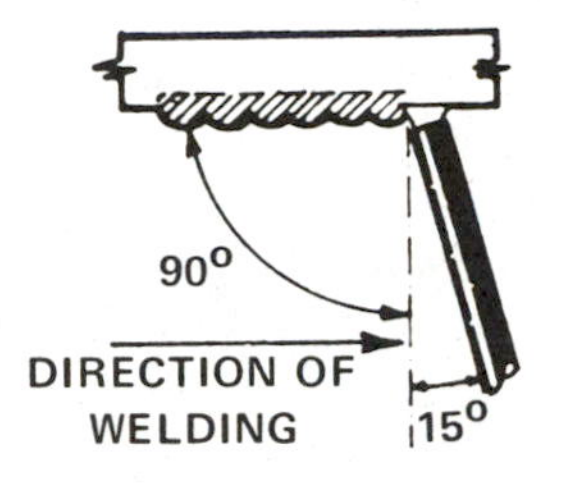

Figure 104. Overhead Groove Joint, Backhand Technique.

Some welders use a technique of overhead welding that "whips" the arc in a small elipse ahead of the weld puddle. I do not recommend this technique because it preheats the base metal and defeats the purpose of rapid solidification in the weld puddle. A steady welding travel speed with good fusion is most important and can be done with small diameter electrodes and patience.

Welding Power Sources for the SMAW Process

There are a variety of welding power sources that are used with the shielded metal arc welding (SMAW) process. They include the electric motor-generator units, the static transformer, the fuel powered engine driven generator power sources, and the transformer-rectifier power sources. All the power sources designed for the shielded metal arc welding process are constant current units. Constant current type power sources have a severe negative volt-ampere curve with a limited maximum short circuit current. This output characteristic is often called the "drooping" volt-ampere curve. Figure 105 illustrates typical volt-ampere curves for both AC and DC welding power output.

You can see that the AC volt-ampere curve is slightly convex while the DC volt-ampere curve shows a concave shape. The apparent loss of power is explained by the fact that there is a voltage drop across the main power rectifier in the welding power source. The power source circuitry is designed to provide full rated amperage at rated load voltage for either AC or DC power output.

AC Welding Power Sources

Alternating current power sources are successfully used for many SMAW applications in industry. AC power sources are preferred where magnetic arc blow is a problem to the welder. There is no magnetic arc blow with AC welding power.

There have been significant advances in the development of welding electrodes for AC applications. Various flux covering formulas are now produced that may be used for either AC or DC welding. The most important addition to the flux coverings designed for AC welding is potassium which is an ionizer. The fractional percentage of potassium added to AC electrode flux coverings helps sustain the ionized condition of the developed CO_2 shielding gas through the arc outages each half-cycle of AC power.

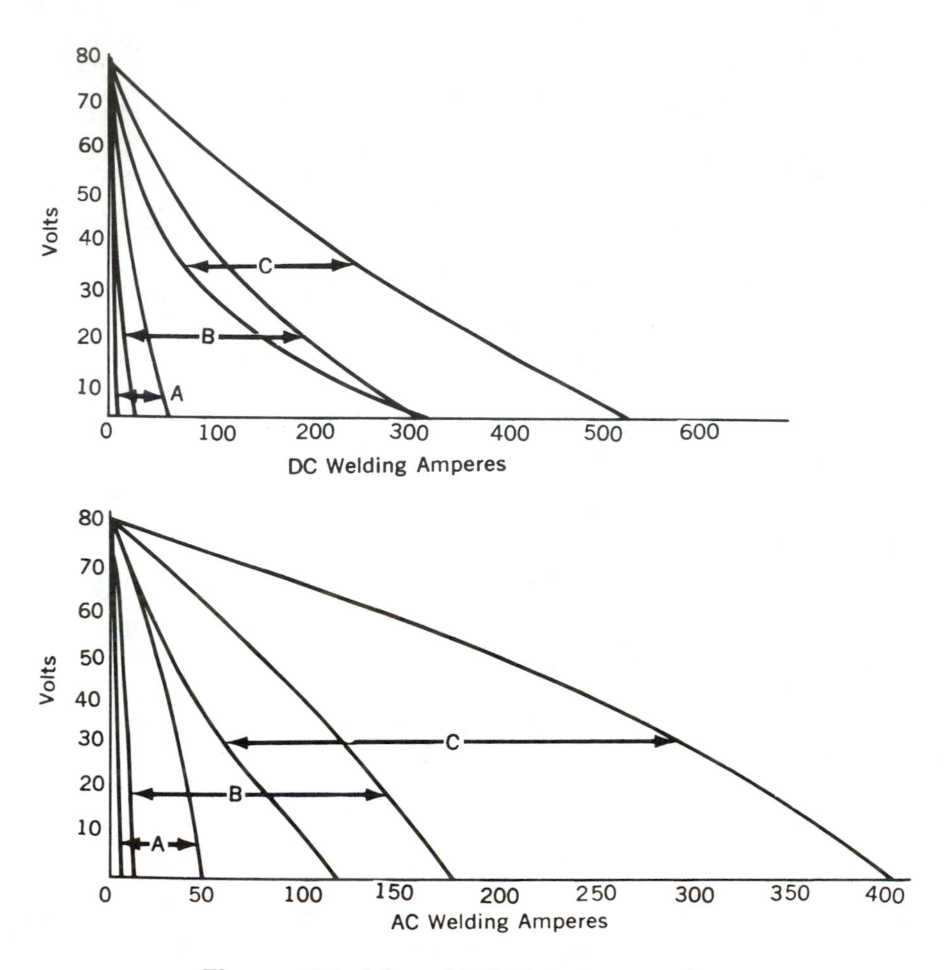

Figure 105. AC and DC Volt-Ampere Curves.

AC/DC Welding Power Sources

AC/DC welding power sources are constant current units that operate from single phase primary electrical power similar to AC power sources. The main difference between AC and AC/DC power sources for SMAW applications is the addition of a main power rectifier with allied DC circuitry to the AC/DC unit. The rectified single phase welding power is not as stable as DC welding power from a three phase based system.

DC Welding Power Sources

Direct current welding power flows in one direction only. There is no change in the direction of current flow as there is with alternating current. DC welding power sources usually have polarity switches. When the power source is set for direct current, reverse polarity (DCRP) the electrode is positive. DCRP could also be written as DCEP (electrode positive). The electrode is negative when using DC straight polarity. DCSP could also be written as DCEN (electrode negative).

DC welding power sources for the SMAW process may operate from either single phase, or three phase, primary electrical power. The most stable DC welding power will be provided by units with three phase transformers and full wave bridge rectifiers.

Electrical Safety With the SMAW Process

There have been many discussions of electrical power safety and, in particular, which type of power is safer, alternating current (AC) or direct current (DC). The clear cut answer is that neither is safer! Any type of electrical power can be deadly if it is handled carelessly. The greatest asset to anyone working around, and with, electrical power is common sense. Following good electrical safety practices when welding will minimize the chance of electrical shock. It is possible, however, to receive an electrical shock from any exposed electrode that is electrically energized.

It is certainly true that AC will give more of an electrical shock than will DC. The reason for this is that alternating current changes direction of flow each half-cycle (each 1/120 second for 60 hertz power). The constantly changing voltage and amperage of AC, coupled with the changing polarity (direction of current flow) each half-cycle, is the cause of the electrical shock with AC.

It has been said that electrical power will "reach out and grab" an individual. That is not true. As you may recall from high school biology, nerves and muscles of dead tissue may be stimulated by an electrical current and caused to move or quiver. The same type of reaction occurs when the human body is stimulated by electrical current and voltage.

If a human body becomes part of an electrical circuit of even modest power, the muscles and tendons of the body will tense. The muscles and tendons will stay in the tensed condition until the electrical power is removed from the circuit.

NEVER TOUCH ANYONE WHO IS IN CONTACT WITH ANY KIND OF ELECTRICAL POWER LINE. The power will be transmitted to you and serious injury or death could result.

The best rule to follow is this: If you are working with electrical voltage in excess of 6 volts, keep one hand in your pocket!

CHAPTER 8

Constant Current Welding Power Sources

Welding power sources are manufactured for use with all the electric arc welding processes. The power sources are available in two basic types:

1. electro-mechanical systems
2. static electrical systems

The power sources discussed in this chapter are the constant current units used for the shielded metal arc process (SMAW), the gas tungsten arc process (GTAW), the submerged arc process (SAW) and sometimes for the air carbon-arc process (AAC).

The specific types of constant current power sources are explained here to inform and instruct the reader concerning their characteristics and uses. Please understand that the applications discussed in this chapter do not limit the abilities of the various power sources to perform in other areas as well.

Electro-Mechanical Generators

The welding power **generator** portion of this class of power source is called a DC generator. The **drive method** is normally either an electric motor or some type of internal combustion engine. This portion of the text is concerned with the electric motor-generator classification of welding power source.

The electric motor-generator welding power source was the first individual source of generated electrical power for welding. The welding power produced by these units is direct current (DC) with the capability of either polarity (DCRP = DCEP; DCSP = DCEN). To define:

DCRP = direct current reverse polarity, or
DCEP = direct current electrode positive.
DCSP = direct current straight polarity, or
DCEN = direct current electrode negative.

The electric driving motors used for motor-generator type welding power sources are normally **AC induction motors.** The NEMA Standard EW-1, 1971, states that, "alternating current induction motors driving DC generator welding power sources shall be three phase and shall have voltages and frequencies in accordance with the following:

60 hertz = 200, 230, 460, and 575 volts.
50 hertz = 220, 380, and 440 volts."

It is apparent that motor-generator welding power sources must operate from three phase primary power. This type of power source, commonly called an "MG set" has excellent shielded metal arc welding output characteristics.

In the operation of a motor-generator power source the three phase primary electrical power is brought to the electric drive motor. The sole function of the primary electrical power is to operate the electric drive motor.

When the drive motor operates, it provides the energy to turn the rotor shaft and assembly. The primary electrical power energy is changed to mechanical energy in the rotor assembly as the motor turns the rotor shaft. The rotor is often called the armature although this is not correct. The rotor is actually the total shaft assembly which includes the armature iron core material, the armature coils, the commutator and the shaft. In most cases, there is also a small DC generator to provide excitation of the magnetic field coils in the generator stator assembly.

The block diagram shown in Figure 106 is for a typical motor-generator welding power source. As illustrated, the three phase primary power is brought to the electric drive motor. The induction type electric motor causes the rotor shaft to turn. The **armature** is a large mass of iron core material and copper conductor coils located on the rotor shaft. The rotor assembly armature rotates within the **stator,** or stationary portion, of the welding power source generator. The magnetic field coils, which are necessary for power generation, are located in the stator portion of the circuit.

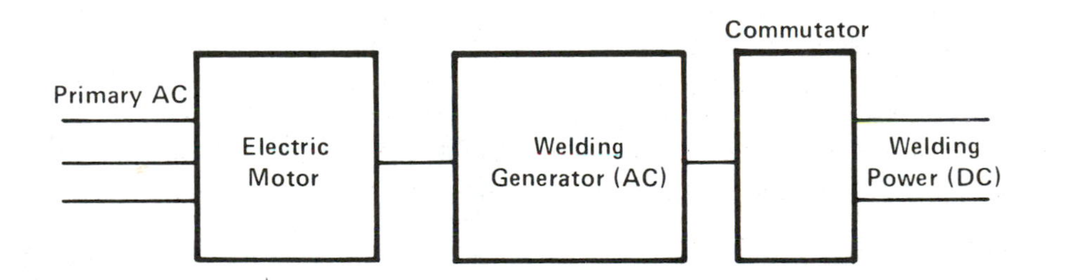

Figure 106. Motor-Generator Block Diagram.

The rotation of the armature within the stator generates alternating current. The current is actually generated in the armature coils. It flows through the armature conductors to the commutator.

The commutator is a system of copper collector bars placed concentric to the shaft centerline. (Concentric means the commutator bars are all an equal radius distance from the centerline of the shaft). The conductor wires from the armature are soft-soldered to the individual commutator bars. In a sense, the copper collector bars of the commutator collect the generated AC welding power and present it for pickup by carbon-type brushes which are placed at discrete intervals around the commutator.

For a four pole generator system there are four contact brushes positioned at specfic intervals around the commutator. The brushes are so arranged that they pick up specific half-cycles of the generated current and direct it down a single conductor as **direct current.** It is at the brush-commutator assembly that the generated AC welding power is changed to DC welding power. To use an analogy, it may be said that the commutator-brush arrangement is a type of mechanical rectifier.

It is at the commutator-brush assembly that the mechanical energy is changed back to electrical energy. Anytime there is a transfer from one type of energy to another there will be energy losses. The energy losses decrease the electrical efficiency of the welding power source considerably.

This helps to explain why the maximum electrical efficiency of the best set up and balanced motor-generator set is approximately 55% at rated load output. The overall electrical efficiency of a motor-generator welding power source is considerably less. A brief comparison of a typical motor-generator power source and a typical transformer-rectifier power source is made in Figure 107.

Power Source Type	Energy Type	Approximate Efficiency at Rated Output	Idle Power Required	Total Electrical Efficiency
Motor Generator	E-M-E	55%	3,000 watts	15-18%
Static Rectifier	E	73%	600 watts	65-68%

Figure 107. Motor-Generator and Transformer-Rectifier Comparison.

A three phase transformer-rectifier type welding power source takes the primary power from the input distribution system. The primary power is changed to usable welding voltage and amperage by the transformer. The power may go through certain control circuitry before it comes to the rectifier where the AC is changed to DC. There is no change of energy type since the transformer-rectifier power source is totally electric. Electrical efficiency of transformer-rectifier power sources is normally about 73% at rated voltage and amperage load.

For the comparison and evaluation shown in Figure 107, we selected 300 ampere rated constant current power sources with the old NEMA 40 load volt rating. Both power sources are rated at 60% duty cycle. This type and rating of welding power source has been a standard unit in the welding industry and is probably the most popular class and rating of power source manufactured.

Keep in mind that the calculation of electrical efficiency for any electric welding power source is the amount of primary electrical power used, in KW, divided into the welding power secondary output, in KW. The percentage of electrical efficiency is one method of determining the power losses in electrical equipment.

The complete story of any welding power source is not told by the electrical efficiency at rated load output. Analysis of arc welding operations using the SMAW process has shown the average welder's arc-on time as about 23% of the working day. The balance of the work time is spent on setting up, chipping slag, changing electrodes, etc.

In a typical eight hour workday the welding power source is operating at rated load, or less, for about two hours. The rest of the time the unit is idling and cooling. As long as it is energized, the welding power source is using current from the primary power system. The amount of primary power drawn but not used for productive purposes becomes a vital factor in computing net welding costs. It certainly has an effect on the overall electrical efficiency of the welding power source.

The table in Figure 107 shows the approximate electrical efficiencies of the two power sources at rated load output. Since approximately 75% of the working day is spent in the "idle" condition, an examination of the idle time efficiencies and costs is necessary.

When idling under no welding load, the welding power sources have a substantially lower power demand from the primary power supply system than they do when operating at rated load output.

As shown in Figure 107, the motor-generator power source requires approximately 3,000 watts per hour to maintain the electric motor at its proper speed in rpm. If there is less than the required amount of primary power, due to a 10% or more voltage drop over a period of time, the electric motor can overheat or stall. A stalled electric motor will not necessarily damage the welding power generator but it could cause serious damage to the electric motor coils and circuitry because of overheating.

Remember: It takes a certain amount of power in kilowatts to keep the electric motor operating. If there is low primary voltage value the primary amperage has to increase to make up the difference in power (volts times amperes equals watts). If excessive amperage is used over a period of time, it will cause the electrical conductors in the electric motor to overheat. This will destroy the conductor insulation and cause the motor to have catastrophic failure by burning.

Figure 107 shows the primary power requirement for the static rectifier power source is approximately 600 watts per hour. Some of the electrical power is used to operate the fan motor for cooling the power source components. Part of the power is used to provide excitation current for the electric control circuit and to compensate for the small electrical losses through the transformer coils and iron core.

Considering the total primary power input to the two types of welding power sources in an eight hour day, it is apparent that the total electrical efficiencies of the power sources will be radically different. The motor-generator set, idling for about six hours each work day, has an approximate electrical efficiency of 15-18%. The static rectifier unit, working the same hours and idling the same hours, has an electrical efficiency of approximately 65%. The total primary current drawn by the static rectifier unit is considerably less than that drawn by the motor-generator unit.

In many areas of the world, utility companies operate on a demand meter basis where the customer pays a rate per KW hour based on the highest power demanded at any specific time. This is referred to as "instantaneous demand power". In some cases, the instantaneous demand power is substantially higher than the average power used by the customer.

The problem is that the utility company is usually required by law to maintain at least 75% of the demand power available to the customer at all times. As a result, customers with poor electrical efficiency in their shops are penalized for the excessive demand power.

Any inductive electric motor, such as those used on motor-generator power sources, will demand three to ten times its operating amperage for starting purposes. The demand is for more primary current since primary voltage is considered to be constant.

For example, if the MG welding power source uses 100 amperes when putting out its rated load, the starting current requirement would be from 300 amperes to 1,000 amperes. This places an extremely heavy short time load on the primary power system as well as the shop primary wiring. It is this very high "instantaneous demand power" load that creates the penalty factor per kilowatt-hour charged.

The static rectifier is a transformer-rectifier power source which has a relatively small primary current demand. The idling current is a very low value until the power source is actually used for welding. The amount of primary current used is regulated by the amount of output current needed. A transformer type power source only demands, and uses, the primary current necessary to supply the welding arc with power.

Static AC Transformer Welding Power Sources

AC transformer type welding power sources are basically simple electrical devices. The purpose of a welding transformer is to change voltage and amperage without changing frequency. The main transformer of a welding power source is designed to take the relatively high voltage, low amperage of the primary power and change, or transform, it into usable welding power.

Another type of transformer often used in welding power sources is the "isolation transformer". The isolation transformer is normally used for a control circuit of some type.

AC transformer power sources consist basically of an iron core, a primary coil, and a secondary coil. The control system may be either mechanical or electric although most AC units have mechanical control. All AC and AC/DC transformer type welding power sources operate from single phase primary power.

Duty Cycle and Service Use

The National Electrical Manufacturers Association (NEMA) publishes the governing documents and standards for electric welding power sources in the Standard EW-1. Part of the requirements for various classifications of power sources is a prescribed duty cycle. Although not all welding power sources conform to the NEMA Standard EW-1, most manufacturers use the NEMA Standard as a guideline for their products.

Until 1962, industrial class welding power sources were arbitrarily rated at some specific amperage and 40 load volts. The NEMA Standard EW-1, revised and published in 1962, and re-affirmed in 1972, proclaimed a graduated load voltage chart based on the power source amperage ratings and the duty cycle. The present load voltage and amperage ratings for NEMA Class I constant current welding power sources are shown in Figure 108. This chart applies to all constant current electric welding power sources, either AC or DC output.

Duty Cycle

The duty cycle of a welding power source is the actual operating time it may be used at its rated load without exceeding the temperature limits of the insulation of the component

Ampere Rating	Voltage Rating
200	28
300	32
400	36
500	40
600 and up	44

Figure 108. NEMA Class 1 Voltage Ratings.

parts. In the United States, duty cycle is based on a ten minute period of time. In some other areas, notably Europe, duty cycle is based on a five minute period of time.

To explain the ten minute duty cycle period the following example is used. Suppose a welding power source is designed to operate at 60% duty cycle, 300 amperes, 32 load volts. This means that it has been designed and built to provide the rated amperage (300), at the rated load voltage (32), for six minutes out of every ten minute period. The other four minutes of the ten minute period the unit must idle and cool. It is mandatory that the full ten minute period be completed for proper duty cycle operation.

Duty cycle is not accumulative. To operate a welding power source for 36 minutes at rated load and then let it idle and cool for 24 minutes is not 60% duty cycle. **REMEMBER:** DUTY CYCLE IS BASED ON A TEN MINUTE PERIOD OF TIME. It is important that the full ten minute period be observed.

In the example provided, it is not 60% duty cycle to operate the power source for six minutes at rated amperes (300) and then turn off the power! The residual heat retained in the electrical conductors and iron core of the transformer would cause severe heat damage to the insulating materials. The cooling fan must run so the heat of the power source components can be removed.

Those power sources rated at 100% duty cycle may be operated continuously at, or below, their rated amperage. If such power sources are operated with current output greater than their rated amperage, they no longer have 100% duty cycle. For example, a power source rated at 300 amperes, 32 load volts, 100% duty cycle, would normally have about 55% duty cycle at 400 amperes output.

The reason for the differences in duty cycle is the conductor cross sectional area used in welding power sources with specific amperage ratings. If the amperage used is above the normal rated capacity of the electrical conductors, excessive heat will develop due to the electrical resistance of the conductors. The greater the temperature of a conductor, of course, the higher the electrical resistance. The result would be possible overheating of the conductor and burning of the insulation.

In recent years, the NEMA Standard EW-1 has re-classified welding power sources by their duty cycle. This makes it relatively easy for the purchaser of a power source to determine exactly what he is buying.

Types of Service Use

There are three general types of constant current welding power sources from the service standpoint. These classifications of power sources are most often used for the shielded metal arc welding (SMAW) and gas tungsten arc welding (GTAW) processes. They may be used for other welding and arc cutting processes as well.

NEMA Class 1. The NEMA Class 1 welding power sources have 60%, 80%, or 100% duty cycle. The welding output may be AC, AC/DC or DC. Formerly called the "industrial" class power sources, they are considered to be the work horses of the welding industry.

NEMA Class 2. NEMA Class 2 welding power sources will have 30%, 40%, or 50% duty cycle. Such welding power sources are normally rated at less than 300 amperes. The welding output may be AC, AC/DC, or DC. This group of power sources was known as the "limited duty" classification at one time.

NEMA Class 3. The NEMA Class 3 power sources are limited in the amount of primary amperage they can draw. They are always rated at 20% duty cycle with maximum amperage output ratings of 230 amperes for the most popular size units. These utility power sources are inexpensive and very useful. NEMA Class 3 power sources were known as "limited input" power sources before 1962.

Welding Output Control Methods

The welding power developed in the secondary portion of the power source circuitry must be controlled so that the welder can have whatever amperage the job calls for. Commonly called "amperage control", the output of welding power sources can be controlled in several ways. The actual method of control depends on the cost effectiveness of the control, the necessity for remote control capability, the economics of manufacturing, and certainly the welding process requirements. Some of the methods used for controlling AC constant current power sources include:

Movable Coil Control
Movable Shunt Control
Tapped Secondary Coil
Tapped Reactor Control
Saturable Reactor Control
Magnetic Amplifier Control
Solid State Electronic Control

Of these various control methods, this section will only consider the movable coil and movable shunt systems. Some electric controls will be discussed in later sections of this book.

One of the most important features of a well designed welding power source is its capability for infinite amperage control within the minimum-maximum amperage output range. If this capabililty is provided in two or more output current ranges, there must be a generous overlap in the maximum current of the lower range and the minimum current of the higher range.

Movable Coil AC Welding Power Sources

The movable coil design is a classic example of an AC welding power source transformer. The primary and secondary coils of the main transformer have loose magnetic coupling to provide the negative ("drooping") volt-ampere output characteristic required for the SMAW process. A drawing of a typical movable coil transformer is shown in Figure 109. The secondary

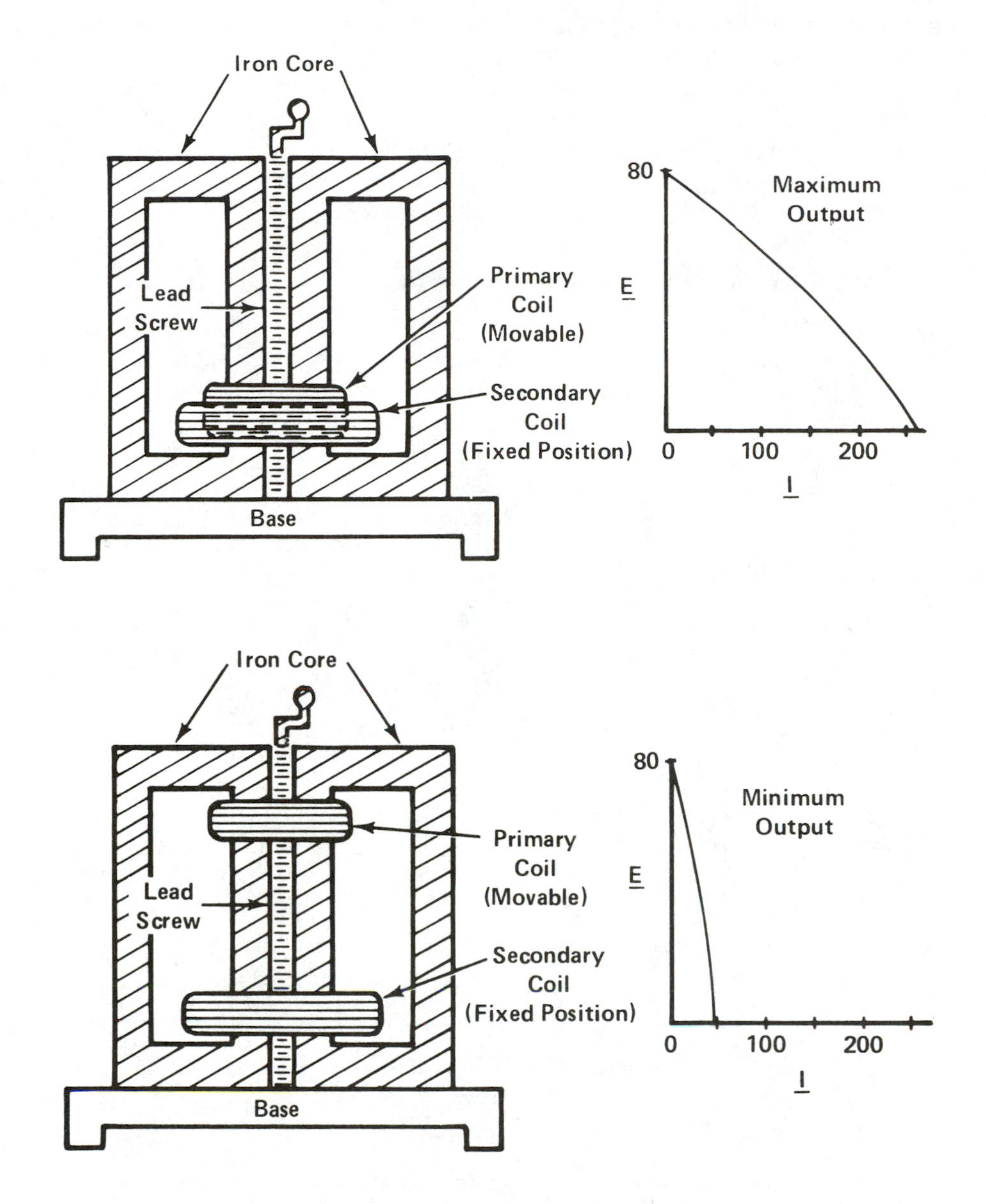

Figure 109. Movable Coil AC Power Source.

coil of the transformer is fixed in position in this power source design. The primary coil is smaller in physical size. It is the primary coil that moves on the lead screw which positions the coil according to the welding amperage requirements.

There is no mechanical connection between the primary and secondary coils. As we previously discussed, voltage is induced into the secondary coil. The actual current output of the AC power source is regulated by the distance the two coils are apart. The closer together the coils, the higher the amperage output. Conversely, as the primary coil is moved away from the secondary coil, the output amperage decreases. This is shown in the two volt-amperes curves in the illustration, Figure 109.

If mechanical controls, such as the handwheel and lead screw illustrated, become difficult to turn, apply some silicone grease to lubricate the threads of the lead screw. It is always best to thoroughly clean the lead screw threads of all dirt and old grease before applying new lubricant.

Movable Shunt AC Welding Power Sources

Movable shunt AC transformer type welding power sources also have loose magnetic coupling. In the movable shunt design, however, the primary and secondary coils are **fixed** in position and do not move. Instead, a **laminated iron shunt** in a shunt holder assembly is moved between the primary and secondary coils. The iron shunt materials are the same as those used in the main transformer iron core. The pieces are rectangular in shape and are film insulated on each side of each piece. The iron shunt acts as a magnetic "flux" diverter. (The term "flux" is the same as saying "magnetic lines of force" in a magnetic field.)

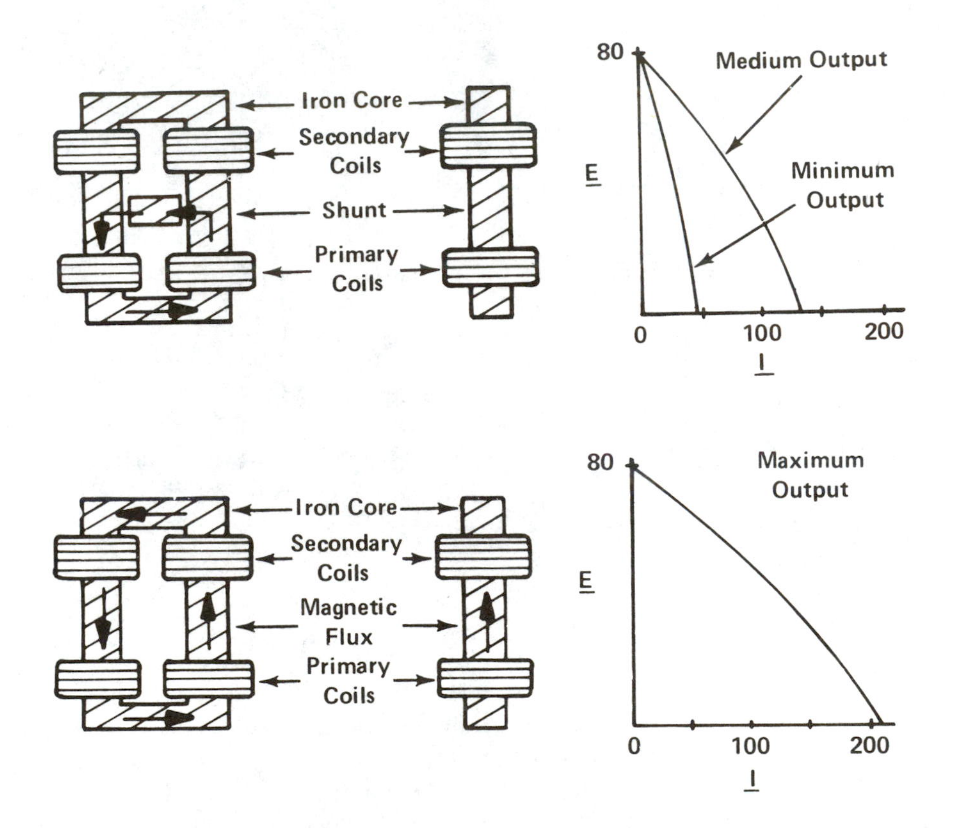

Figure 110. Movable Shunt AC Power Source.

As illustrated in the lower portion of Figure 110, the movement of magnetic lines of force are unobstructed when the iron shunt is not between the primary and secondary coils of the main transformer. When the iron shunt is caused to move between the primary and secondary coils (upper portion of Figure 110), the magnetic lines of force are literally diverted to the iron shunt rather than having free access to the secondary coil. The volt-ampere curves show graphically the output from the shunt type power source in various conditions. In this manner, welding output amperage from the power source can be adjusted from minimum to maximum within the range of the power source.

When the iron shunt is moved in between the primary and secondary coils the output welding current decreases. When the iron shunt is moved out from between the two coils the output welding power is increased. Current output adjustment is infinite within the minimum-maximum amperage range of the power source.

Shunt type power sources may be NEMA Class 1, Class 2, or Class 3. Preventative maintenance is usually confined to periodically blowing out any accumulated dirt with compressed air. If there is any other problem, such as shunt rattle, check the power source operation manual for troubleshooting tips.

The welding power cables used with AC welding power sources should always be uncoiled while welding. If they are coiled, the alternating current will create a reactance and minimize the current output from the power source.

Troubleshooting AC Welding Power Sources

Always have the manufacturers Installation and Operation manual for the power source available for troubleshooting problems. It should be read and understood by all who will use or work on the welding power source.

Every welding power source installation and operation manual has a section on the recommended installation procedures. Another section tells the welder how to make secondary connections and how to use the unit. In all cases there should be a troubleshooting section in the book which describes some of the more common malfunctions which could occur with the power source and its related circuitry. Suggested repair methods are also located in this section. A circuit diagram is usually part of the troubleshooting section of the manual. The circuit diagram shows the electrical relationship of the power source parts.

AC/DC Static Welding Power Sources

The AC/DC static welding power source was first developed right after World War II. Initially designed for the gas tungsten arc welding (GTAW) process, it was soon evident that there was considerable application for such a power source with the SMAW process. Although this type of welding power source is a compromise between an AC power source and a DC power source it does have specific advantages for some welding process applications.

An AC/DC transformer-rectifier welding power source is basically an AC power source operating from single phase primary power. The DC capability is added by the inclusion of a main power rectifier and allied DC circuitry in the secondary portion of the power source. The main power rectifier changes the AC to DC.

The AC/DC type power sources operate from single phase primary power which, when rectified and changed to DC power, creates a substantial ripple factor percentage in the output welding current. This simply means there are peaks and valleys in the output current which can cause arc instability. The reason for the high ripple factor percentage is that the single phase-based DC power theoretically goes to zero current each half-cycle of AC input. The current actually does not go to zero each half-cycle nor does it reach the maximum output. Rather, there is what is termed the "RMS", or effective, value. This is shown in the equation:

$$\text{RMS value} = 0.707 \times (\text{maximum circuit current value})$$

To provide continuity of current flow, and maximum stability in the welding arc, a heavy duty "stabilizer"is normally added to the DC portion of the power source circuitry. (The term "stabilizer" and "inductor"are synonymous in electrical terms. A stabilizer is an iron core with a current carrying coil wrapped around some part of its mass. It is always in the DC portion of the power source circuitry). The function of the stabilizer (inductor) is to slow down the rate of response of the power source to changing arc conditions. This smooths out the DC welding output and provides a more stable welding current value at the arc.

All AC/DC welding power sources are designed to provide constant current output with limited maximum short circuit current. They are especially designed for use with the shielded metal arc welding (SMAW) and gas tungsten arc welding (GTAW) processes. For other DC applications, such as air carbon-arc cutting and gouging (AAC), single phase AC/DC power sources do not have the total output power level that three phase transformer-rectifiers have.

The AC volt-ampere curve shown in Figure 111 has a definite convex shape while the DC volt-ampere curve is slightly concave. The difference is because of the voltage drop that is inherent through the main power rectifier of the power source. This is characteristic of all DC transformer-rectifier power sources.

Most output control methods used with present NEMA Class 1 AC/DC power sources are either electric control or electronic control. The basic reason for using electric controls such as the saturable reactor or magnetic amplifier are for remote amperage control by the welder. This permits the welder to adjust welding power source amperage output from the work station without having to return to the power source.

The transformer design and function is the same for AC/DC power sources as it is for AC transformer power sources. The addition of the rectifier, stabilizer, and other DC circuitry makes it possible for the welder to have AC, DCSP, or DCRP at the flick of a switch. To have this convenience, users will tend to sacrifice the better arc characteristics obtainable with standard three phase transformer-rectifier DC welding power sources.

A number of manufacturers have developed NEMA Class 2 AC/DC power sources for both the SMAW and GTAW processes. These power sources provide limited duty cycle for AC gas tungsten arc welding. They do give the user who has only single phase primary power available the opportunity to use certain SMAW electrodes that will only operate with DC welding power.

Some models of AC/DC welding power sources are designed especially for the gas tungsten arc welding process. Most of these units are NEMA Class 1 which normally have some of the following components as standard equipment:

A built-in high frequency system, gas valve and solenoid, water valve and solenoid, postflow timer for gas and water, primary contactor, and start control circuit with rheostat. In addition, some electronic control systems have line voltage compensation, amperage upslope and downslope, pulsing of current, and other electronic functions.

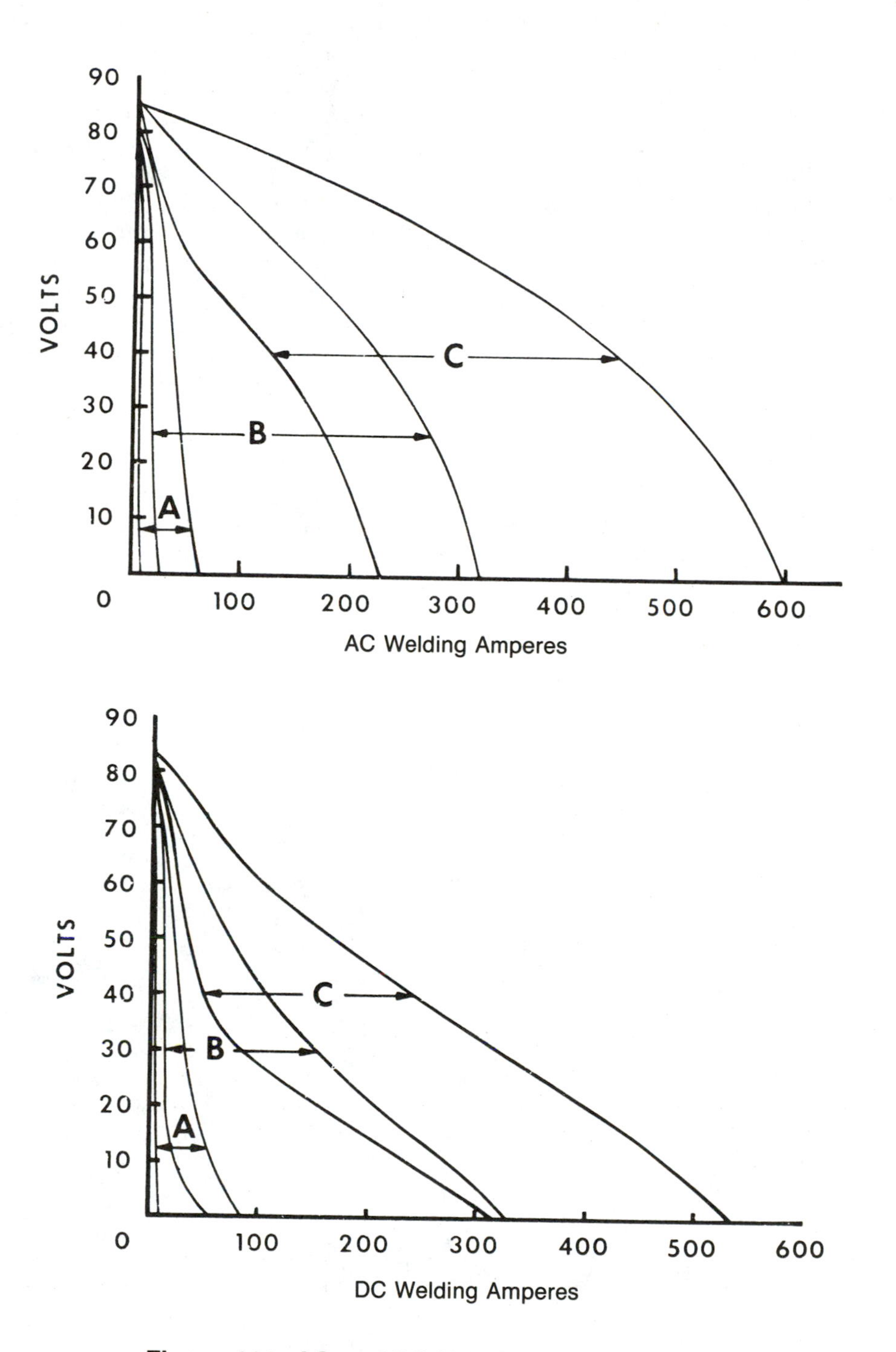

Figure 111. AC and DC Volt-Ampere Curves.

One of the specific reasons for developing the AC/DC welding power source classification is to have the capability to weld aluminum and magnesium with alternating current and the gas tungsten arc welding process. When welding either aluminum or magnesium with AC and the GTAW process there is developed the phenomenon called "DC component". Some method of dissipating the energy of DC component is usually incorporated as part of NEMA Class 1 power source circuitry. This may take the form of ni-chrome resistor bands, capacitors in series with the welding arc, or some other medium. The subject of DC component will be discussed further in the portion of this book relating to gas tungsten arc welding.

In summary, it is evident that AC/DC transformer-rectifier welding power sources have had wide acceptance in the welding industry despite the fact that they are a form of compromise power source. The fact that both AC and DC welding power is available appears to offset any disadvantages that may occur due to single phase primary power. As a matter of fact, the AC/DC power source is an advantage in areas where the only primary power available is single phase power.

DC Transformer-Rectifier Welding Power Sources

The initial arc welding power came from battery banks. As illustrated in Figure 112, the batteries were continuously charged by large dynamos (generators).

Figure 112. Battery Powered DC Arc Welding.

In the early days of generated welding power, three phase generators were used. They were either electric motor driven or internal combustion engine driven. The only real criteria for the engine or electric motor horsepower and size was its ability to permit the generator to reach full power output.

With the inception of the three phase transformer-rectifier power source in 1950, there has been a predictable trend toward greater use of this type of welding unit. With the increasing world-wide energy problems, the use of generators has decreased substantially due to their poor electrical efficiency and higher maintenance costs. Noise levels in factories are another reason motor-generator welding power sources are being replaced by quieter, more energy efficient transformer-rectifiers.

Although transformer-rectifier power sources may be designed to operate from either single phase or three phase primary power, those that supply **DC welding power only** usually are three phase transformers. This type of transformer-rectifier power source is considered the most electrically efficient of the constant current systems. Three phase transformer-rectifier power sources will always have lower ripple factor percentages than single phase transformer-rectifier power sources. This is shown in Figure 113 which shows the relative characteristics of rectified single phase and rectified three phase power.

Rectified DC

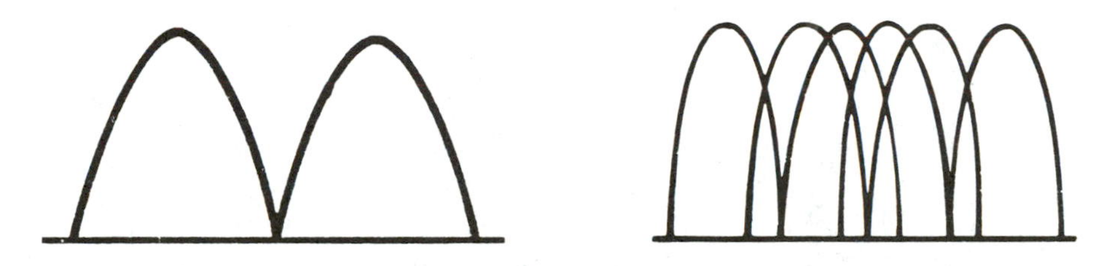

Figure 113. Single Phase and Three Phase Rectified DC Power.

Several manufacturers have produced engine driven welding power sources with revolving field coils and static rectifier design. This type of engine driven power source is electrically more efficient than the DC generator type. Another good point is the fact that the rectifier will not permit the power source to change polarity while welding. This is a relatively common occurrance with DC generators having revolving armature design.

Constant Voltage/Constant Potential type power sources for the gas metal arc (GMAW) and flux cored arc (FCAW) welding processes will be discussed in detail in a subsequent chapter.

Inverter/Converter Power Sources

A relatively new type of welding power source has been presented to the welding public recently. This type of power source is called by a variety of names including ''Inverter'', ''Converter'', ''High Frequency'', and ''Transformer-Rectifier''. The interesting part is that all the names used are essentially correct. We use the term ''Inverter/Converter'' since the design of the unit requires both types of power conversion.

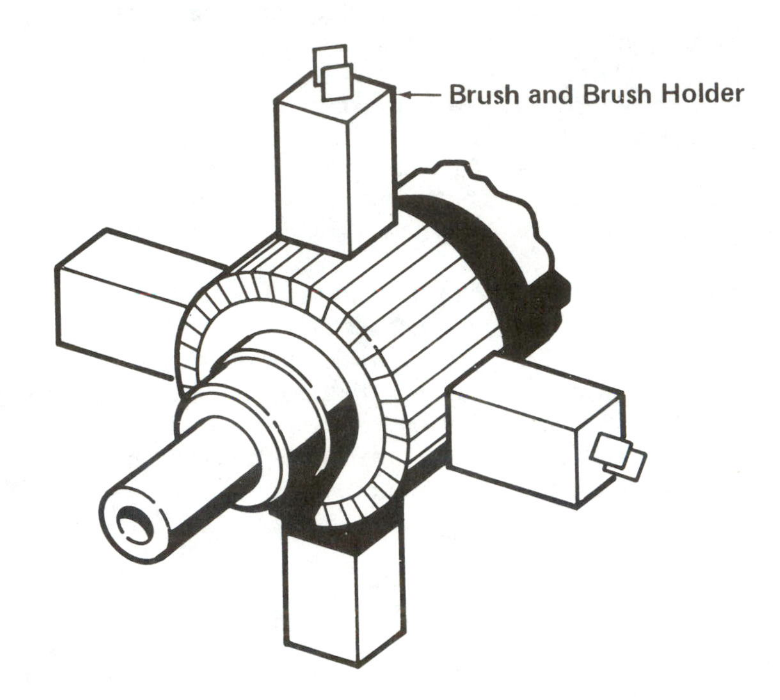

Figure 114. Typical Brush-Commutator Arrangement.

A typical inverter/converter type power source will have primary alternating current (AC) input voltage of either 230 or 460 volts in the United States. Some have connections for either voltage.

Other primary voltages such as 380 or 575 volts are available from some manufacturers. The 575 volt systems are used in some parts of Canada and 380 volt primary is common in Europe.

Power Flow Sequence

The alternating current primary power is brought directly to an "On-Off" switch or circuit breaker in the power source. When the switch is turned "On", AC power energizes the fan motor, the control circuit and the main power rectifier circuit. The components work as follows:

The **fan motor** provides cooling air flow over the component parts of the welding power source. This is important to the duty cycle rating of the unit.

The **control circuit** is mostly contained within a printed circuit board. The printed circuit (PC) board is the "brain" of the power source.

The **main power rectifier circuit** is a full wave bridge rectifier system that converts the primary AC to direct current (DC). As you can see, the converter portion of the power source goes right to work!

Figure 115. Typical Inverter/Converter Power Source.

The DC power is applied alternately to two silicon controlled rectifiers (SCR's). The SCR's alternately charge and discharge two power capacitors in the circuit. The rate, or frequency, of the alternations of the two SCR's may range from about 800Hz to 25,000Hz. This is the "high frequency" that is referred to in this type of welding power source. The actual frequency is controlled by the electronic control circuit. The power coming from the two SCR's is a new, single phase alternating current.

At this point, the circuit power is still at the primary voltage (230 or 460 volts). The *single phase AC* primary voltage is applied to a transformer which changes the relatively high primary voltage to a usable welding voltage. The frequency of the secondary AC power is exactly the same as the frequency of the primary AC. Remember, a transformer can change voltage and amperage but it cannot change frequency.

The single phase, low voltage AC power is rectified by two silicon diodes to provide DC welding power at the output terminals of the power source. These silicon diodes are NOT part of the main power rectifier circuit. It is usual to have some form of electronic filtering circuit to absorb the "high frequency" ripple current from the inverter/ converter system. Inductors and capacitors may be used in the filtering circuit. The output welding power is DC.

Observations and Comments

The information available at this time indicates some advantages and disadvantages to the inverter/converter type welding power sources. This data is derived from several sources so what is true for one manufacturer may not be true for another. It is the responsibility of the individual to examine and test the equipment before purchasing. Be sure it will actually perform the welding operations you need!

Advantages

1. **Compact size.** Most inverter/converter power sources are relatively small in physical size compared to standard transformer-rectifiers having the same welding power output. They use less floor space and are highly portable.
2. **Light Weight.** The inverter/converter design permits a total weight of about 1/10th the weight of a standard transformer-rectifier power source having the same welding power output.
3. **Energy efficient.** The inverter/converter design requires considerably less primary power input for a specific amount of welding amperage output. Smaller primary cables may be used.
4. **Portability.** Many welding applications must be done in hard-to-reach locations. The compact inverter/converter units can be taken any place they are required where there is primary electrical power available.

Disadvantages

1. Inverter/converter units are not normally designed to operate at voltages above 50 volts except on an intermittent basis. Damage to the thyristors (SCR's) could result. This would certainly limit its use with the AAC process. (This limitation could be removed at any time due to the fast moving advances of the electronics field.)
2. The volt-ampere curves submitted by manufacturers indicate some distortion in certain voltage-amperage ranges.
3. The output DC welding power is derived from single phase AC. This is offset, however, by the high cycling frequency of the base AC power.
4. Spare parts inventory must include extra PC boards for the solid state control circuit. Replacement or exchange of the damaged PC boards may increase operating costs for the units. This will depend on the manufacturers exchange policy.

Applications

Manufacturers of inverter/converter power sources have models that may be used with SMAW only. Other models are designed for use with either the SMAW, PAW or GTAW processes. Still other inverter/converter power sources are designed to operate with either SMAW or GMAW.

There are a variety of obvious applications for the inverter/converter type power sources. Any welding job that requires mobility of equipment, such as shipbuilding, construction work, and machinery repair can use this type of welding power source. The correct primary voltage (230 or 460 volts) must be available. These units may not be re-connectable for two or more primary voltages.

It is important that the specific model(s) of inverter/converter type welding power sources be tested for approved operation on any specific application. This means that actual comparative welding tests should be made under field welding conditions. The tests should be made under the supervision of personnel competent to judge the power source performance properly. The same joint designs, welding processes, metal thicknesses, and personnel should be used in making the tests. The tests would be considered satisfactory:

1. If the welds applied meet the minimum testing requirements of the governing Code document (ASME, AWS, or API).
2. If the operating characteristics of the power source provides minimum spatter, good welder appeal, and minimum noise levels.
3. If the power source meets the manufacturers minimum specification claims for both primary input power and secondary output welding power and duty cycle.
4. If the unit is price competitive with other welding power sources of similar output ratings and characteristics.

It is important to test the power source under working conditions with your own welders. Then you can evaluate correctly the applications for the inverter/converter power source with your company.

CHAPTER 9

Rectifiers for Welding Power Sources

The function of any rectifier is to change alternating current to direct current. The rectifier may be any of a number of materials which will, under the proper circumstances, provide the rectifying action.

Rectifiers had been used by industry for some time before they were applied to welding power sources. While the use of AC welding transformers was not new to the welding industry, the use of rectifier elements to change AC welding power to DC welding power was a novel idea. Several methods of rectification were known at the time but only two of them provided the current carrying capacity required for welding operations. The two types of rectifiers were the **vapor-arc type,** commonly known as ignitrons and thyrotrons, and the **metallic types** using such elements as cuprous oxide, selenium, magnesium-cupric-sulphide, and germanium.

The vapor-arc rectifiers never achieved commercial popularity for use in welding power sources. This type of rectifier functions at relatively high voltages and, of course, arc welding processes are essentially low voltage applications. In addition, the original cost factor is high for vapor-arc rectifiers and maintenance is expensive.

The metallic group of rectifiers have excellent characteristics for welding power source applications. Of the several metallic elements used, selenium has been the most successful rectifying element. Selenium rectifiers have a higher voltage rating per plate than many other metallic elements considered. The rectifier plate voltage rating, when properly considered in the design of rectifier stacks, makes it relatively simple to determine the exact rectifier requirements for any welding power source with a known open circuit voltage.

In the mid-1960's the element silicon was introduced as part of a two element rectifier, or "diode". Subsequently, the silicon diode has become the most widely used rectifier in welding power sources. Electronic controlled silicon rectifiers (SCR's) are presently in use in not only welding main power rectifiers but in other control circuitry also.

To provide a sound basis of understanding for welding power source rectifiers, we will consider the selenium rectifier in one section and the silicon rectifier in another section. Remember, a rectifier does only one thing which is to change AC to DC. Both selenium and silicon have specific advantages that should be considered when applying them to welding power sources.

It is important to realize that the power source circuit design must be different for silicon rectifiers and selenium rectifiers. When silicon rectifiers were placed in DC output welding power sources designed for selenium rectifiers, the welding output characteristics were very harsh. In this situation, the manufacturer was required to totally re-design the DC power sources for silicon rectifiers.

Selenium Rectifiers

One of the most well known selenium rectifiers built for welding power source applications was the Gold Star selenium rectifier manufactured by the Miller Electric Manufacturing Company, Appleton, Wisconsin. The following data is basically a modified outline of the selenium rectifier manufacturing process used by that company.

The selenium rectifier single plate "cell" originates with an aluminum base plate designated the **back electrode.** The aluminum plate, which is about 0.040" thick, is preferred because of light weight, good radiation heat loss characteristics, and good thermal transfer.

In preparation for the selenium deposit the surface of the aluminum plates are dry grit blasted to remove the tenacious aluminum surface oxides. The roughened surface provides a greater area for the selenium deposit. The selenium will be deposited by the vapor deposition method. This operation is performed in a controlled vacuum system.

The high purity selenium is boiled in the vacuum chamber as the aluminum plate is being passed over the molten metal. The selenium vapors condense on the surface of the aluminum plate to a thickness of approximately 0.004". To achieve this thickness of selenium deposit the aluminum plate is passed over the boiling selenium 150 times! The selenium layer covers the entire surface of one side of the aluminum base plate.

During and after the vapor deposition the selenium layer is processed in such a manner as to establish proper electrical conductivity. Figure 116 shows one step in the manufacture of the selenium rectifier cells.

After removing the selenium covered aluminum plates from the vacuum system, they are placed in a masking device. The masking device covers a circular portion of the center of the plate as well as approximately 1/4" around the periphery of each cell plate.

The selenium cell is then metal sprayed with a low melting alloy of metals complementary to selenium. The application technique is called "thermal spraying". The low temperature alloy spray deposit is called the **front electrode** of the selenium cell. The front electrode materials are usually an alloy of cadmium, bismuth, tin, and other elements. The front electrode alloy has a very low melting temperature. A view of the selenium cell components is illustrated in Figure 117.

When the masking device is removed there is no front electrode coating on those areas that were "masked off". These include the central circular area and the 1/4" outer edge of the selenium cell. The purpose in leaving these areas uncoated is to provide insulation between the front electrode and the back electrode.

The front electrode, sometimes referred to as the "counter electrode", is the cathode element in selenium rectifier plates. (The cathode is always the negative pole in an electrical circuit). The front electrode alloy will have a melting temperature range of approximately 230° F. to 338° F. The element selenium has a melting temperature of about 422° F. This fact shows that selenium rectifiers are temperature sensitive and must operate at lower temperatures than those shown to maintain rectifier integrity.

At this point in the rectifier manufacturing process, the reverse resistance of the rectifier cell falls off too rapidly to permit the cell to be put into practical service. To make full use of the

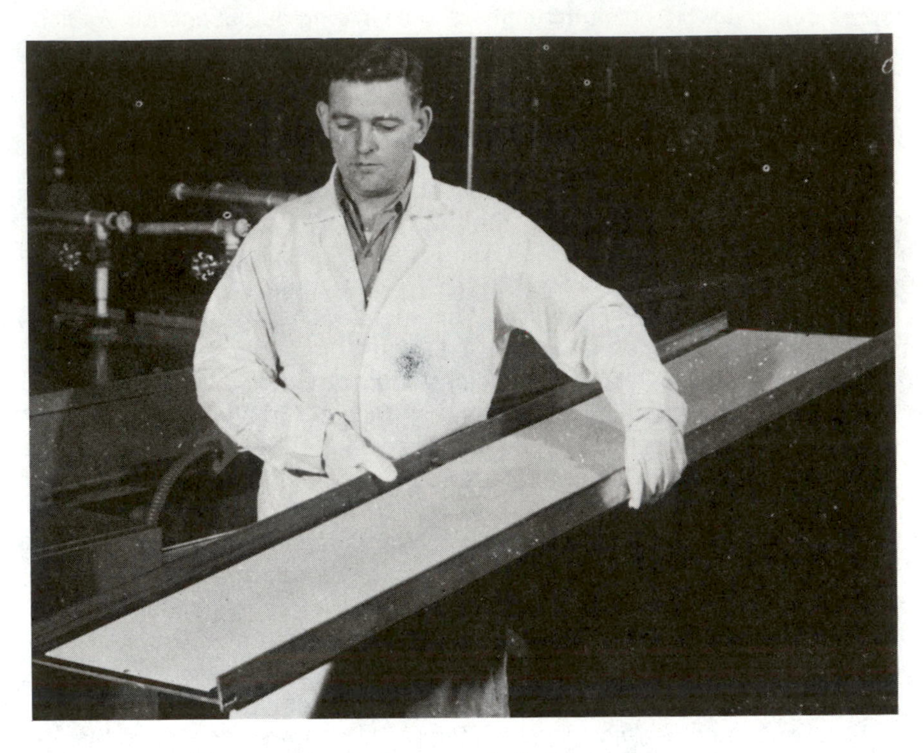

Figure 116. Selenium Cell Back Electrode.

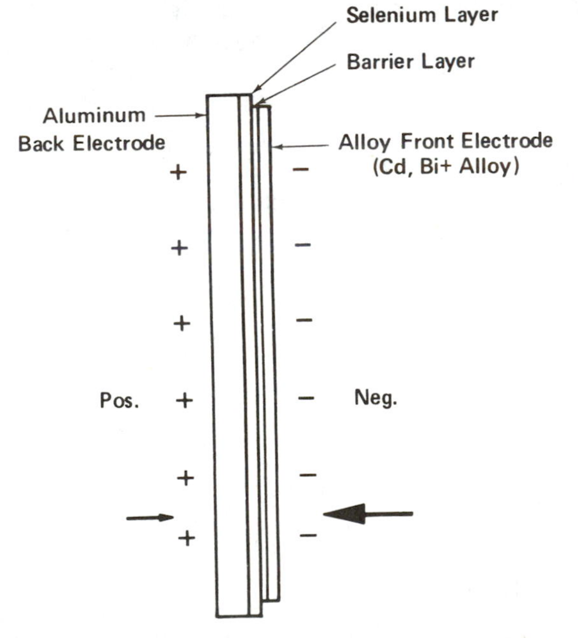

Figure 117. Selenium Cell Component Parts.

selenium rectifier cell it must be processed further by electro-forming. This is done by placing the selenium cells on electro-forming racks where DC voltage is applied to each cell.

The DC voltage is applied to the selenium cells in the reverse direction; that is, positive (+) to negative (−). A "barrier layer" of alloy is formed between the selenium layer and the front electrode alloy. The barrier layer is an alloy of both materials. As the reverse voltage is applied, the cell temperature is allowed to rise under controlled conditions. Additional treatment and "seasoning" completes the formation of, and stabilizes, the barrier layer. The actual barrier layer is only a few molecules thick.

At the beginning of the electro-forming operation, the applied DC voltage is low in value to prevent overheating the selenium cell. As the reverse resistance to current flow builds up in the cell, the applied voltage is increased until its maximum value is well above the rated voltage of the selenium cell.

The theory is that the electro-forming process increases the effectiveness of the barrier layer and aligns the excess electrons in the front electrode immediately adjacent to the barrier layer. There is now an excess of electrons at the front electrode and a deficiency of electrons at the back electrode. This produces a **high negative electrical charge at the front electrode** and a **high positive electrical charge at the back electrode.** Since electron movement is from negative to positive, and electron movement constitutes current flow, current flow is excellent from the front electrode to the back electrode of the selenium cell.

Electrical connections made to the back surface of the back electrode (in the aluminum base plate) are positive (+). Connections made to the front surface of the front electrode (the low melting alloy) are negative (−). The base plate acts as the total assembly support and one electrode (+). The low melting alloy of the front electrode acts as the other electrode (−). The combination presents a one piece rectifier plate requiring no high pressure contact for rectification.

When a specific polarity DC voltage (or one-half of an AC sine wave) is applied to the selenium cell, **the back electrode is polarized and electrically charged positive** and **the front electrode is polarized and electrically charged negative.** In this situation, a current of much higher value will pass through the selenium cell than when the polarity of the cell is reversed. The amount of current flow depends directly on the effective rectifying area of the cell and the applied voltage.

Rectifiers are used to change AC to DC. Each half-cycle of AC power the current flow direction is reversed. Now current is trying to flow from the back electrode to the front electrode. This is not a good situation since we are trying to take electrons from an area of electron deficiency and move them to an area having a surplus of electrons. It is obvious that very little current will flow in the reverse direction. This is the function of a rectifier; that is, to permit current to flow in one direction easily and to minimize current flow in the reverse direction.

Selenium rectifiers have the advantage of being able to accept high voltage surges (voltage spikes) without breaking down. The very large rectifying surface area tends to spread the voltage over many square inches of rectifier surface. The heat transfer rate of the aluminum back electrode dissipates the heat over a wide area. The natural shape of the selenium rectifier stack shown in Figure 118 makes cooling of the recitifer relatively easy.

It is simple to add more rectifying area to selenium rectifiers if it is required for a specific application. Just add more selenium cells to the existing rectifier assembly.

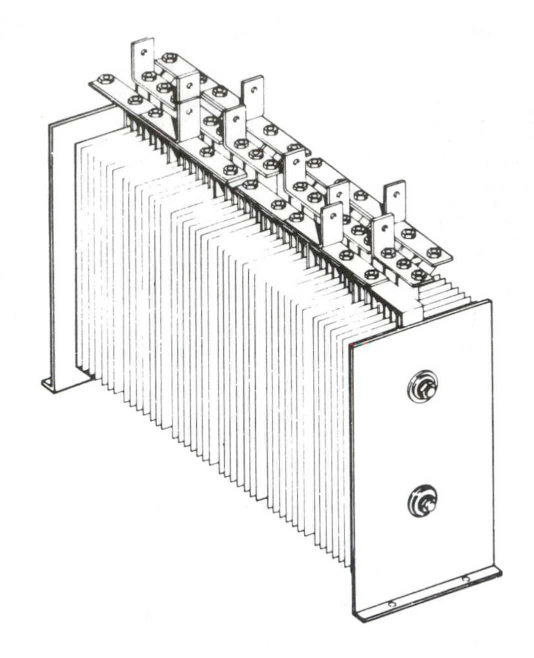

Figure 118. Complete Selenium Rectifier.

Silicon Rectifiers

The silicon diode is a two element rectifier that is used for main power rectifiers by many manufacturers of welding power sources. Silicon diode rectifiers perform the same basic function as selenium rectifiers which is to change AC to DC. Both selenium and silicon rectifiers are classed as semi-conductors.

Silicon is the most common natural element on earth. It is found on every continent and under every sea. The basic element is, of course, sand.

When sand is melted and refined to a certain state it is known as glass. In this form it is one of the best non-conductors of electrical current known. When used with other materials it can be converted to a semi-conductor. (A semi-conductor is an element or material that exhibits almost perfect electrical resistance at 0° Kelvin but which increases in electrical conductivity as its temperature increases.)

Before becoming a semi-conductor the basic materials must be refined to an ultra-pure state. There are several methods of refining silicon and often several methods are used, stage by stage, to achieve the purest end item silicon.

One of the ultimate methods of refinement is known as "zone refining". In this process the ingot of silicon, which is approximately one inch in diameter and twelve inches in length, is placed in an inert gas atmosphere. A wave of heat is applied at one end and caused to move the length of the ingot. Any impurities in the ingot are driven out ahead of the applied heat.

Pure silicon must be transformed into a state that will be suitable for silicon diode rectifiers. This is done by placing the silicon material in a crucible and bringing it to a temperature of approximately 1450° C. (2642° F.). At this temperature the mass is molten. During the time the material is being brought to the correct temperature, desired impurities are added to convert the pure silicon from a non-conductor to a semi-conductor of the proper type. A small amount of gallium or arsenic may be added for this purpose. The impurity addition is usually only a few parts per million (ppm).

The crucible of molten silicon is placed in a crystal growing furnace for the purpose of growing a Czochralski crystal. The temperature of the glowing red mass is accurately maintained within one-quarter of a degree celsius and a "seed" is planted.

The seed is a specially selected crystal of super-silicon which will start the crystal growth phenomenon. The seed is partially immersed in the molten silicon and then very slowly withdrawn at a controlled rate. One rectifier engineer compares the silicon seed withdrawal to "pulling your foot out of a mudhole over a period of two hours!". As the silicon seed is withdrawn from the surface of the molten silicon, a **single crystal** forms that eventually reaches approximately twelve inches in length and about one inch in diameter. It may take as long as four hours to complete a single crystal growth.

The result is a single crystal of **either "N" or "P"type silicon.** The "N" or "P" designation is based on the type of impurity that was used for the silicon semi-conductor. It also designates the basic polarity that will be assigned to the finished silicon diode. The total silicon ingot crystal will weigh approximately one to two pounds and will provide enough material for several thousand silicon diode rectifiers.

The silicon ingot is sliced into thin wafers with a diamond saw. The wafers are only a few thousandths of an inch thick. The next step is the formation of a "junction" where the actual rectification takes place. Without this junction, there would be equal electrical conductivity in both directions through the silicon wafer.

There are several methods that might be used to create the junction. For this discussion we will consider only the "alloy method". The silicon wafers, often called "chips", are placed on the surface of an inert plate. A solution containing another element, usually arsenic or aluminum, is applied to the upper surface of the wafer by silk-screening, painting, or some other suitable method. The entire tray of silicon chips is then sent through a diffusion furnace.

The temperature of the silicon wafers is brought to a critical point and maintained for a specific period of time in the diffusion furnace. During this time the elements in the applied impurity solution diffuse through the silicon wafers. At a precisely timed moment, when the impurities have diffused approximately half-way through the silicon wafer thickness, the diffusion process is stopped.

If the total process has been successful, and it isn't always, the result is a batch of silicon wafers with an excess of electrons on one side (N) and a deficiency of electrons on the other, or (P) side. The actual designations are the "N" and "P" sides of the silicon wafer. After the silicon wafers have been removed from the diffusion furnace the residue of the upper alloying element is removed in preparation for making the ohmic contact.

The silicon wafer must be mounted on the necessary hardware. The wafer, by nature of its thin cross section and brittleness, is a fragile component and must be treated with the utmost care. Figure 119 shows the various component parts of a silicon diode in cross section.

When put into service as a rectifier there will be considerable heat generated in the silicon wafer. The heat must be dissipated to prevent fracture or melting of the wafer. Since

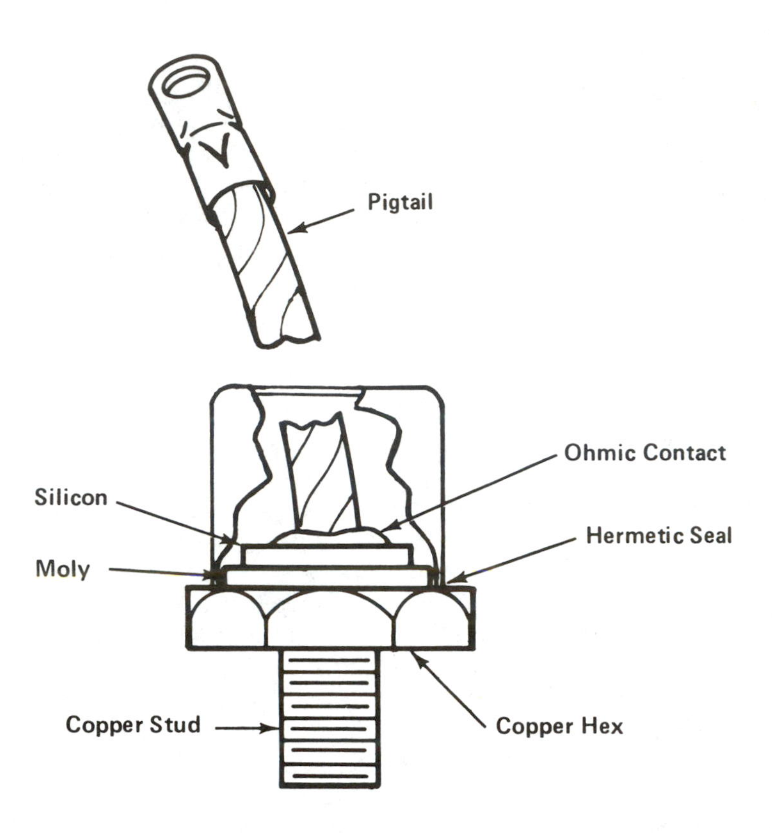

Figure 119. Silicon Diode Components.

copper has excellent thermal conductivity, as well as excellent electrical conductivity, copper stud heat sinks are used. The copper used is normally an oxygen-free grade of high purity.

The copper stud and hex-head arrangement is made suitable for installation on thick aluminum plate heat sinks. The aluminum plate heat sinks host only a single silicon diode per plate in the main power rectifier. Any heat generated in the silicon diode is quickly dispersed by the heat transfer capabilities of the copper stud and the aluminum plate.

On top of the copper hex-head is placed a material which has a coefficient of expansion similar to that of the silicon wafer. Molybdenum is normally used for this purpose. The "moly" is secured to the copper hex-head with some type of hard solder (not a tin-lead alloy) which has a relatively high melting temperature. The silicon wafer is then placed on the molybdenum and hard soldered into place. This is a very critical operation in the manufacturing sequence of silicon diodes. Many parts are rejected at this point because of the high standards of quality control imposed by the manufacturers of silicon diode semi-conductors.

The next step in the construction of the silicon diode is to connect the ohmic, or low resistance, contact to the upper side of the silicon wafer. It is to this component part that the flexible copper pig-tail is attached.

Quality control is very important in the manufacturing of silicon diode rectifiers. All through the manufacturing operations there are cleaning procedures for the hardware, the copper heat sinks, the silicon wafers, the ohmic contact area, etc., because it is imperative that no foreign object be in the finished product. Even the clothing worn by manufacturing personnel must be lint-free.

Water vapor must also be excluded and, to accomplish this, most of the manufacturing and assembly operations are performed in a dry, inert gas shielded atmosphere.

After the connection of the ohmic contact to the silicon wafer, the ceramic seal top cap is applied over the entire device and cold welded to the copper hex-head. Finally, the copper pig-tail is crimped into place. This completes the manufacturing operations except for testing. Completed silicon diode rectifiers are illustrated in Figure 120.

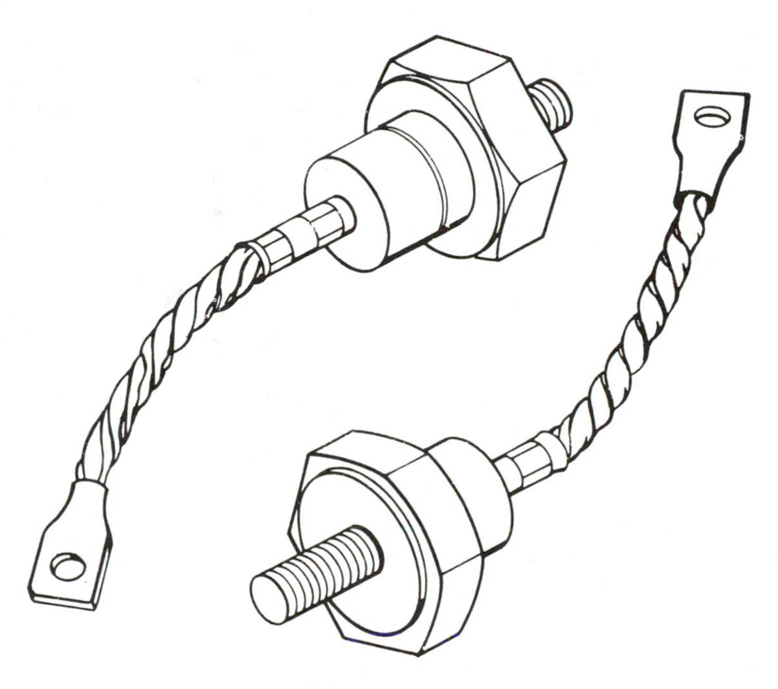

Figure 120. Completed Silicon Diode Rectifiers.

As part of the quality control, electrical tests are made at every step of the manufacturing process. This is done to eliminate faulty parts as early as possible in the manufacturing process. After the final assembly each silicon diode is tested for voltage and amperage rating.

To test the silicon diode for **voltage** capability, a pre-programmed increasing DC voltage is impressed in the reverse direction and the leakage current observed on an oscilloscope. When the leakage current reaches a certain value, the voltage is noted and this becomes the **peak inverse voltage rating** of the silicon diode.

To test the silicon diode for **current** rating, a known forward current value is forced through the rectifier and the voltage drop across the diode is observed. The diode is then rated according to its losses. Forward voltage drop should be as low in value as possible. It is this voltage drop that creates the concave shape to the volt-ampere output curves of constant current power sources.

The result of the total manufacturing effort is to produce an electrical device which, by nature of the rectifying junction, will conduct current in one direction and not the other. The silicon diodes used for main power rectifiers in NEMA Class 1 power sources have a minimum peak inverse voltage rating of 250 volts with a non-repetitive transient over-voltage rating of 350 volts. In actual manufacturing practice, the voltage ratings range from 600 to 900 volts.

The two most commonly used silicon diode rectifiers in welding power source main power rectifiers are the 150 ampere unit with a 3/8'' or 1/2'' diameter stud, and the 275 ampere rated diode with a 3/4'' diameter stud. The silicon diodes are stacked into single phase and three phase bridge rectifiers in much the same manner that selenium cells are used except that fewer diodes are needed for a given rectifier amperage rating.

The aluminum plates used as heat sinks with silicon diodes are large in surface area and thick in depth. These aluminum heat sinks are used to conduct the heat generated in the silicon diode away from the diode and to the air stream created by the welding power source fan. Only one silicon diode is mounted on each aluminum heat sink plate. The mass of the aluminum heat sink will remove any heat generated almost instantaneously.

When installing a silicon diode on the aluminum heat sink, a heat-conductive compound should be used. This specially formulated material provides excellent heat conduction between the silicon diode copper hex-head stud and the aluminum heat sink plate.

Although a torque wrench was required for installing silicon diodes in early rectifier stacks, it is no longer necessary in most applications. Instead, a special Belville washer (a combination washer-compression nut) is used to control the tension on the mounting stud.

The installation of a silicon diode, with Belville washer, is accomplished by putting the diode in place in the aluminum heat sink plate and bringing the Belville washer up to firm finger tightness. An ordinary wrench of the proper jaw size is applied to the nut and one-half turn (180 degrees), no more, no less, is added. The additional half-turn compresses the calibrated Belville washer the exact amount necessary to insure the proper connection. A typical installation diagram for silicon diodes is illustrated in Figure 121.

Care should be taken not to strain the flexible copper pig-tail when installing a silicon diode. A stress could be introduced within the structure of the diode assembly which could

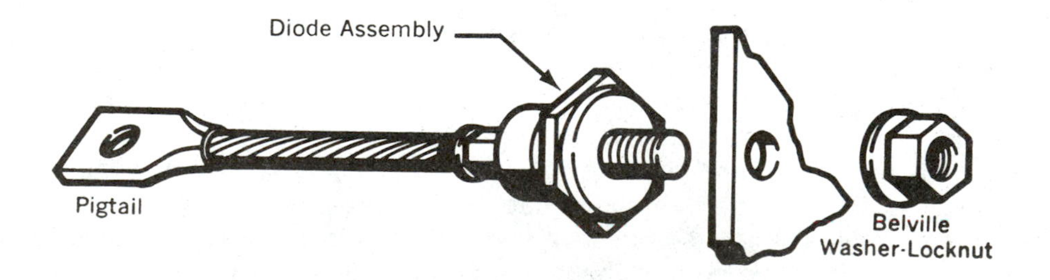

Figure 121. Typical Silicon Diode Installation.

possibly cause the silicon wafer to fracture. Proper installation procedures dictate that the diode should be placed in the heat sink plate, firmly connected by hand, and then the pig-tail should be attached to the proper connection. Only after this has been done can the diode copper hex-head be brought up tight and secured. Before the final tightening of the copper hex-head and Belville washer, the stud should be rotated so no strain is apparent on the pig-tail.

Care should be taken not to strain the flexible copper pig-tail when installing a silicon diode. A stress could be introduced within the structure of the diode assembly which could possibly cause the silicon wafer to fracture. Proper installation procedures dictate that the diode should be placed in the heat sink plate, firmly connected by hand, and then the pig-tail should be attached to the proper connection. Only after this has been done can the diode copper hex-head be brought up tight and secured. Before the final tightening of the copper hex-head and Belville washer, the stud should be rotated so no strain is apparent on the pig-tail.

It is true that rectifiers may fail for a variety of reasons. Silicon rectifiers are no exception to this rule. Diodes that are suspected of malfunctioning may be tested in the field using an ohm meter. THE OHM METER SHOULD BE SET ON THE "X1" SCALE ONLY. Otherwise it will give a false reading of continuity.

Put one probe of the ohm meter on the pig-tail portion of the diode and the other ohm meter probe on the stud portion of the diode. The meter should either read no continuity or some level of continuity. By reversing the probes on the diode and pig-tail you should obtain an opposite reading on the meter to the one noted previously. For example, if you had no continuity shown on the meter in the first placement of the probes, you should have some level of continuity when reversing the probes.

In the rare event that a silicon diode does fail in service it will normally fail in the short circuit position. For this reason it is important to shut off the primary power to the power source as soon as possible after a suspected diode failure.

When a diode fails in the short circuit position a full short circuit load is placed on the secondary AC in that phase of circuitry. This would affect only one-half cycle of AC power, and one diode, so the load on the other diode in the same phase would soon cause it to malfunction. We are talking about time in seconds! The total short circuit current would flow through the secondary coils of the main transformer which could cause them to overheat and possibly damage the insulation. The result would be catastrophic failure of the transformer.

Figure 122 shows a typical silicon diode main power rectifier for three phase power. Each silicon diode is protected from momentary voltage spikes by a wafer capacitor. The configuration shown is for a three phase full wave bridge rectifier.

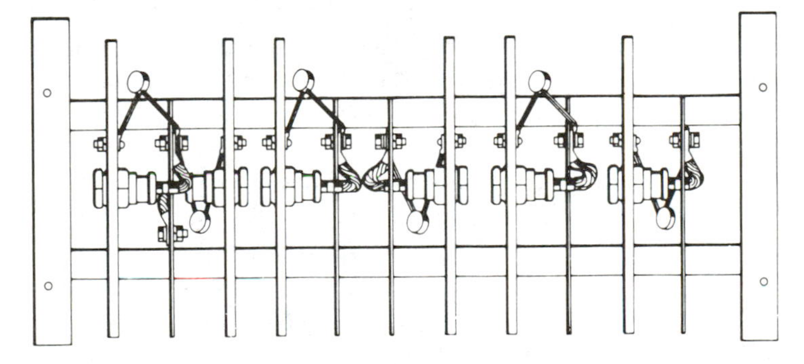

Figure 122. Silicon Diode Main Power Rectifier.

CHAPTER 10

Key Circuits Used in Various Welding Power Sources

The term "key circuits" is defined as "those electrical circuits that are important to the operation of more than one class and type of welding power source". Key circuits are usually some form of control system for regulating welding current output from a power source. Some key circuits that are described in this chapter include:

1. SATURABLE REACTOR AMPERAGE CONTROL.
2. CURRENT FEED-BACK CONTROL.
3. STANDARD 30 VOLT CONTROL.
4. ELECTRONIC SOLID STATE CONTROL.

This is written to help the reader understand the concept of these various control systems as they are used for welding power sources. Those who desire more information on the electrical aspects of these controls should contact the manufacturers for more resource information. The controls discussed are normally used in NEMA Class 1 welding power sources.

Saturable Reactor

The term "saturable reactor" usually causes anyone not familiar with its concepts to veer around the problem and continue in ignorance. Although the term is formidable, the saturable reactor is a simple electrical device used to regulate the output amperage flow in the secondary circuit of certain transformer and transformer-rectifier type welding power sources.

The illustrations shown in the next few pages of the text are designed to explain the functions of the various component parts that make up the saturable reactor control system. The drawings are intended for illustration only. They do not represent any particular electrical circuitry or welding power source model.

We will start with the basic transformer circuit shown in Figure 123. At the left side of the drawing is shown the **primary coil symbol.** Just to the right of the primary coil are three straight lines which represent the transformer **iron core.** Further to the right is the **secondary coil symbol.** The total group of electrical symbols together represents a welding transformer. The balance of the circuit illustrated shows the secondary leads going to the electrode holder (upper right) and the work connection (lower right).

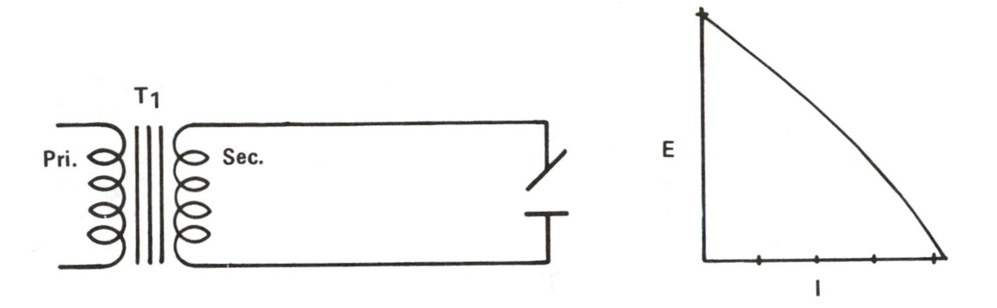

Figure 123. AC Welding Transformer With One Output.

The drawing and volt-ampere curve show the power source has only one output which is both minimum and maximum.

If a coil is added to the secondary portion of the circuit, as shown in Figure 124, the drawing has this added symbol:

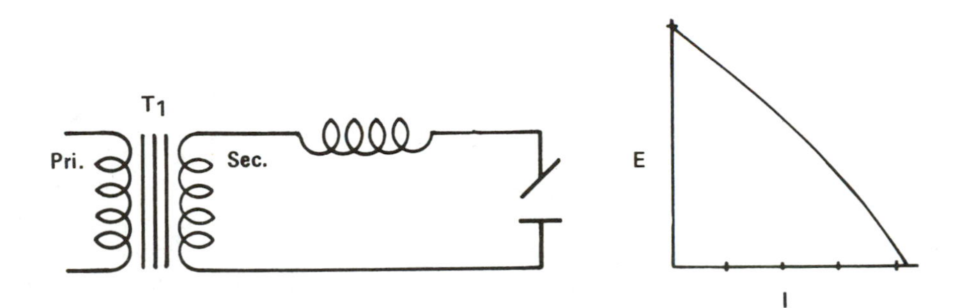

Figure 124. AC Transformer Circuit With Coil.

The addition of the coil to the electrical circuit does create more electrical resistance but only to the value of the added conductor length. The welding current output will be essentially the same as it was before the coil was added. This is shown in the volt-ampere curve illustrated.

The next electrical component to be added to the circuit is an iron core. This is indicated by the three straight lines placed under the coil in Figure 125. Note that we have incorporated the letter "Z" near the coil and iron core electrical symbols. The total electrical symbol shown is the American National Standards Institute (ANSI) symbol for a **reactor.** It is important to know, and understand, that the same electrical symbol also is used to designate a **stabilizer (inductor).**

Some of the electrical symbols approved by the American National Standard Institute (ANSI) are shown in Data Chart 11 located in the appendix of this text.

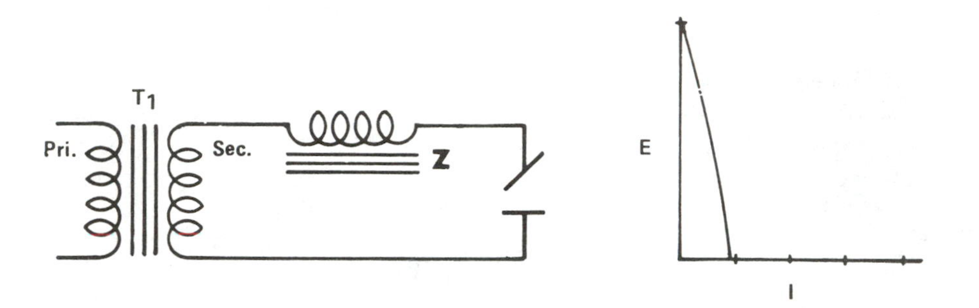

Figure 125. AC Transformer With Coil and Iron Core.

The rule of thumb that is used to determine whether the circuit diagram is showing a reactor or stabilizer is as follows:

a. A reactor is normally located in the secondary AC portion of the welding power source circuitry.
b. A stabilizer (inductor) is normally located in the DC portion of the welding power source circuitry.

The creation of a reactor, as shown in Figure 125, in the welding power source circuitry alters the amperage output characteristics of the power source considerably.

The letter "Z" is the electrical symbol for impedance. Impedance is a combination of resistance and reactance in the electrical circuits of a power source. While amperage flow could essentially be stopped by impedance this is not the design purpose in a welding power source. Instead, impedance is used to limit, or slow down, amperage flow in a welding power source. Impedance does not just "happen". It is the result of several separate occurrences that happen almost simultaneously.

The electrical reactor, although the name has interesting connotations in this nuclear age, is a very simple electrical device. It consists of a conductor, usually copper wire, which is wound as a coil and placed around one leg of the reactor iron core. The two components that make up the reactor are an iron core and a coil of some type of electrical conductor material. When the conductor wire circuit is not electrically energized the two components are electrically inactive. It is not until amperage is caused to flow in the power source circuitry that the reactor comes to life.

Perhaps a bit of review concerning the relationship of current carrying coils, iron cores and magnetic fields is in order. You will recall that when current, or amperage, is caused to flow in a coil wrapped around an iron core, a magnetic field will be created. The strength of the magnetic field depends on three factors:

1. The mass and type of iron in the core.
2. The number of effective electrical turns in the coil.
3. The numerical value of the alternating current in the coil.

Varying any one of these factors will cause the strength of the magnetic field to change.

Again you are reminded that voltage is the electrical force that causes current to flow in an electrical circuit. Sometimes called "electro-motive force" or "electrical pressure", **voltage does not flow** but is the pressure behind current flow.

There are actually several important steps that take place when current is caused to flow in the welding power circuit that has a reactor as part of the system. For clarity, we will number them in the sequence in which they occur. Please keep in mind that everything listed here takes place within 1/120th of a second, or one-half cycle of 60 hertz power.

1. Voltage is induced into the secondary coil and circuit of the welding power source from the main transformer.
2. An arc is initiated at the electrode-work area.
3. Voltage causes current to flow in the secondary electrical circuit, including the reactor coil.
4. As the current level increases each half-cycle, a magnetic field is created in the reactor.
5. Energy is used to create the magnetic field.
6. The energy is not "used up" but is momentarily stored in the magnetic field.
7. The magnetic field strength increases as the current level increases until it reaches either 90 or 270 electrical degrees. (See Figure 126.)

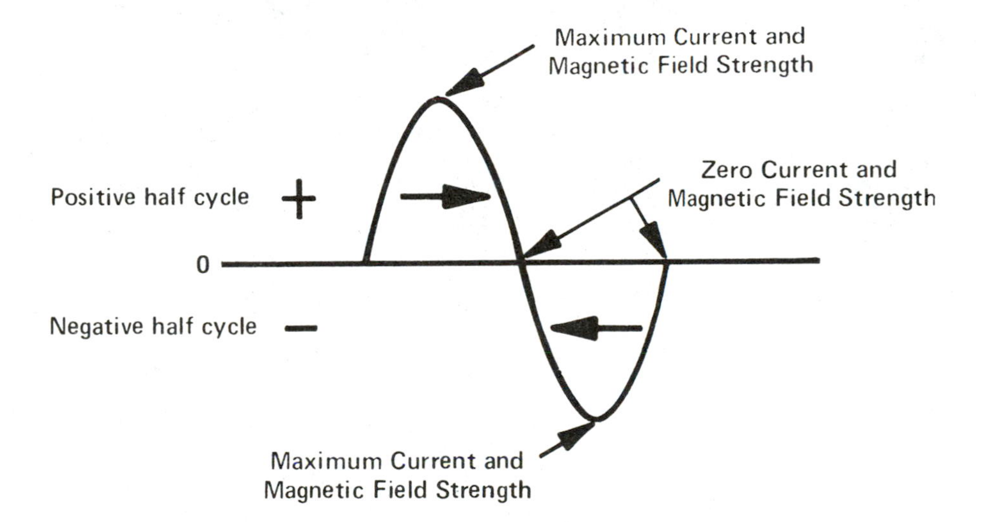

Figure 126. AC Sine Wave and Magnetic Field Trace.

8. Immediately the current level begins to decrease and the magnetic field strength decreases also. (The decrease in magnetic field strength is called the "collapsing" of the magnetic field).
9. The energy stored momentarily in the magnetic field is returned to the circuit, and the reactor, where it originated.

10. The energy from the magnetic field is apparent in the circuit as a counter-voltage with force direction exactly opposite that of the impressed AC voltage in the circuit. This is a classic example of Lenz's Law in action.
11. The impedance factor, or counter-voltage value, will determine how much welding current is allowed to pass through the reactor coil.

The electrical circuit in Figure 125 shows the full reactor in the welding power source circuit. In this situation, the maximum amount of AC magnetic field strength will be created in the circuit and, therefore, the maximum amount of counter-voltage, or impedance. The result is that current flow will be minimum in the welding circuit.

At this point there has been some significant data assembled. The maximum output current level has been determined. The minimum output current level has been determined. By adding a reactor to the welding power circuit a certain amount of output current control has been achieved.

In this circuitry the reactor seems to be a "key" component. From what we have learned it appears that we may be able to obtain even finer output current control if the reactance of the welding circuit could be varied in some manner.

There are at least three methods by which the output welding current could be controlled in the circuit illustrated in Figure 125. One way would be to physically move the iron core in or out of the confines of the reactor coil. This method is cumbersome and rather awkward. Being a mechanical control method, it would certainly limit the use of remote control for welding current output.

Another method would be to saturate the iron core with magnetic lines of force. Ideally, the magnetic field would be created by current flowing in a DC coil wrapped around some portion of the iron core. This would alter the effects of the iron core. This method would require substantial alteration of the existing circuit shown in Figure 125.

The third method would be to change the number of effective turns in the series power coil (reactor coil) around the iron core. This may be accomplished by physically tapping the reactor coil at specific electrical turns of the coil. Use of the various taps is provided by either a tapped range switch or plug-in receptacles. A typical tapped reactor is illustrated in Figure 127.

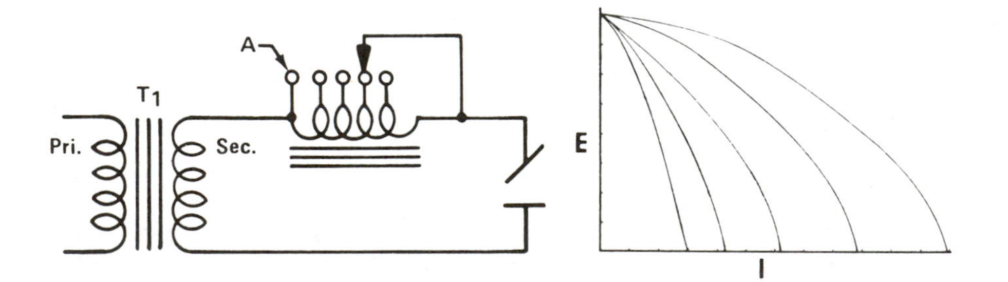

Figure 127. Tapped Reactor Current Control.

The illustration shows that tap "A" will permit the highest current output from the welding power source. Analysis will show that under this condition the entire reactor coil is out of the welding power circuit. There is no impedance to the flow of welding current in the secondary circuit.

The circuit shown in Figure 127 shows there are several amperage ranges that may be obtained with this method of output current control. It is still not possible, however, to achieve the fine current control necessary for most welding applications.

It is true that infinite amperage adjustment is highly desirable in any welding power source. This is especially true for welding applications on foil gauge thicknesses of metals with the gas tungsten arc welding process. It is necessary, then, to devise some method for obtaining infinite current adjustment in the welding circuit with which we are working.

Probably the best way to accomplish our purpose is to change the effective amount of iron in the reactor core. This may be done either mechanically or electrically. Physical movement of the reactor core would be troublesome from the design standpoint. The alternative is to consider the electrical control concept. The necessary modifications to the electrical circuit are shown in Figure 128.

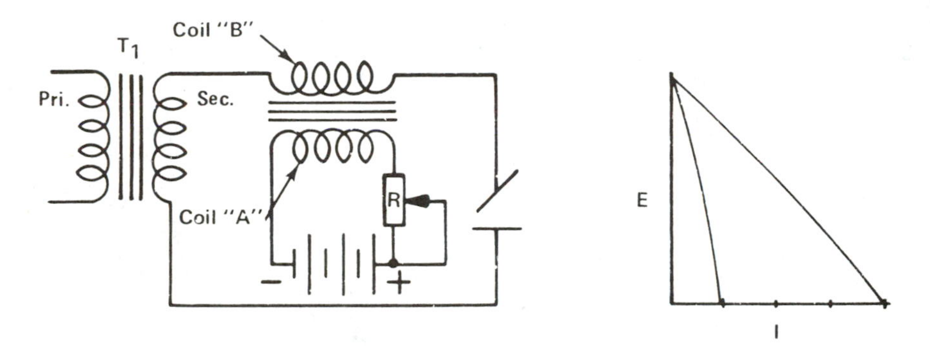

Figure 128. Modified AC Transformer Circuit.

The circuit diagram is now considerably different from those previously illustrated. The basic components are still the same but several more items have been added to the circuit. The equipment additions include a DC control coil (coil "A"), a potentiometer connected as a rheostat (R), and a source of DC power. (In this case, the DC power source is shown as a battery. A battery always provides direct current and is used here only to show the circuit is DC). The DC control coil (coil "A") introduces the electrical control concept of welding output amperage regulation.

Using "R", a potentiometer connected as a rheostat, DC voltage can be impressed on coil "A" which is the DC control coil. DC power is used because direct current flows in one direction only. A magnetic field established with DC power will maintain its strength until the DC power is removed from the circuit.

The full use of the saturable reactor control can now be explained. We have discussed the plain reactor and its limiting effect on the welding output current. Figure 128 shows the reactor with the added DC control circuitry including the DC control coil.

The DC control circuit is an isolated circuit that is common to all phases in the secondary AC welding power circuit of the power source. There is, therefore, no possibility of the DC control current impinging on the alternating current welding power circuit. By the same reasoning, there is no possibility of the AC welding power impinging on the direct current control circuit.

The term "saturable" comes from the word "saturate". To define the electrical meaning:

Saturate = "a state of maximum magnetization".

A "saturable reactor" is a reactor that can be brought from zero magnetization to a state of maximum magnetization by varying values of either AC or DC continuing electrical energy input.

The iron reactor core is common to both the AC welding power circuit (coil "B") and the DC control circuit (coil "A"). It has the ability to absorb a certain number of magnetic lines of force from magnetic fields. It is important to understand that the magnetic lines of force can be provided by either alternating current (AC) or direct current (DC). The impedance factor, or counter-voltage value, is totally dependent on the strength of the AC magnetic field. It is the counter-voltage value that determines how much welding current is available at the output terminals of the welding power source.

We will look at Figure 129 to further illustrate the electric control concept for welding amperage output. This illustration shows a saturation curve at the left side, a volt-ampere curve and a rheostat knob at the right side.

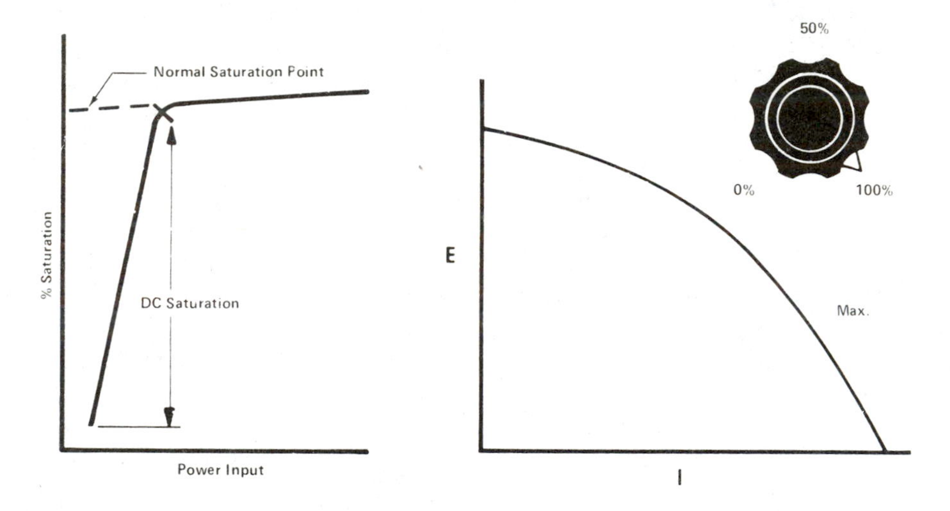

Figure 129. Reactor Iron Core Saturation Curve Data.

The purpose of the saturation curve is to show the amount of reactor iron core saturation by both AC and DC electrical power. Be aware of this fact: the iron core of the reactor is always completely saturated with magnetic lines of force, either AC or DC or a combination of the two, whenever the power source is producing welding current output. Complete saturation of the reactor iron core means that the iron has accepted all the magnetic lines of force that it is capable of carrying per square inch of iron core lamination surface area. Figure 129 shows the total saturation of the iron core with DC power.

The rheostat (R) controls the amount of DC power permitted to flow in the DC control circuit, including the DC control coil. When the rheostat is placed in the maximum output position the reactor iron core is fully saturated with DC power. As long as the DC power is maintained in the DC control circuit the magnetic field created will continue as a steady force.

The **AC welding power,** coming through the welding power circuit and coil "B" at the same instant of time, finds the reactor iron core already completely saturated and unable to accept any more magnetic lines of force. In this situation, there can be no AC magnetic field created and, therefore, no counter-voltage impedance factor. The result is maximum welding current at the output terminals of the welding power source.

At this point we can develop a useful rule for the saturable reactor method of welding current output control:

"Maximum DC power in the reactor control circuit provides maximum welding current at the output terminals of the welding power source".

It is possible to adjust the power source current output by setting the rheostat (R) to approximately 50% of maximum power output. At this setting the reactor iron core would be approximately half saturated with DC power. This is illustrated in Figure 130.

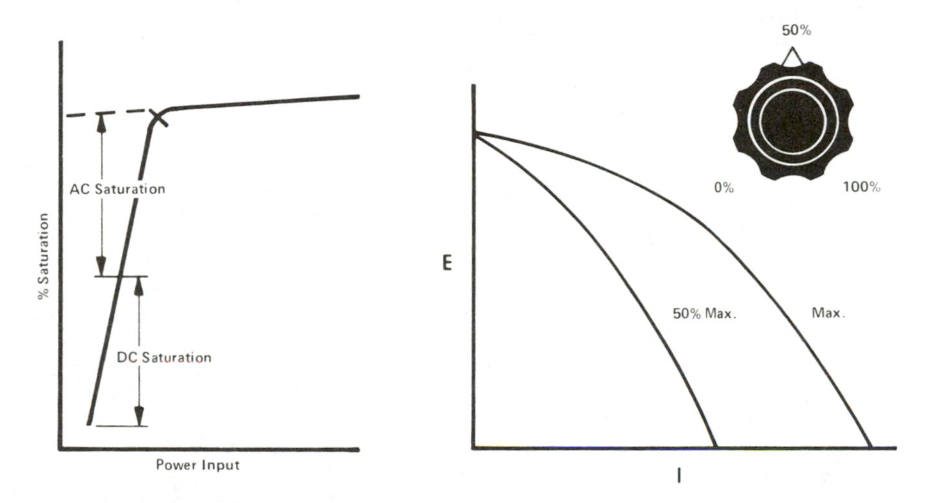

Figure 130. Reactor Saturation Curve With Both AC and DC Power.

With the reactor core only half saturated with DC power, the AC welding current flowing in the series power coil (coil "B") now has the opportunity to create an AC magnetic field. The strength of the AC magnetic field will be limited by the amount of iron available in the reactor iron core. The saturation curve shown in Figure 130 indicates that approxmately half the iron core saturation is accomplished with AC power. Remember this important point: the reactor iron core is always completely saturated when welding regardless of the welding amperage used or the position of the rheostat control.

The energy stored momentarily in the AC magnetic field is returned to the saturable reactor circuit each half-cycle as counter-voltage. The counter-voltage is the impedance factor to current flow in the welding power circuit. For the example cited, the welding current output would be approximately 50% of the amperage range set on the welding power source.

It is apparent that by varying the amount of DC power in the control circuit it is possible to also vary the amount of the AC magnetic field strength. As the AC magnetic field strength is varied so is the induced counter-voltage value (impedance factor). The affect of the impedance factor on the output amperage will cause the power source to stabilize at any amperage value within the capability of the power source. The control of welding amperage is now infinite within the minimum-maximum amperage range of the power source.

In summary, the saturable reactor electric control provides full range amperage output control. One other advantage to this method of current control is the fact that it may be operated from a remote rheostat amperage control unit. Either a remote hand control, or foot control, rheostat may be used by the welder. This eliminates the problem of the welder having to go back to the power source each time he wants to change power source settings.

Three Phase Transformer-Rectifier Circuit Diagram

The circuit diagram in Figure 131 is relatively easy to read and understand if you consider the following:

1. A welding power source is made up of several types of electrical circuits.
2. Each type of electrical circuitry has a specific function to perform.

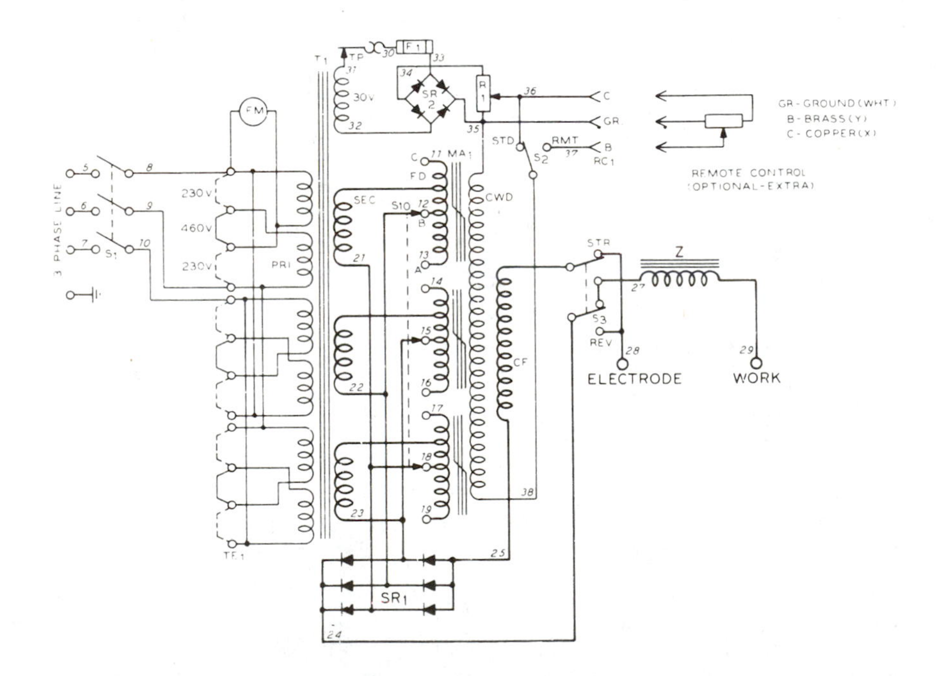

Figure 131. Standard Three Phase Transformer-Rectifier Power Source Circuit Diagram.

3. Various component parts may be used to build transformer-rectifier power sources. They include primary and secondary coils, iron cores, switches, rectifiers, receptacles, fan motors and contactors.
4. A welding power source must be designed so that electrical power follows a logical path through the circuitry to produce the desired result.

An excellent method of tracing out welding power source electrical circuits is to use different colored pencils for each separate part of the circuitry. For example, use red for the primary AC, blue for the secondary AC, etc. This will help you to remember each of the parts of the diagram.

In tracing out the circuits of Figure 131, we will first follow the welding power circuit from primary input to the output DC terminals.

Beginning at the upper left side of the circuit diagram we find the words "3 phase line". Note the lines numbered 5, 6, and 7. This is the primary power coming into the power source from the utility company. Just below those lines is a single line with two cross lines and a dot at the base. This is the primary power safety ground to the power source frame.

Just to the left is S_1, the primary power switch which turns the power source on or off. Proceeding to the right on lines 8, 9, and 10 there are terminals marked "230 V"and "460 V". This means that the power source primary power system is re-connectable for either 230 or 460 volt primary power. The circle, with the letters "FM", extending toward the top of the drawing indicates the power source fan motor. Note that the fan motor is connected across a single phase of the 230 volt connections regardless of the primary voltage used with the power source.

Continuing to the right as we trace the primary AC circuit we find the **primary coils** indicated by the letters "Pri.". There are three complete sets of primary coils shown so we can be sure this is a three phase transformer. This completes the description of the primary AC portion of the power source circuitry.

Just to the right of the primary coils are three straight lines. This is the electrical symbol for an iron core. In this case, it is **the main transformer iron core.** Immediately adjacent at the right of the transformer iron core are three coil symbols labelled with the letters "Sec." They are the **secondary coils** of the main transformer.

This is also the beginning of the secondary AC portion of the welding power source circuitry. (You should use a different color pencil to identify this part of the circuit). The coil and related circuitry at the top of the drawing is another circuit which will be discussed in detail later. It is a separate and isolated circuit called the **control circuit.**

The secondary coils have conductors numbered 21, 22, and 23 leading to another set of electrical coils labelled "FD". The FD means **flux diverter.** In the circuit diagram they are shown as the **AC reactor control coils.** These coils are part of the welding power circuit.

The AC reactor control coils are each wrapped around one leg of separate **reactor iron cores.** The reactor iron cores are shown as three separate units. The reason for three reactor iron cores in the circuit is that each of the three electrical phases of secondary AC has a separate reactor.

The AC reactor coils have three taps indicated by the letters "A", "B", and "C". The arrows pointing to tap "B" are connected by a dotted line which indicates that, when one moves, they all move. This is known as a "gang" switch. The switch is labeled S_{10}. It is called the **amperage range control switch.**

Following the circuitry from the AC reactor control coils, the lines go to SR_1 which is the main power rectifier. The rectifier changes alternating current to direct current. The main power rectifier may be either selenium or silicon. At the point where the three conductors attach to the main power rectifier the secondary AC part of the circuit ends. Everything in the welding circuit from this point on is direct current.

Using another color of pencil, we will now trace the DC portion of the welding power circuit. In the DC portion of the circuit there are two conductors attached to the main power rectifier. The conductors are numbered 24 and 25. The number 24 conductor leads to S_3 which is the polarity switch. This switch changes the DC welding power from one polarity to another without having to change the welding cables from one power source terminal to another. At this point the electrical conductor becomes number 27 (as indicated in the illustration). The unit in the circuit diagram is set up for DCSP (electrode negative).

Line 27 goes to an electrical symbol showing a coil, iron core, and the letter "Z". You may recall the rule of thumb that explained this symbol as being either a reactor or a stabilizer. If it is in the AC portion of the circuit it is a reactor. If it is in the DC portion of the circuit it is a stabilizer. The purpose of the stabilizer is to smooth the ripple factor that is always present in rectified DC welding power output. The result is a more stable welding arc. The conductor goes from the stabilizer as line number 29 to the "work" terminal.

Beginning at the electrode terminal and following line number 28, we come to S_3, the polarity switch. The conductor then goes to a coil labelled "CF". This is the **current feed-back coil.** It's position in the circuit diagram indicates that it is common to all three reactor iron cores. The purpose and function of the current feed-back coil is discussed in subsequent paragraphs of the text.

Following line number 25 leads back to the main power rectifier, SR_1. This completes the tracing of the welding power circuit from primary AC to DC welding power at the output terminals of the power source.

The electric control circuit is the key to amperage output regulation in this type of welding power source. The next step is to trace the control circuit and relate it to the entire circuit diagram.

The coil located near the top of the circuit diagram and next to the main transformer iron core is a **standard 30 volt AC coil** designed to provide power for the entire control circuit. Conductor 32 goes directly to SR_2, the control rectifier. Conductor 31 goes to the symbol "TP", a **bi-metal thermostat.** In this circuit diagram, the thermostat is physically located on the selenium rectifier as a protective device. It will open at about 175° F. and will automatically re-set and close at about 150° F.

F_1 is a ten ampere plug type fuse used for protection of the control rectifier shown as SR_2. It is at SR_2 that the AC control voltage and amperage are changed to DC. Conductors 34 and 35 extending from the SR_2 control rectifier carry direct current. Conductor 34 leads to R_1 which is the main welding rheostat for controlling welding current output. Lines 35, 36, and 37 are conductors leading to an electrical symbol shown as RC_1. This is the remote control receptacle into which a remote control device, such as a foot control or hand controlled rheostat, may be inserted.

The switch indicated as S_2 is the remote-standard switch for control of the remote amperage control receptacle. When this switch is in the standard position, welding power output is controlled by the main welding rheostat on the power source. When the switch is in the remote position, welding power may be controlled at the work site by either a remote hand rheostat or a remote foot rheostat.

Proceed along line number 38 to coil "CWD". The CWD coil indicates the **DC control winding,** better known as the **DC control coil.** The DC control coil is common to all three reactor iron cores. Proceed to the black dot which is an indication that line number 35 connects to the CWD coil. Follow line number 35 to SR_2. This completes the examination of the standard 30 volt AC control circuit typical of a three phase transformer-rectifier power source having DC output.

Current Feed-Back System

The current feed-back circuit is used on some three phase transformer-rectifier power sources with DC output only. The current feed-back coil is wound on the same basic coil form as the DC control coil and is common to the same reactor iron cores. As previously noted, the control circuit carries low voltage, low amperage DC power. The current feed-back coil is in the DC portion of the welding power circuit and reflects the DC amperage output of the power source.

The circuit diagram illustrated in Figure 131 is for a standard three phase transformer-rectifier power source with DC welding output. An examination of the circuit diagram shows the current feed-back coil, labelled CF in the drawing, in the DC portion of the welding power circuit. Of course, DC welding amperage is flowing in the current feed-back coil when welding is in progress.

The current feed-back coil is physically wound around the outside of the DC control coil. The greatest difference in the two coils is that the DC control coil is made of many electrical turns of light gauge conductor wire (about 550 turns) while the current feed-back coil is made of very few turns of heavy gauge conductor wire (about 15 turns). Keep in mind that the current feed-back coil is in the DC welding power circuit and must be made of conductor wire capable of carrying the rated amperage of the power source.

In the discussion of the saturable reactor there was a thorough explanation of the meaning of saturation. The diagram in Figure 132 shows the saturation curve of the iron in the reactor cores.

The vertical axis indicates the percentage of iron core saturation. The horizontal axis shows the amount of power applied. The normal operating point of saturation is shown with the cross line at the "knee" of the curve. The saturation point is located here because the reactor iron cores are always fully saturated when the power source is being used for welding.

Remember: Saturation of the reactor iron cores may be accomplished with direct current (DC) from the DC control circuit, or, by alternating current (AC) from the AC reactor coils (series power coils) in the welding power circuit.

DC welding current is passing through the current feed-back coil wrapped around the reactor iron cores. This helps stabilize the output welding current in the arc.

If there is a sudden increase in DC amperage at the electrode, such as would be caused by "sticking" the electrode to the work surface, the current feed-back coil would reflect the increased amperage at the saturable reactors. The greater DC power input to the reactor iron cores would cause additional magnetic lines of force to super-saturate the reactor iron cores.

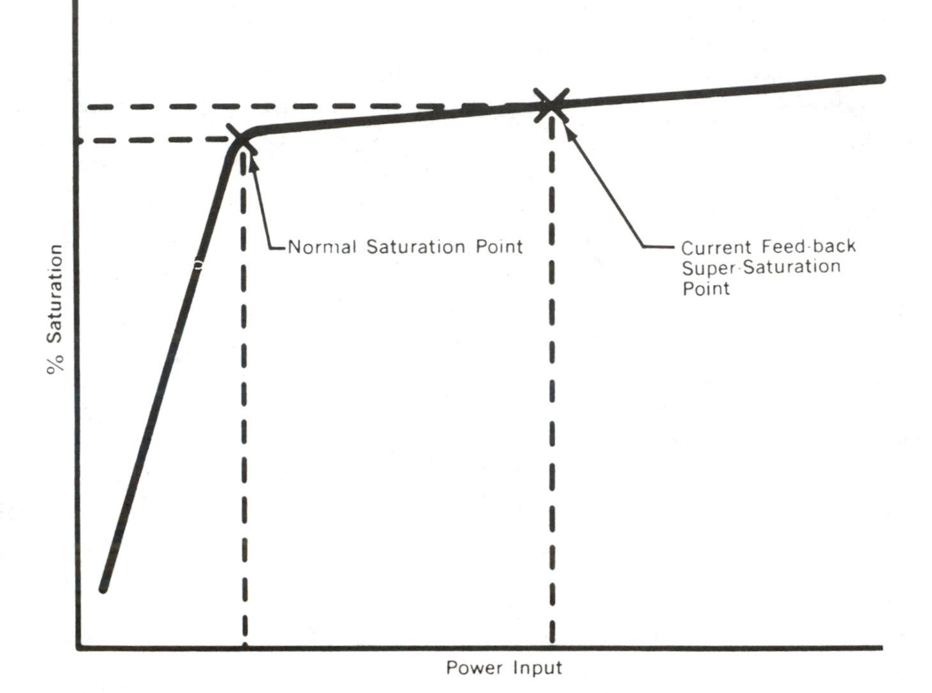

Figure 132. Reactor Iron Core Super-Saturation Curve.

This may be called super-saturation of the reactor iron cores because a much greater amount of DC power is now applied to the saturable reactors. In the illustration, Figure 132, the increased saturation level is shown by the dotted lines at the right side of the curve. Although there is substantially increased power input, there is very little increase in the saturation level of the iron cores.

Standard 30 Volt Control Circuit

The electric control circuit, and its relationship to the rest of the welding power source circuitry, has been well illustrated in Figure 131. A more simplified drawing of the control circuit only, with conductor wires numbered as in the regular circuit diagram, is illustrated in Figure 133. This drawing shows specific voltage values for different parts of the standard 30 volt control circuit. Careful review of the voltage check points in the DC control circuit will assist in making actual checks on this type of control circuit.

Between lines 31 and 32 there is a coil symbol shown. This coil supplies 30 volts AC to the basic control circuit. Line 31 leads directly to SR_2 which is the control rectifier.

Line 32 shows two safety devices in the control circuit: a thermostat (TP) and a fuse (F_1). Note that line 32 becomes line 30 after the thermostat and then changes to line 33 just prior

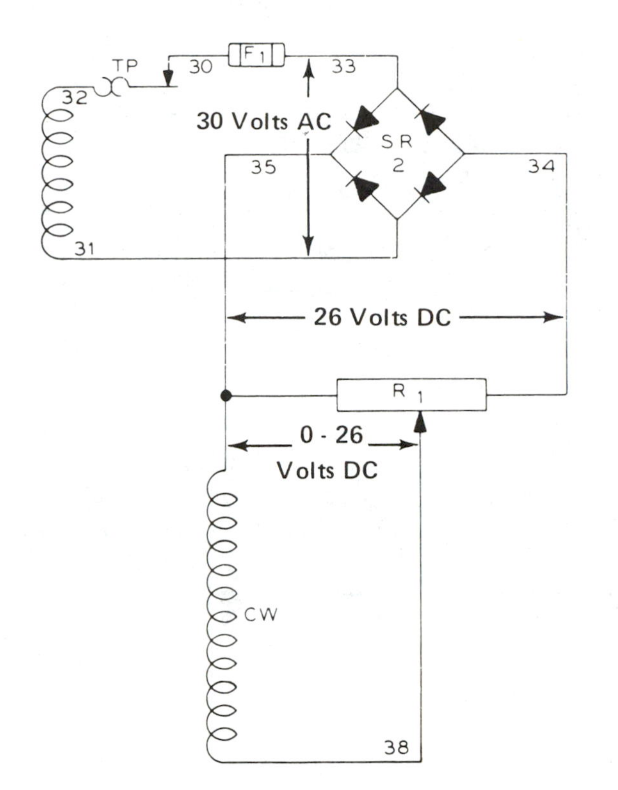

Figure 133. DC Control Circuit Diagram.

to entering the control rectifier (SR_2). There are 30 volts AC between lines 31 and 33. This completes the AC portion of the control circuit.

Lines 34 and 35 are the DC output conductors from SR_2, the control rectifier. There are approximately 26 volts DC between lines 34 and 35. The small voltage loss is typical of the energy loss through a rectifier of any type.

Following line 35 vertically we find a black dot which indicates a connecting junction between two or more wires. Below the connection is a coil, labelled CW, which represents the DC control coil, or "winding". Line number 38 terminates at R_1 which is the main welding rheostat on the power source. The voltage between lines 35 and 38 should read about 0-26 volts DC as the rheostat is rotated through its entire range.

Some electric control circuits have higher voltages than those shown here. Check the circuit diagram for the power source being tested if the voltage values are substantially different from those discussed here.

Electronic Solid State Circuitry

The term "solid state" has been applied to a lot of different electronic and electrical devices in recent years. Unfortunately, there is seldom an explanation defining the meaning of the term "solid state". According to Webster's Dictionary, the term is defined as follows:

Solid State = "Any electronic device which has solid material such as transistors, SCR thyristors, etc. substituted for movable parts, gaseous elements, vacuum tubes or filaments."

The silicon diode, discussed in a previous chapter about "rectifiers", is a solid state device. It is, as you know, simply a one way valve that allows current to flow easily in one direction but not the reverse direction. Subsequent to the introduction of the silicon diode for rectification of welding power, the silicon controlled rectifier (SCR) was developed for direct welding output control.

Silicon Controlled Rectifiers (SCR)

An SCR is a silicon diode rectifier with an electronic "switch" called a gate. The gate is activated by an electrical signal. This permits current to flow as long as the anode is positive (+) and the cathode is negative (–). The SCR gate is simply an "off-on" switch for the diode. It does not control the current flow and it cannot shut the current flow off.

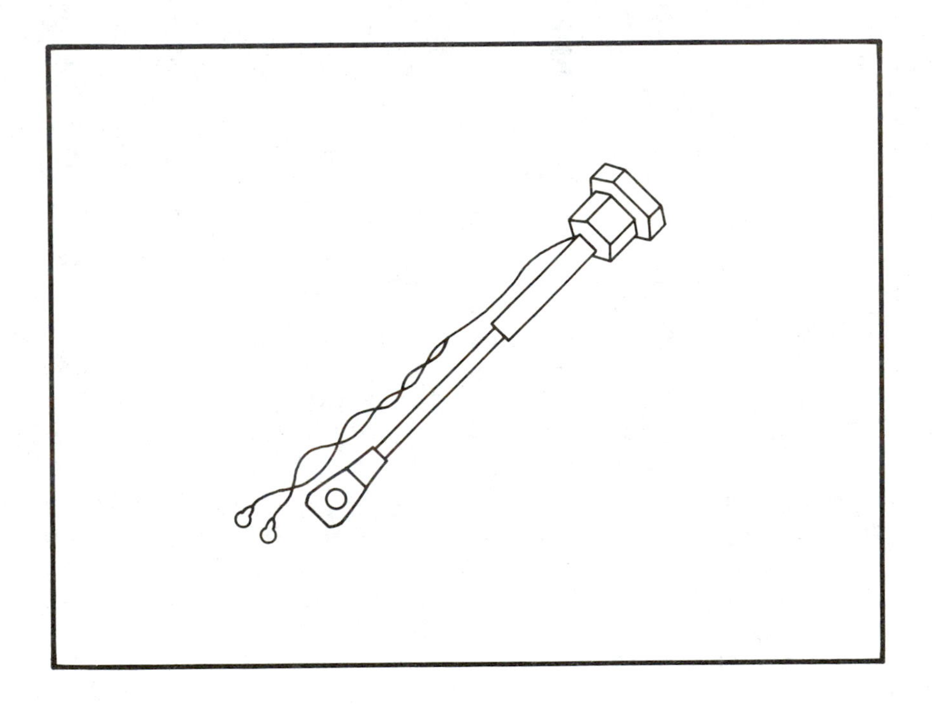

Figure 134. Silicon Controlled Rectifier (SCR).

When the direction of alternating current changes, as it does each half-cycle, the anode becomes negative to the positive cathode. This stops current flow in the SCR diode. Another electrical pulse to the gate is required before current can flow again.

Printed Circuit Boards

Much of the solid state electronics used with welding equipment is supplied on printed circuit (PC) boards. In most cases, these are not shown by individual components in circuit diagrams. Rather, they are simply shown as a "printed circuit board" with inlet and outlet receptacles. There is no indication of the electronic devices on the PC board or their values. When a PC board malfunctions, most manufacturers recommend an exchange PC board with the defective unit returned to the manufacturer for repair. In most cases, only one of the electronic devices on the board is defective and easily replaced. In this situation, the PC board would be tested and put back in stock as a replacement unit.

Transistors

Another solid state device that is used in many types of welding power sources is the transistor. Most of us are familiar with the term "transistor radio" but seldom do we know exactly what it means. Many people, in fact, equate the term transistor radio with all radios of small size. The term "transistor" is defined as follows:

Transistor = "A device using a semiconductor, such as germanium or silicon, which performs many of the functions of a vacuum tube but without the physical size and power requirements." It therefore allows for miniaturization of electrical and electronic equipment.

Transistors offer essentially proportional relationship of input to output. When there is no input signal there is no electrical conduction. An input of small value causes a proportional small value of conduction. A large input signal causes a proportionally large conduction. An advantage of transistors is that they can be switched off at any time.

Integrated Circuits

The development of integrated circuits has created a situation where welding equipment controls can be computerized. Essentially low voltage, low current devices, integrated circuits are used mainly in control circuits on solid state controlled equipment. To define:

Integrated Circuit = "A miniaturized electronic circuit comprised of transistors, diodes, resistors, and other electronic devices; the IC is formed by processing a semiconductor 'chip', usually silicon."

Additional data on specific solid state circuits may be obtained from the manufacturers of welding power sources and equipment.

CHAPTER 11

Some Power Source Troubleshooting Techniques

Troubleshooting welding power sources requires some knowledge of electricity, good safety practices, and lots of common sense. It is wise to seek all the assistance available from the manufacturer before beginning to check out any power source internally.

For example, all major manufacturers of welding power sources provide an Installation, Maintenance and Operation Manual with each new power source. The ''M & O'' manual, as it is often called, should be read and understood by the welder and the maintenance person who works on the unit. Very often the source of trouble can be found by reading the correct installation procedures for the power source. Most M & O manuals also provide a circuit diagram for the power source as well as a ''troubleshooting guide''.

The troubleshooting data provided in this chapter will be helpful in working with any brand of transformer-rectifier welding power source. All of the component parts discussed may not be on your particular power source. Not to worry; just check out those parts that are present.

Useful Tools

Certain tools will be useful in working on welding power sources. For example, the outer covering is normally installed with self-tapping machine bolts and screws. Adjustable wrenches and large size screw drivers may be used to manually remove these fasteners. Pliers, wire cutters, and a volt-ohm meter will be useful at certain times. All electrical hand tools should be well insulated to protect the worker from electrical shock.

The senses of smell and sight will certainly be useful when you first remove the cover of the power source. In many cases, welding power source problems are caused by overloading the power source output system when drawing excessive amperage. This may cause the conductors to overheat and burn the insulation. The resulting damage would be visible and would probably have some odor.

A check list, with each step of a typical welding power source checkout, would be very helpful. The check points should be correlated to the circuit diagram and the troubleshooting guide provided by the manufacturer.

The data supplied in this text is presented as a guide for determining the proper methods for troubleshooting the power source. This includes the power source internal circuitry as well as the external circuit comprised of the welding cables, electrode holder and ground clamp.

Each of the following listed items will be discussed and explanations given as to what you should look for and what you should look out for!

1. Primary voltage and phase.
2. Primary power input terminal panel linkage.
3. Check open circuit voltage (OCV).
4. Welding power source front panel components.

It is true that a professional craftsman values his working tools above almost anything else in his possession. In the welding industry, the least understood tool is the one that is most important to the welding process, the **welding power source.**

Primary Voltage and Phase

This is an area of concern that is often overlooked when troubleshooting welding power sources and other electrical equipment. Primary voltage and phase are considered to be constant values. **Primary amperage** is a **variable value** which is dependent on the welding power source secondary output rating and requirements.

It is well to check out the primary power distribution system for stable voltage if there are fluctuations in the welding power output of a power source. Primary voltage may be any one of the common values such as 208, 230, 380, 460, or 575 volts. A voltmeter is used to measure the actual line voltage in the primary system. Remember that voltage is always measured across two electrical conductors.

For any specific primary voltage rating in an electrical system it is not unusual to find the actual primary voltage either high or low in value. The result would be either high or low amperage output values for a particular power source setting.

The reason for this is quite simple. A welding transformer is a totally electrical device and it can only produce, in welding power, a percentage of the total primary electrical power drawn. If the input voltage is low, therefore, the output power must be low for a specific control rheostat setting.

It is important that the primary electrical service be the correct phase for the welding power source. For example, if the power source has AC or AC/DC output, the primary electrical service must offer single phase or use one phase of a three phase system. If the power source produces DC output only, it will probably require three phase primary power.

Be sure and check all primary line fuses for continuity. This should be done by opening the line disconnect switch BEFORE removing the fuses. Remove the fuses with a fuse puller and check them with an ohm meter. Defective fuses will have no continuity indication on the ohm meter.

Should one fuse be defective in a three phase system it would be apparent as single phase power input to the unit. The result would be a slow running fan and low welding current output from the power source.

Terminal Linkage

All transformer-rectifier welding power sources have a primary terminal panel for connection of the primary power leads to the welding power source. It is necessary that the primary terminal linkage and the primary voltage be compatible for safe, efficient operation.

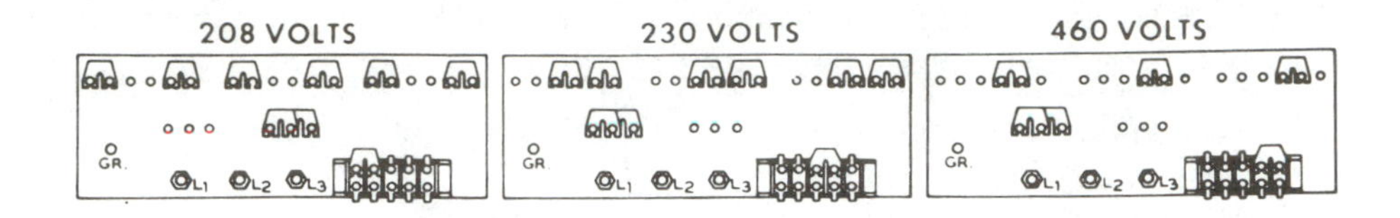

Figure 135. Terminal Linkage at the Primary Terminal Panel.

A typical terminal linkage setup for several different voltages is illustrated in Figure 135. Note the lower right hand terminal strip in each voltage illustration. This would be the typical linkage connection for an auxiliary transformer such as a control transformer. It would not appear on all welding power sources.

Welding power sources are shipped from the manufacturers plant set for the highest primary voltage at which they will operate. If the power source is a standard unit with 230/460 volt reconnectable linkage it should come to the user set for 460 volts primary power.

A welding power source that is set for 460 volt primary power but connected to 230 volt power will not produce the welding power, for a given setting of controls, that it should under normal conditions. Conversely, a welding power source set up for 230 volt power but connected to 460 will be seriously damaged by the excess voltage.

Test Open Circuit Voltage

Most commercial welding power sources have both the primary and secondary electrical ratings listed on a front control panel where the output controls are located. Most NEMA Class 1 power sources have 80 volts or less open circuit. Certain NEMA Class 1 power sources with DC output may have up to 100 volts open circuit.

NEMA Classes 2 and 3 power sources, often either AC/DC or AC output, may have two open circuit voltages in different output amperage ranges. The low amperage range usually has 80 volts open circuit. Often the high amperage range will have much less than 80 volts open circuit. Trying to weld with AC output power sources having less than 60 volts can cause arc pop-outs with certain SMAW electrodes.

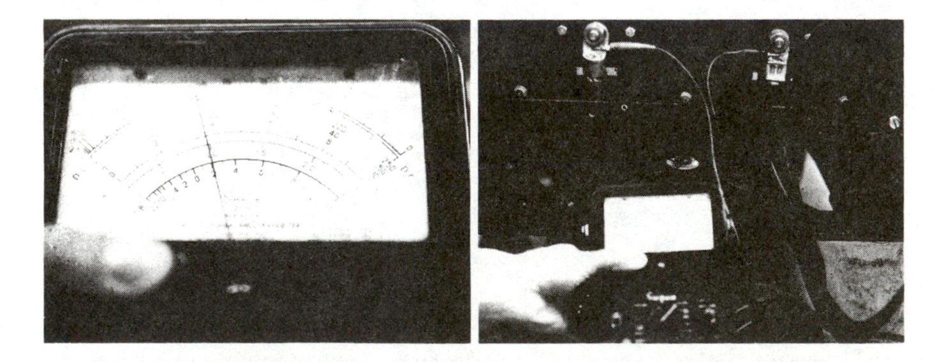

Figure 136. Checking Open Circuit Voltage.

As shown in Figure 136, open circuit voltage may be checked by placing voltmeter probes on the positive and negative, or electrode and work, terminals of the power source. The welding power source is energized for this simple test. It is a safe practice to remove all welding cables from the power source when testing with the power source energized.

The open circuit voltage meter reading should be within five percent (5%) of the rated open circuit voltage of the power source for good electric welding operations.

Welding Power Source Front Control Panel

Many constant current type welding power sources have similar control parts and systems. All electrical control units, for example, provide some form of remote control access to the power source output. Figure 137 shows a typical control panel for a three phase transformer-rectifier power source. This section of the text will discuss the various controls that affect the operation of the power source.

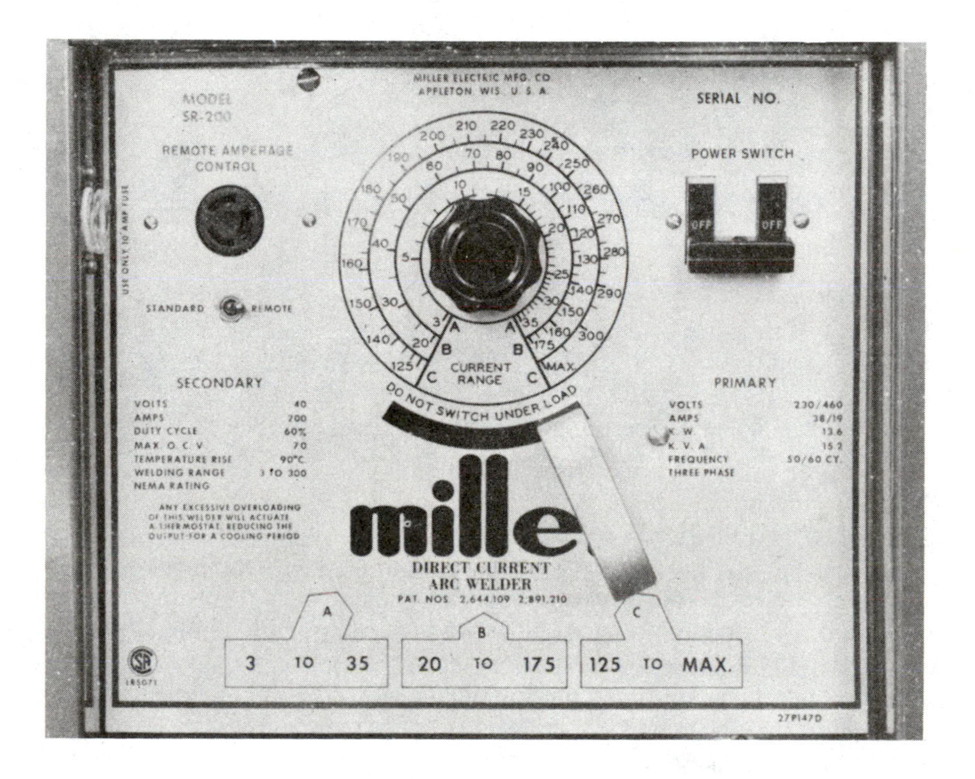

Figure 137. Constant Current Power Source Control Panel.

Standard-Remote Switch and Receptacle

The three-pronged **receptacle** (Figure 138) is where a remote foot or hand amperage control is plugged into the power source. When the standard-remote switch is in the **standard position** all welding current output control is accomplished at the main welding rheostat

Figure 138. Standard-Remote Switch and Receptacle.

on the front panel of the power source. When the switch is placed in the **remote position,** amperage output adjustment may be controlled at the welder's work station with either a remote foot control rheostat or a remote hand controlled rheostat.

CAUTION: DO NOT SWITCH OUT OF THE STANDARD POSITION UNLESS A REMOTE CONTROL DEVICE IS PLUGGED INTO THE REMOTE AMPERAGE CONTROL RECEPTACLE. Such action could cause damage to the toggle switch.

Ten Ampere Fuse

The ten ampere fuse is a plug type, screw based unit located on the front panel of some welding power sources. It is used to protect the remote control circuit as well as the control rectifier.

This fuse link is visible and may appear to be in good condition yet still be defective. If there is any doubt in your mind, check the fuse for electrical continuity with an ohm meter. An operating fuse will show continuity. A defective fuse will show no reading on the ohm meter.

In some instances people have placed a piece of paper behind the fuse base in the fuse receptacle. The paper acts as an insulator and the fuse does not function in the control circuit. Another favorite trick is to place a small piece of plastic tape over the fuse base contact point. Be alert to this type of horseplay because it can, and does, occur.

In connection with the ten ampere fuse there is a ceramic fuse block located just behind the fuse and the control panel. The ten ampere fuse is threaded into the fuse block. Since it is a ceramic block it can be broken fairly easily. If the fuse has tested out satisfactorily it would be wise to check the fuse block contacts for continuity. This may be done by placing the ohm meter probes on the input and output contacts of the fuse block. No continuity means a defective fuse block.

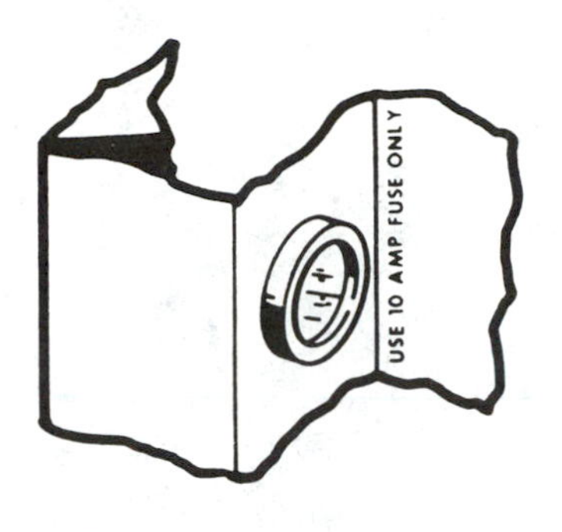

Figure 139. Ten Ampere Protective Fuse.

Current Selector Rheostat

The current selector rheostat may be tested by setting it at maximum in any range and attempting to weld. The object is to see if the power source is producing proper amperage values for the setting used. The next step is to set the welding rheostat at minimum in the range and again attempt to weld. There should be a radical difference in the welding output values of the two settings. As the rheostat is then adjusted through its entire range, the welding current should vary accordingly.

If there is little, or no, difference between the minimum and maximum settings for welding current, check the rheostat wiring for breaks or disconnected terminals. Be sure to check the small contact carbon brush that is part of the rheostat control. The carbon could be worn or chipped and broken. In such a case, the brush would not make firm contact with the rheostat coil windings and welding current output would be erratic.

If the rheostat wiring and carbon contact brush test out satisfactorily check the actual coil windings on the rheostat body. In operations where there is a highly corrosive atmosphere, such as salt water or chemical vapors, it is possible that the rheostat windings may have developed a layer of oxides on their surfaces. This condition would cause the rheostat carbon brush to have poor contact with the rheostat windings. The result would be erratic amperage output when the rheostat is rotated through its range.

When installing any welding power source with rheostat control it is good practice to instruct the welder to rotate the rheostat control through its entire range at least twice each day. This will assist in keeping the rheostat coil windings polished and bright which will promote better welding amperage output control.

IMPORTANT: ON MOST ELECTRICALLY CONTROLLED WELDING POWER SOURCES THE REMOTE CURRENT CONTROL IS LIMITED BY THE MAXIMUM OUTPUT SETTING ON THE MAIN WELDING RHEOSTAT. This is an excellent feature that has great application when the power source is used with the gas tungsten arc welding (GTAW) process.

A remote control rheostat located at the work station will have current control variation from the minimum of the range set on the welding power source to the maximum set on the

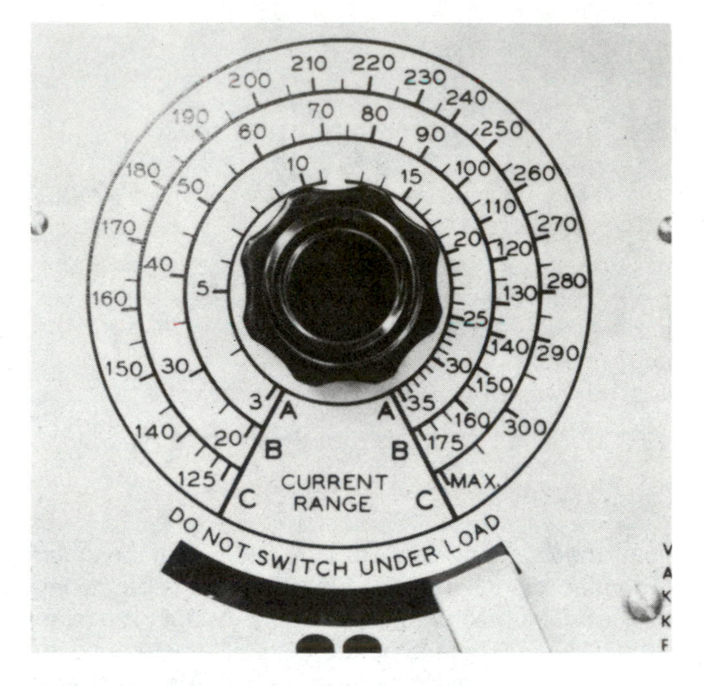

Figure 140. Welding Current Selector Rheostat.

main welding current rheostat on the power source front panel. A remote current control device cannot override the main rheostat setting in this type circuitry. For example, if the main rheostat is set at 50% of the range, as shown in Figure 141, the maximum available amperage at the remote work station would be the value set at the main rheostat. The remote-standard switch would be in the "remote" position on the control panel.

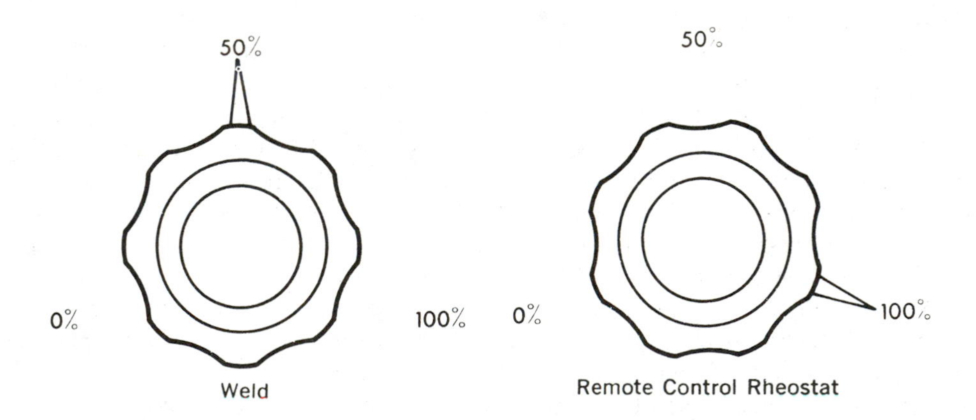

Figure 141. Electric Control Rheostat Settings.

The benefits of such a control system are substantial. The ability to limit amperage output when working with the GTAW process can help to maintain the welding current levels within the current carrying capacity of the tungsten electrode. Excessive welding amperage will cause the tungsten electrode to deteriorate rapidly. This type of control is also very helpful to the welder who can make substantial physical changes in the remote rheostat setting with relatively small amperage changes at the arc.

Start Current Control

Although not all welding power sources have a start control amperage rheostat and allied circuitry, such a circuit is standard on almost all NEMA Class 1 welding power sources designed for the gas tungsten arc welding (GTAW) process. The start control circuit provides initial amperage at the welding arc that may be either higher or lower than the actual welding current. The specific welding application will determine how the start rheostat will be set or, indeed, if it is used at all.

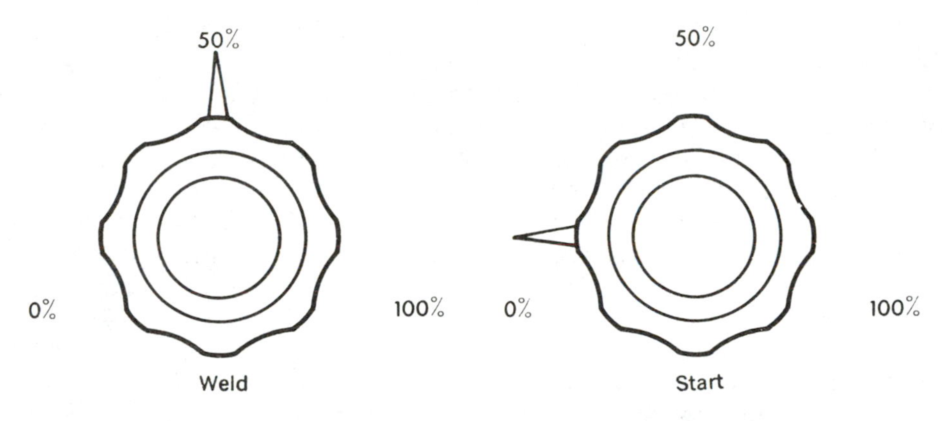

Figure 142. Low Amperage Start Rheostat Position.

The low amperage start, sometimes referred to as a "soft start", is usually used when the welder wants to begin a GTAW weld with a hot tungsten electrode and a stabilized welding arc. A weld begun in this manner is excellent for relatively thin gage sheet metal where arc stability is a prime requirement. In addition, a pre-heated tungsten electrode has better electron emission characteristics and, therefore, provides a more stable arc column when the weld is actually started. Figure 142 illustrates the relative positions of the main weld rheostat and the start control rheostat for a low amperage weld start.

The high amperage "hot start" is used primarily for heavier thicknesses of metals or thinner sections of metals which have good thermal conductivity. In this manner, a large amount of heat energy is given to the metal at the beginning of the weld. The intent, of course, is to pre-heat the base metal arc start area. Since heat will travel through hot metal much more slowly than it will through cold metal, the pre-heat helps to keep the welding heat energy in a smaller area for better penetration at weld starts. This technique will help to minimize the formation of cold laps at the beginning of weld beads. Figure 143 shows the relative positions of the start rheostat and the main welding rheostat for a typical high amperage weld start.

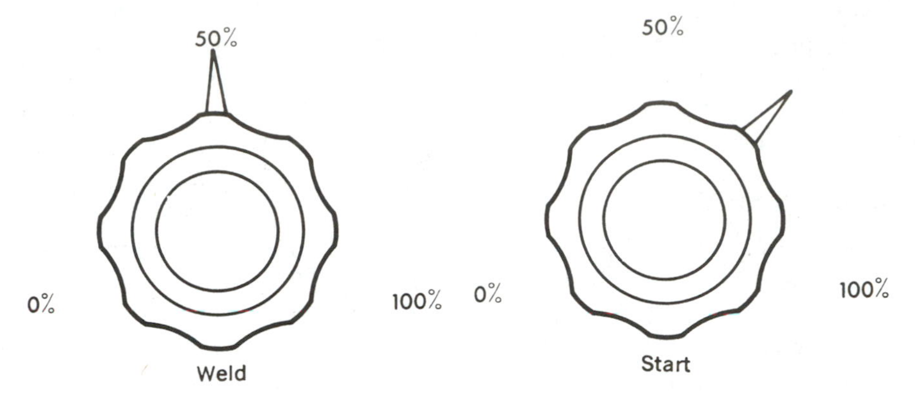

Figure 143. High Amperage Start Rheostat Position.

This discussion of troubleshooting is obviously incomplete if all types of welding power sources are to be considered. The information given is applicable, however, to many standard power sources regardless of name or origin. The importance of the manufacturers Installation, Operation and Maintenance manual (the "M & O manual") cannot be stressed too much. There should be such a manual available for every power source with which you work. Read it carefully, heed it thoroughly, and USE THE INFORMATION IT CONTAINS!

A parts list showing the various component parts of the welding power source is usually incorporated as part of the information furnished with a welding power source. Use the parts list when tracing out a circuit diagram to help you identify the various parts of the unit shown with symbols and letters. For example, "T_1" shown on the circuit diagram may be found listed in the parts list as the main transformer of the power source.

It is important to realize that information of any kind published by the various welding equipment manufacturers costs money! The intent in publishing such data is to assist you in using the welding equipment to its fullest potential. No individual can retain in his memory all the information about every welding process and power source. It is important for the welding person to ask questions, to continue to be inquisitive about a welding process and to ask what it can do for his application. The person who loses interest in new welding methods and new welding products will soon be out of the welding business.

Power Source Preventative Maintenance (PM)

There are many different actions that can be listed as preventative maintenance for welding power sources. Some of the principle concepts include the following:

1. Remove the cover of a welding power source and remove accumulated dust and dirt. This may be done with an industrial vacuum cleaner, by blowing it out with compressed air, or by wiping the unit down with solvent. This latter effort would be applicable where the power source is in a very oily atmosphere. Internal cleaning should be done as often as necessary to prevent buildup of dust, dirt, and metal dust.

2. Check all electrical connections and cables for tightness.
3. Position all power sources so that fan exhaust openings have a minimum of 18" of open space behind them. This permits them to cool properly.
4. NEVER USE ANY TYPE OF FILTER ON POWER SOURCE INLET OR OUTLET AIR PORTS.
5. Examine the outer cases and electrical connections of welding power sources periodically. Repair as required.
6. Compare voltage and amperage meter readings with other meters of known calibration at least every ninety days.

A professional person takes care of the equipment that helps make his living. Be a professional!

CHAPTER 12

Rotating Type Welding Power Sources

Rotating type welding power sources, such as motor-generators and various types of engine driven units, have been used by the welding industry for many years. As a matter of fact, motor-generator welding power sources were some of the first electric arc welding power sources built. The engine driven welding power sources were instrumental in freeing the welding profession from the shop and giving mobility to this most useful craft.

An important thing to remember is that all rotating type welding power sources are some type of **electro-mechanical device.** This means they are a composite unit which uses mechanical energy to generate electrical energy for welding. This is the same basic principle used in power generating plants the world over.

All *generator and alternator units produce a maximum total amount of kilowatts.* If you recall that volts times amperes equals watts, and that number divided by 1,000 equals kilowatts, the rest is easy. The rule, as given earlier in this text, is that for a specific amount of total electrical power: "As the voltage increases the amperage will decrease proportionately. Conversely, as the voltage decreases the amperage will increase proportionately".

This information may be tied in with the operation of some rotating type welding power sources. If there is a relatively high open circuit voltage at some particular setting on a power source, there must be a relatively limited amount of maximum short circuit current at the same time. This is typical of a number of models of motor-generator and engine driven welding power sources on the market today.

As a matter of fact, when the power sources are used with small diameter SMAW electrodes such as 1/16" and 3/32", the open circuit voltage can usually be read in a range of 100-105 volts. The open circuit voltage measurement is made at the output terminals of the welding power source. For normal operations with 1/8", 5/32", and 3/16" diameter SMAW electrodes, the open circuit voltage is usually lower (in a range of 75-80 volts) but the maximum short circuit current of the power source is considerably greater than before. This is shown on the several volt-ampere curves illustrated in Figure 144.

Setting the rotating type welding power source for even larger diameter SMAW electrodes will produce open circuit voltage that is further decreased in value (possibly in the range of 55-65 volts). Maximum short circuit current will be quite high at these settings.

The logical question is: "Why the different open circuit voltages and maximum short circuit current levels for different diameters of electrodes used for shielded metal arc welding?" The answer also follows logic or, if you like, plain common sense.

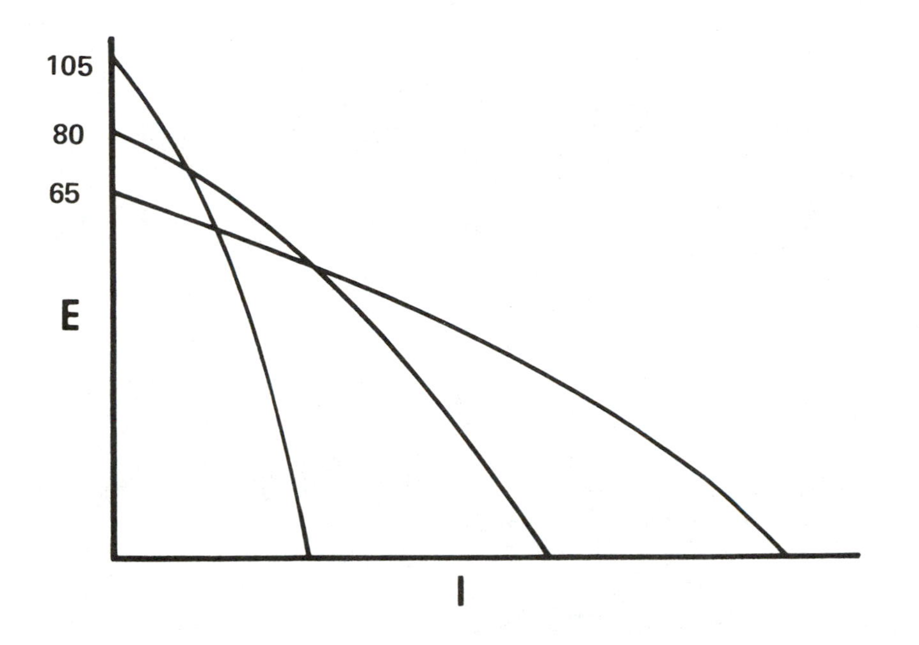

Figure 144. Typical Volt-Ampere Curves
For a Variable Open Circuit Voltage Constant Current Welding Generator.

Consider, for example, the small electrode diameters of 1/16'' and 3/32''. First determine the cross sectional area of the **electrode core wire** in square inches. Then determine the cross sectional area of the **electrode flux covering** in square inches. Now compare the two values and you will find that usually there is more flux cross sectional area than core wire cross sectional area.

The greater amount of flux makes it difficult to initiate the welding arc with most small SMAW electrodes unless there is a relatively high open circuit voltage at the welding power source output terminals. This is particularly true of low hydrogen type electrodes.

The normal, most commonly used SMAW electrode diameters are 1/8'', 5/32'', and 3/16''. In most AWS classifications these electrode diameters will have greater core wire cross sectional areas than they have flux cross sectional area. They do not require the extra high open circuit voltage to initiate the welding arc. The larger diameter electrodes do require more welding amperage from the power source to properly maintain the melt-rate of the electrodes. This is illustrated by the volt-ampere curves in Figure 145.

The ''voltage control'' normally located on the front control panel of constant current type rotating power sources is basically an open circuit voltage control. Some manufacturers use other terms to describe the type of voltage being used but the welder can quickly determine that it is indeed open circuit voltage. Just apply the two probes of a voltmeter to the output terminals of the power source at different settings of the voltage control.

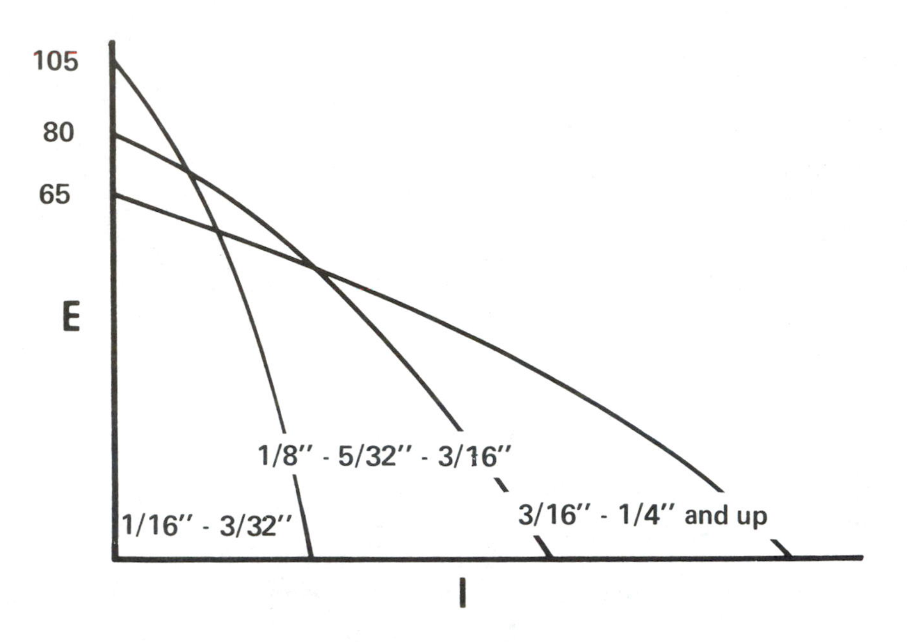

Figure 145. Volt-Ampere Curves and SMAW Electrode Correlation.

Types of Welding Generators

There have been three types of commonly used generator designs available to the welding industry over the years. Two of the three have found widespread use. The third has not because of its inherent lack of adaptability for welding use. The three generator types are:

DC Generator	=	Provides DC welding output only.
AC Generator	=	Normally not used. Provides AC output.
AC Alternator	=	Provides either AC or DC welding output.

Each of these electro-mechanical devices will be explained in this chapter of the text.

Types of Motive Power Used

Both electric motors and fuel-powered engines are used to turn the rotor assemblies of the rotating type welding power source generators. The basic criteria for motor or engine size and horsepower rating is that the electric motor or fuel-powered engine be capable of permitting the generator to reach full power output.

Electric Motors

The electric driving motors used for motor-generator type welding power sources are normally AC induction type motors. The NEMA EW-1 Standard for Electric Arc Welding Apparatus states the following:

"Alternating current induction motors driving DC generator and DC generator-rectifier arc welding power sources shall be three phase and shall have voltage and frequency ratings in accordance with the following:

60 hertz = 200, 230, 460, and 575 volts.
50 hertz = 220, 380, and 440 volts."

Fuel Powered Engines

There are a variety of engine types and sizes that are used for portable engine driven welding power sources. Both liquid cooled and air cooled engines are employed for specific power source applications. Many smaller engine driven units of less than 250 amperes output use air cooled engines satisfactorily. Most larger engine driven power sources have liquid cooled engines.

A variety of fossil fuels may be used for running the engines of welding power source generators. Gasoline is probably the most popular fuel because it is readily available in just about any area. It also is used for a variety of other engine driven equipment.

Diesel fuel is very popular in many areas because of its high flashpoint. There is normally less pilferage loss of diesel oil than there is with gasoline. Some Federal laws will only permit diesel fuel for engines used in specific applications. A good example of this restriction is the use of diesel engines for welding power sources on off-shore drilling rigs.

Propane is used in some applications of engine driven welding power sources. It is less expensive and cleaner burning than gasoline but it does require a special carburation system.

The DC Generator Design Concept

The DC generator design concept considers that the **rotor assembly** is comprised of a through shaft, two end bearings to support the rotor and shaft load, an armature which includes the laminated armature iron core and the current carrying armature coils, and a commutator. It is in the armature coils that welding power is generated.

The **stator** is the stationary portion of the generator within which the rotor assembly turns. In this design the stator holds the magnetic field coils of the generator. The magnetic field coils have a small amount of DC voltage and amperage applied to maintain the necessary continuous magnetic field required for power generation. The DC amperage is normally no more than 10-15 amperes and very often is less.

In electric power generation there must be relative motion between a magnetic field and a current, or electric, field. It makes no difference which type of field is in motion as long as there is relative motion between the two. In the DC generator it is the **armature** that is the current, or electric, field. The magnetic field coils are located in the stator. The armature turns within the stator, and its magnetic field system, and welding current is generated. A typical DC generator rotor assembly is illustrated in Figure 146.

The current that is generated in any welding power generator is alternating current (AC). The alternating current is carried to the copper commutator bars through electrical conductors from the armature coils. The conductors are soft-soldered to the individual commutator bars. The commutator bars may be considered as terminals, or "collector bars", for the generated alternating current from the armature.

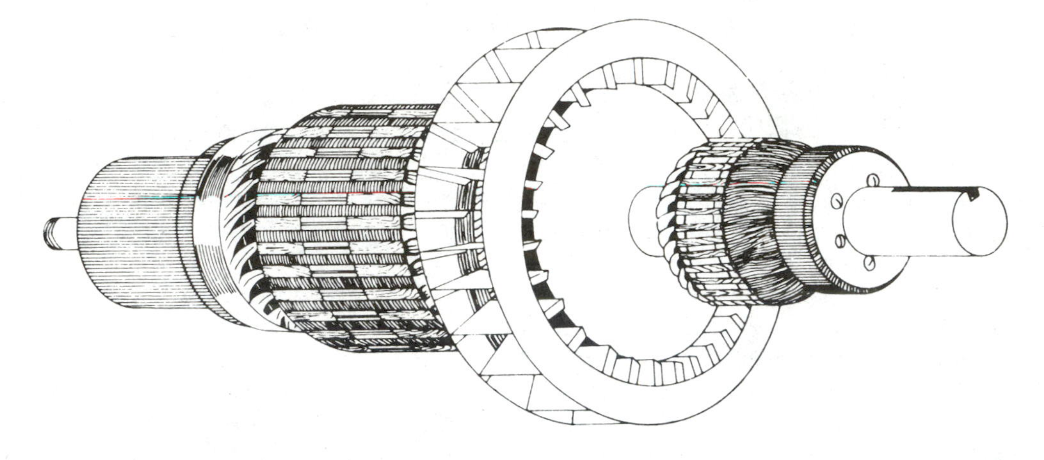

Figure 146. Typical DC Armature Assembly.

The commutator is a system of copper bars that are placed concentric to the centerline of the rotor shaft. Each copper bar of the commutator has a machined and polished top surface upon which carbon type contact brushes ride to pick up each half-cycle of the generated alternating current. The purpose of the commutator is to carry both half-cycles of the generated AC sine wave but on separate copper commutator bars. Each of the copper commutator bars is insulated from all the other copper bars.

The carbon contact brushes actually pick up each half-cycle of generated alternating current and direct it into a conductor as direct current. It may be said that the brush-commutator arrangement is a type of mechanical rectifier since it does change the generated alternating current (AC) to direct current (DC). Most of the carbon type brushes used are an alloy of carbon, graphite and small copper flakes.

The DC generator is so-called because it has the commutator-brush arrangement for changing AC to DC welding power. Normally the DC generator is a three phase electrical device. Three phase systems provide the smoothest welding power of any of the electro-mechanical welding power sources.

The AC Generator Design Concept

The AC generator design concept is similar to the DC generator with the major exception that it has no commutator and brush arrangement for changing the generated AC to DC. AC generators do have a rotor assembly which consists of the armature iron core and armature coils, through shaft and bearings, and something different in place of the commutator, slip rings. (Slip rings are solid brass parts that are machined, polished and fitted concentric to the shaft centerline). It is still in the armature coils, located on the rotor assembly, that welding power is generated.

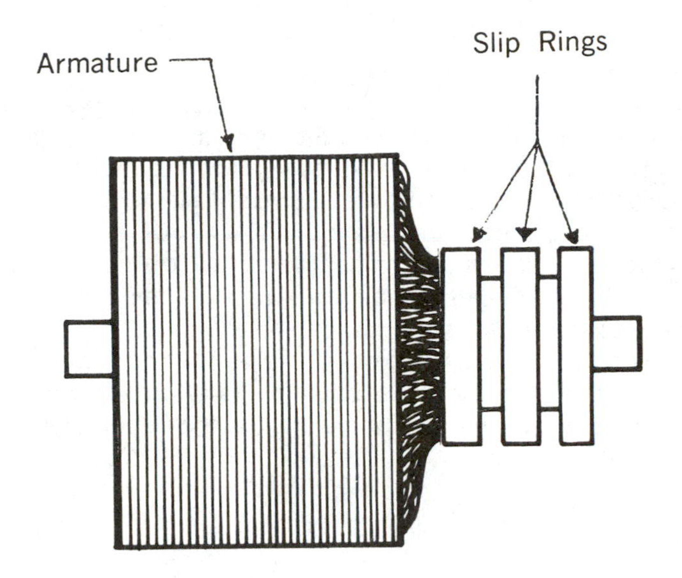

Figure 147. Typical AC Rotor Assembly, Diagrammatic View.

The stator is the stationary portion of the generator. The magnetic field coils are at discrete fixed positions in the stator. As the rotating armature moves within the magnetic field of the stator, welding power is generated. The generated power is alternating current.

The generated electrical power is carried by conductors to the brass slip rings. The alternating current is picked up by carbon type contact brushes from the slip rings and conveyed by conductors to some type of control device such as a reactor. The alternating current is then directed to the output terminals of the welding power source. The welding power output is AC.

At this writing, I do not know of any commercially manufactured AC welding generator. This data is placed here for information only.

The AC Alternator Design Concept

The AC alternator design concept has both the rotor and the stator as do the other two types of electric power generators. There is, however, a marked difference in the design characteristics of this unit.

The rotor consists of a through shaft, end bearings at one end only of the shaft, and a coupling plate for direct connection of the rotor to the engine driving it at the other end. The rotor assembly has the magnetic field coils attached instead of the armature coils as the other generators have. There are brass slip rings on the rotor shaft through which small amounts of DC voltage and amperage are brought to the magnetic field coils for excitation purposes. In many designs there is some portion of the exciter circuit power system located on the rotor.

The stator is the stationary portion of the alternator. It has the armature coils wound in slots in the armature iron core which is also the basic stator. Welding power is generated in the armature coils but it does not have to be picked up and transferred through carbon type brushes.

The power is already in the stator portion of the circuit and is conveyed through electrical conductors to a reactor for division into various amperage output ranges. In those models where the welding output is either AC or DC, the generated welding power goes through a main power rectifier system to obtain the DC power. The rectifier may be either selenium or silicon although most welding rectifiers today are made with silicon rectifiers (diodes). The rectifier does only one thing which is to change the AC to DC.

A rheostat is normally in the output amperage control circuit to obtain fine amperage adjustment within each of the amperage ranges. Ammeters and voltmeters are often standard equipment on engine driven power sources having higher amperage capabilities.

One excellent option that should be on every engine driven power source is the **running hour meter.** This meter keeps track of the actual hours of running time of the engine and generator. For preventative maintenance of the engine, the hour meter is the unit that tells you when to change oil, filters, etc.

MAGNETIC
ROTOR—FIELD COILS

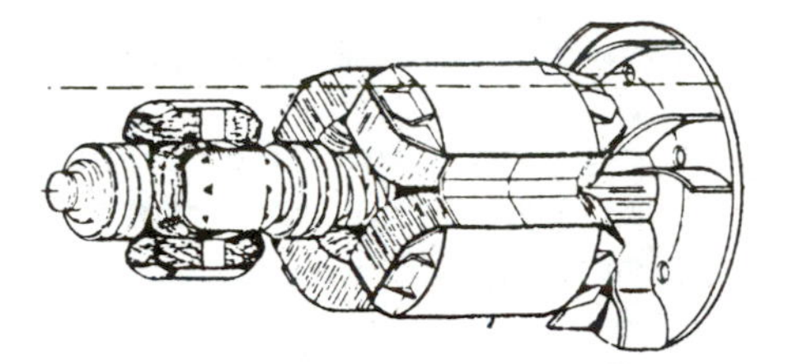

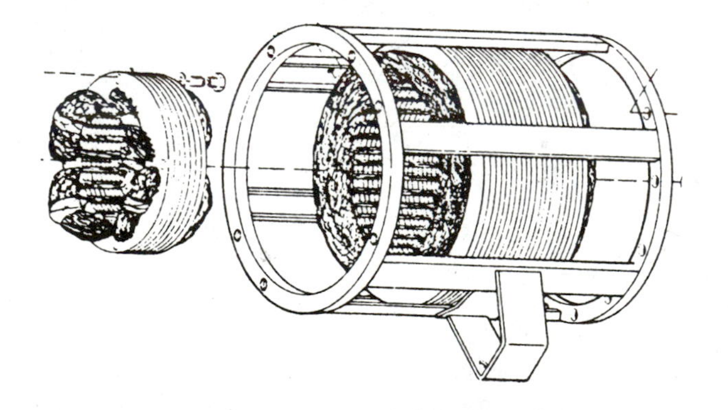

STATOR—ARMATURE COILS

Figure 148. Typical AC Alternator System.

The alternator type of welding power source is normally an engine driven unit and is totally portable for field operations. In most cases the alternator welding power sources also provide auxiliary power for tools and electric lights. Some models of alternator power systems can function as portable electric generators providing 60 hertz power. These units are single phase units.

Most engine driven welding power sources used for auxiliary power generation are four pole systems. If the engine welding speed is above 1800 rpm it must be slowed to 1800 rpm to obtain 60 hertz, 115 or 230 volt electrical power.

Comparison of Rotating Equipment Designs

To better understand the design concepts, and the differences, of both the rotating armature and the rotating magnetic field coil generating systems we will examine the two types of units as shown in Figure 149.

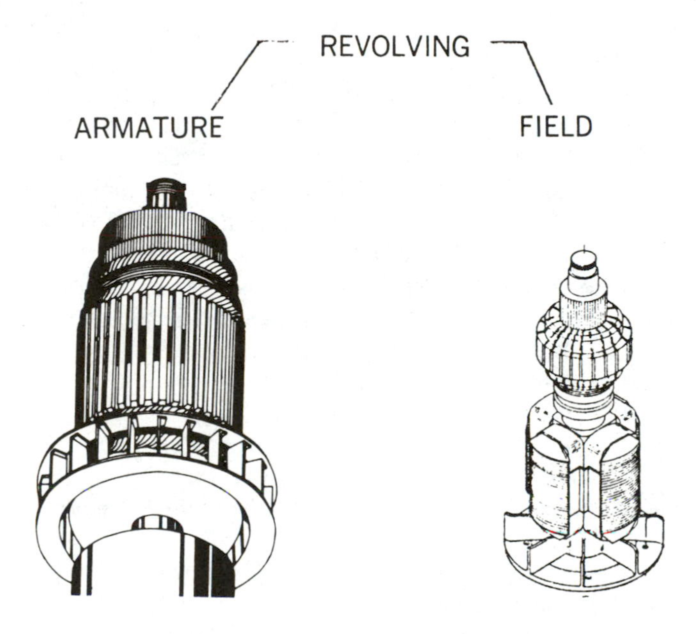

Figure 149. Rotor Comparisons.

In looking at the **rotating armature design,** first you see the massive iron core, the armature coils, and the rather large commutator assembly. The mass and weight of the system is considerable. The stator portion (not illustrated) is also quite massive although its most important function is to house the magnetic field coils of the generator. The stator also carries the weight of the rotor assembly.

By comparison the **rotating magnetic field coil design** rotor looks small. You see the four magnetic field coils and their iron core material. (This is, of course, a four pole system).

The stator (shown in Figure 148) holds the armature coils in their slots. The stator iron is the core material for the armature.

The interesting point is that both units illustrated are rated at 200 amperes DC welding output. The duty cycle rating of the two power sources is, however, considerably different. The rotating armature power source is rated at 200 amperes, **60% duty cycle.** The rotating magnetic field coil unit is rated at 200 amperes, **100% duty cycle.**

Remember: Duty cycle is based on a 10 minute period of time.

At 60% duty cycle the rotating armature type welding power source can operate at its rated 200 amperes for six minutes out of ten. The other four minutes of the ten minute cycle the unit must idle and cool.

The rotating magnetic field coil power source may be operated continuously at, or below, the rated 200 amperes of DC welding current output. There is no need for a cooling period each ten minutes because the unit is designed to operate continuously at or below 200 amperes. There is no overheating of the alternator system which could cause possible damage and welding downtime.

The logical question here is: ''How can this be when the one power source—**the rotating armature type and design**— is so much larger and heavier than the **rotating magnetic field coil system?**'' Fortunately there is a reasonable and logical answer to the question.

Consider that all welding amperage is generated in the armature coils of a generator. It makes no difference to the power source if the armature coils are in the rotor or the stator. They still carry the generated welding amperage. You will recall that all electrical conductor materials have some measure of electrical resistance to electrical current flow.

In the rotating armature design generator (the **DC generator**) the armature coils are positioned on the rotor assembly which rotates within the stator assembly. There is minimal space between the rotor and stator assemblies so cooling air flow is minimum. There is no way the armature coils can be cooled adequately and so, as the generated welding current flows through them, the coil conductor temperature increases due to the electrical resistance of the conductor wires.

The thermal (heat) energy cannot be properly dissipated so it heats the armature iron core material as well as the armature coils. This causes problems even though the DC generator is producing the proper amount of amperage for its rating. Due to the resistance heating of the armature assembly, all of the generated amperage cannot flow through the conductors as output from the unit.

The effects of the resistance heating in the DC generator rotor assembly will show up in the welding output after a very few minutes of welding operation. It is not uncommon for the welder to return to his power source after about 15 minutes of welding and readjust the amperage output control to a higher setting. This permits him to continue welding with the same diameter of SMAW electrode. If you ask him why he is adjusting the output control he will probably tell you, ''My machine got cold and I need more current to weld!''

As a matter of fact, the DC generator armature coils and iron core became hot due to the resistance heating and lack of adequate cooling. The additional electrical resistance in the armature coils, the iron core material, the conductors, the commutator bars and probably the electrode and work leads used up a considerable amount of kilowatts of output from the generator. The energy, lost due to electrical resistance in the circuit, never had a chance to reach the welding arc.

The only electrical current flowing in the rotating magnetic field coil rotor system is a small amount of DC power necessary for the excitation of the magnetic field coils. At maximum this may amount to 12-15 amperes; hardly enough to cause an overheating problem in the magnetic field coils or any other part of the rotor assembly.

The armature coils carrying the generated welding current are located in the stator which is wide open to the cooling air. The result is that the armature coils have a better opportunity to remain relatively cool while under the welding output load. There is less electrical resistance and not nearly the decrease in welding current output that is found with the rotating armature design. Very seldom does a welder have to adjust his rotating magnetic field coil type power source for increased amperage output due to armature heating.

The efficiency of the rotating magnetic field coil design is well known to producers of electrical power around the world. All of the power generating turbines used in power generating plants the world over are the rotating magnetic field coil design. The electrical efficiency of this design is very high.

Paralleling Electro-Mechanical Power Sources

When two or more welding power sources are placed in parallel operation the amperage is additive while the load voltage remains the same as it was for one power source. Paralleling motor-generator type power sources is time consuming and somewhat difficult. The very nature of paralleling the output amperage requires that the excitation of the two units be in electrical phase, and the AC output power generated in the armature must be in phase. This can get to be a little rough unless you are a highly skilled and qualified electrician.

One of the basic problems encountered in paralleling DC generators, either electric motor driven or engine driven, is balancing the output amperage of the two units to be paralleled. If they are not almost exactly the same amperage output there is a tendency for the unit with the higher amperage output to feed current back into the other power source.

This is done in the following manner. The current goes through the parallel connection, back through the second power source output terminals, through the brushes and commutator, and dissipates as heat in the armature coils. The armature coils are already operating at elevated temperatures because they are producing welding current output. Many rotating armature units are burned out each year because precautions were not taken when paralleling two or more power sources.

Those rotating magnetic field coil power sources which have DC welding output normally put the generated AC through some type of rectifier. Of course, a rectifier changes AC to DC. The rectifier may be either silicon or selenium. Most are silicon in present power source designs.

It is a simple matter to put two or more rotating magnetic field type power sources in parallel operation. Proper parallel connections are to connect the two or more negative cables together and the two or more positive connections together. Each set of cables should be the same length from the power source to the connection terminal. Both positive and negative welding cables must be of sufficient cross sectional area to carry the increased amperage level with minimum resistance.

The welding amperage should be reasonably the same from each power source in the parallel connection. It is not necessary that they be exactly the same since there is no danger of current feedback from one power source to another. The rectifiers absolutely stop any reverse flow of current from one power source to another.

Auxiliary Power Plant Operation

Many of the engine driven welding power sources have capability of being operated as an auxiliary power generating plant. This is often a blessing to the contractor in the field who has no other source of electrical power for operating hand tools, lights, or other electrical equipment.

In most cases the generator is a four pole system. The engine should operate at 1800 rpm when the unit is used as a power plant. At 1800 rpm the unit will produce 60 hertz power at either 115 volts or 230 volts. The 60 hertz frequency is required for most AC electrical tools.

If the engine speed is lower than 1800 rpm the cycles per second (hertz) would be lower. This could cause AC motors to stall and possibly burn out due to the lower frequency. If the engine speed were higher, perhaps 2500 rpm, the frequency would also be higher. This would speed the AC motor operation and would probably burn out the electric motor.

Summary

Engine driven welding power sources are a necessity for the many field welding applications presently being performed. There is a difference in the design and electrical efficiency of various engine driven generators. It is the responsibility of the user to determine which type of engine driven equipment is best for a particular application.

Some engine driven power sources may be used for auxiliary power generation where other electrical power is not available. To obtain 60 hertz power it is necessary for the engine and generator to operate at 1800 rpm in a four pole system. The power generated is always single phase power.

CHAPTER 13

Gas Tungsten Arc Welding

The principles of the gas tungsten arc welding (GTAW) process are relatively simple. The objective is to provide welding heat at the work area without the contaminating influence of the surrounding atmosphere. This is accomplished with a shield of externally supplied inert gas. The inert gases used with the GTAW process are argon (Ar) or helium (He) or combinations of the two. The electrodes used are non-consumable tungsten materials.

This welding process is often called the "TIG" process. The term "TIG" stands for **Tungsten Inert Gas.** Seeking a more descriptive name, the AWS has selected the name **Gas Tungsten Arc Welding** as the correct professional name for the welding process.

Welding in an inert gas atmosphere was first considered in the late 1920's. In 1930 a patent was issued to Hobart and Devers covering the use of an electric arc within an inert gas atmosphere. The process was not developed during the 1930's for several reasons, the most important being it was not cost-effective at the time.

It was about 1939 that Russell Meredith was assigned the task of developing a new process for welding aircraft airframes made of magnesium tubing. The problem was that most aluminum and magnesium aircraft materials were gas welded with the oxygen-hydrogen flame. This required use of a fluoride flux while welding. When the weld was finished, the flux residue had to be removed or it would continue to react with the metal. The flux removal was done by vigorously brushing the parts under hot water.

Unfortunately, when the magnesium tubing was welded into aircraft seats, the weld was a closure weld. This meant that any flux inside the tubing could not be washed away. Fluoride fluxes will continue to react with magnesium and aluminum as long as there is contact between the two materials. The result was that eventually the flux would cause separation of the part. This, in turn, probably caused some consternation to pilots who found their aircraft seats suddenly loose from the aircraft floor while performing aerobatics as fighter aircraft pilots must do from time to time.

Mr. Meredith had the job of coming up with a new welding process that was faster, easier and that would provide better metallurgical characteristics in the weld. In 1941 he devised a method of hand-feeding magnesium wire through a capped nozzle in which an inert gas (helium) was introduced through a copper tube. This was not satisfactory because of the extremely high melt-rate of magnesium. It was determined that a non-consumable, refractory electrode would be better for the welding process. Tungsten was the electrode material eventually selected.

A patent was applied for in October, 1941, by Mr. Meredith. The patent was issued to him in February, 1942. The gas tungsten arc welding process that was eventually to make space travel possible was created.

Since that time, the gas tungsten arc welding process has been used for joining most of the weldable metals including refractory materials. The process has developed into a most reliable method of making extremely high quality welds. It is used for the root pass of all Code high pressure piping systems including those in nuclear applications.

The basic equipment used for the gas tungsten arc welding (GTAW) process is different in appearance than that used for the shielded metal arc welding (SMAW) process. Figure 150 shows typical GTAW process manual equipment.

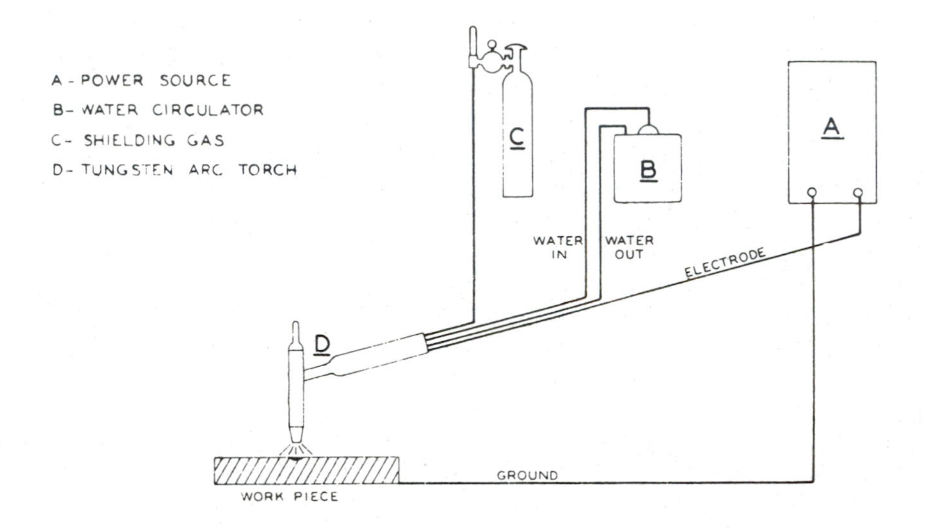

Figure 150. Basic Gas Tungsten Arc Welding Equipment.

The power source used with the GTAW process is normally a NEMA Class 1 constant current type unit. It may have either AC or DC welding power output or it may have a combination of the two (AC/DC). It will usually have special equipment such as gas valves and solenoids as standard GTAW equipment.

The water recirculator equipment permits the use of distilled water instead of "city water". This is obviously for those applications where water-cooled GTAW torches are necessary. Why spend the money for a coolant system when you can get water out of the shop faucets? Simple. The water supplied by almost every water utility in the world has hard particles, such as iron, in its compound. These hard particles will build up in a very short time and clog water hoses, pipes and torch head passages.

Consider the reason for water cooling the torch in the first place. The objective is to make the torch smaller in size and lighter in weight by using water as a heat exchanger. It makes sense that the tubing that carries the water in the torch must be very small in diameter.

If "city water" is used in the cooling system it will very quickly build up chemical deposits in the tubing. Such chemical deposits will slow down, and eventually stop, water flow in the torch. There will be no cooling action and the torch body will overheat and be destroyed. With a water recirculating system of some type, distilled water can be used. Distilled water has none of the contaminants of utility supplied water.

The illustration, Figure 150, shows a compressed gas cylinder and a gas flowmeter. The shielding gas must be either argon or helium or a combination of the two gases. Shielding gases for the GTAW process, and their characteristics, are discussed later in this section of the text.

The tungsten arc welding torch, properly called a "tungsten arc electrode holder", may have a variety of shapes. The principle thing the torch does is to channel shielding gas to the arc area and to hold the tungsten electrode while passing welding current through the torch body and collet to the electrode.

Energy Input Comparisons

The gas tungsten arc electrode holder is often called a "torch". This is because it performs essentially the same function as an oxy-acetylene torch. Both oxy-acetylene and gas tungsten arc welding processes provide heat only at the welding work area.

The oxy-acetylene process uses the chemical combination of oxygen and acetylene to provide an open flame to produce heat. The gas tungsten arc welding process uses electrical energy for this purpose. Any filler metal that might be added with either welding process is non-energized electrically and would be external to the welding torch.

HEAT ONLY	HEAT + MOLTEN METAL
OXY-ACETYLENE	SHIELDED METAL ARC
GAS TUNGSTEN ARC	GAS METAL ARC

Figure 151. Thermal Energy Input Comparisons.

The welding processes that provide heat only at the weld joint are shown in Figure 151 as the oxy-acetylene, and gas tungsten arc, welding processes. These processes may be contrasted with the shielded metal arc, and gas metal arc, welding processes which provide both **heat** and **molten metal** to the weld puddle simultaneously. The differences in the weld deposits of the various welding processes are discussed in other sections of this text.

The Gas Tungsten Arc Electrode Holder Assembly

The typical gas tungsten arc electrode holder ("torch") is made up of several component parts. Each part serves a specific purpose. As illustrated in Figure 152, a typical manual gas tungsten arc welding electrode holder assembly contains the following: a torch body; a highly machined collet; a collet cap; a gas nozzle (cup); the gas and water hoses; the welding current cable.

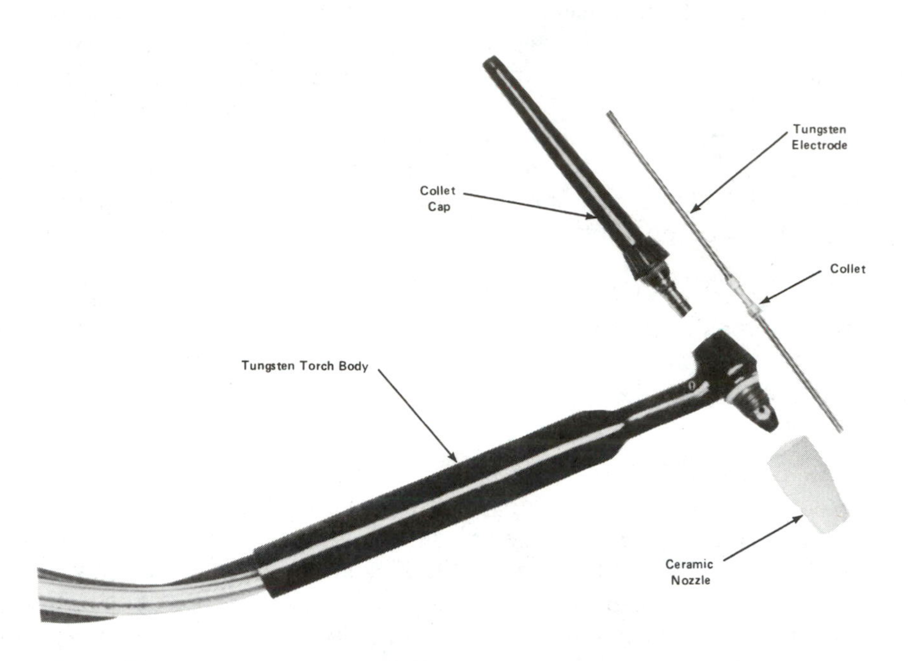

Figure 152. A Typical Gas Tungsten Arc Welding Torch Assembly.

The tungsten electrode diameter will vary depending on the welding application and the current requirements. Each tungsten electrode diameter requires a different collet size.

The hose and cable assembly attaches to the electrode holder at one end and the welding power source, the source of shielding gas and the water source at the other end.

The collet cap will normally have "O" rings for sealing the collet cap to the torch body. This is to prevent aspiration, or sucking in, of outside atmosphere into the shielding gas. Aspiration of outside air would contaminate the shielding gas, the tungsten electrode, and the weld deposit.

The welding power sources used for the gas tungsten arc welding process may have AC, DC or AC/DC welding output. The type of welding current needed will depend on the type of metal to be welded. For example, alternating current (AC) is normally used for aluminum and magnesium and their alloys. Direct current straight polarity (DCSP) is usually used for all other metals such as steel, stainless steel, copper, etc.

Shielding Gases

The shielding gases that may be used with the gas tungsten arc welding process must have certain specific characteristics. They must be inert to the products of the weld zone. They must protect the tungsten electrode from contamination and oxidation. They must protect the molten weld puddle and the weld deposit from atmospheric contamination. The two inert gases that are available in commercial quantities and purities are argon and helium.

Both argon and helium are inert monatomic gases. This means they each have one atom per molecule. There are a total of six known inert gases. They are: argon, helium, krypton, xenon, radon, and neon. Only argon and helium are suitable for welding applications.

Argon (Ar)

Argon is a chemically inert gas that will not combine with the products of the weld zone. As a matter of fact, argon will not combine with any other element involved with welding. Even when it is put into the same compressed gas cylinder with another gas, such as helium, there is only a gas mixture, not a compound gas. This means there are argon gas atoms and helium gas atoms in the same cylinder as a mixed gas.

Argon has an ionization potential of 15.7 electron volts (eV). To define:

Ionization potential = "The voltage necessary to remove an electron from the gas atom, thereby making it an ion".

An ion is an electrically charged atom. The ionized gas creates a column of electrically charged (positive) particles between the tungsten electrode and the base metal surface. This ionized gas column provides a preferential electrical path for the welding current to follow from the electrode to the surface of the base metal.

Argon has low thermal conductivity. The arc column is somewhat constricted with the result that high current densities are present. Gas tungsten arc welding with argon shielding gas will result in weld deposits with a medium width top bead. The total weld deposit resembles a partial parabola configuration. This is especially true in welding aluminum with the GTAW process. The shape of the deposit is due to the heat transfer characteristics of the metal and the shielding gas. Any filler metal that would be added would act as a quench to the weld.

The illustration shown in Figure 153 diagrams the weld cross section under the arc. The arrows indicate the direction of heat flow through the base metal and the weld deposit. The actual arc column impingement on the base metal surface covers a small area at the center of the weld puddle.

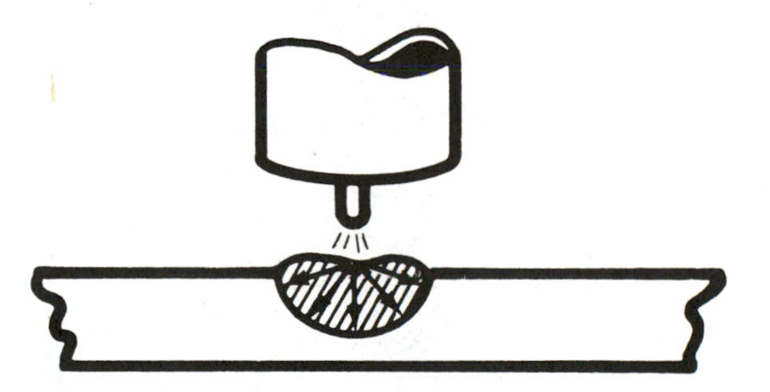

Figure 153. Typical Argon Shielded GTAW Weld in Aluminum.

It is a fact that heat will move more rapidly through cold metal than it will through hot metal. The argon shielded arc column impingement on the base metal surface is quite small. The heat energy is distributed 360 degrees from the arc column center at the work surface. It is used to heat, and melt, the upper area of the weld cross section illustrated.

The welding heat will transfer equally well radially and vertically through the base metal to be welded. Literally, the heat flows into the base metal from the molten metal of the weld puddle. The interface of the weld deposit and the base metal heat-affected zone (HAZ) is created at the place where the heat input is exceeded by the heat dissipation in the metal.

Argon is in abundant supply in the atmosphere since it comprises approximately 8/10 of 1% of the earth's atmosphere. Argon gas is obtained as a by-product of the oxygen reduction process and is heavier than air. Welding grade argon purity is 99.995% minimum. In most cases, actual argon purity is at least 99.998%.

Helium

Helium is another shielding gas that is inert to the products of the weld zone. It is a low density gas that is lighter than air. Helium has an ionization potential of 24.5 electron volts (eV).

Helium has excellent thermal conductivity. The helium arc column will expand under heat, causing thermal ionization of the gas and reducing the arc density. Remember that arc density is calculated by dividing the arc column cross-sectional area, in square inches, into the arc amperage value.

When using helium as a shielding gas there is a simultaneous change in arc voltage where the arc voltage gradient of the arc length is increased by the discharge of energy from the arc column. This means that arc voltage will be higher, and heat energy losses greater, from a helium shielded welding arc than would be true with argon. Some of the lost arc energy is used to heat the helium gas in the arc area.

The nominal arc area impingement, and a typical cross section of a weld deposit configuration, are shown in Figure 154. The effective arc area is larger than that of argon because helium transfers heat much more readily. Note the deep central penetration pattern achieved with helium shielding gas. This is typical of a GTAW weld in aluminum.

Figure 154. Helium Shielded GTAW Weld in Aluminum.

As previously noted, helium has excellent thermal conductivity. It may be assumed, correctly, that a helium arc plasma will have a very low temperature gradient from the center of the arc column to the periphery of the arc area. It is apparent that a greater work surface area is being heated by the arc column when welding with helium shielding gas.

It makes sense that, if the heat energy input is equal over the entire surface of the weld puddle, the heat at the center of the puddle cannot move radially. Remember, heat will travel more easily through cold metal. The only ''cold'' metal is vertically down from the center of the weld puddle. This is where the heat travels from the weld puddle and this causes the deep central penetration achieved with helium shielding gas and the gas tungsten arc welding process.

Helium is a product of natural gas deposits in the earth. Until recently, helium has been in limited supply. Within the past few years new gas wells have been located which have considerable helium content. Some helium bearing gas wells have been located in the central United States, in Canada, and in the North Sea reserves. It is estimated that sufficient helium has been found to last well into the 21st century. At one time the Helium Gas Producers Association speculated that there would be no more helium available by the year 2,000.

Gas Flow Rates

Very often the question is asked, ''What is the correct gas flow rate, in cubic feet per hour (CFH), for welding a specific application with the gas tungsten arc welding process''? Unfortunately, there is no single correct answer that can be written in a book. The flow rates that are provided in most technical literature come from the companies who sell the gas. In most instances, the recommendations for flow rates are on the high side and they should be used as guides only.

A good rule of thumb is, ''gas flow rates should be of sufficient volume to provide adequate coverage of the molten weld puddle deposit and the tungsten electrode''. Any shielding gas flow in excess of the necessary requirement is wasteful and costly. Excess shielding gas flow can also cause defects in the weld because of the inability of the weld deposit to release the gas held in solution with the molten metal before the metal solidifies. This would be apparent as porosity.

A good method to use in setting correct flow rates is to set more than adequate gas flow on the flowmeter. Have the welder begin welding on a scrap metal specimen and reduce the gas flow rate until smoke is visible from the arc. There will also be some crackling sounds. Immediately increase the gas flow by 2-3 CFH and hold there. The gas flow will stabilize within a minute and should be sufficient for the application.

Gas flow rates that are inadequate to protect the weld deposit and the tungsten electrode tip will cause oxidation of the weld surface and the deposited metal. The tungsten electrode tip will show a dull grey-green color and will undoubtedly erode into the weld deposit as contaminating particles. Figure 155 shows two tungsten electrodes (left) that were oxidized by inadequate shielding gas coverage. For comparison, the tungsten electrodes at the right were properly gas shielded.

Current Density

The gas tungsten arc welding (GTAW) process creates high current density at the electrode tip. The amperage per square inch of electrode cross sectional area is much higher

Figure 155. Improperly, and Properly, Gas Shielded Tungsten Electrodes.

than it is with the shielded metal arc welding (SMAW) process. This becomes readily apparent when the electrode diameters for the two welding processes are compared at the same welding amperage.

For example, using amperage as the constant factor, there is this relationship at 120 amperes welding current.

120 amperes = 1/8'' dia. E6010 mild steel electrode.
120 amperes = 1/16'' dia. tungsten electrode.

Current density is determined by dividing the cross sectional area of the electrode, in square inches, into the welding amperage value. For the examples mentioned above the calculations would be as follows:

$$1/8'' \text{ dia.} = \frac{120}{0.01227 \text{ inch}^2} = 9{,}780 \text{ amperes per inch}^2.$$

$$1/16'' \text{ dia.} = \frac{120}{0.00307 \text{ inch}^2} = 39{,}088 \text{ amperes per inch}^2.$$

The cross sectional area of the 1/8'' dia. electrode, in square inches, is four times greater than the 1/16'' dia. electrode. The current density of the 1/8'' dia. electrode, however, is only one-fourth that of the 1/16'' dia. electrode for the specific amperage used. It is evident that the smaller diameter electrode will have a greater heat input to the base metal, in a smaller area, than the larger diameter electrode.

Current density would certainly be one of the factors to consider when selecting a welding process for a specific application. It may also influence the joint design of the base metal. This would be especially true of the metals having good thermal conductivity.

Some of the industrial metals that have good thermal conductivity compared to iron are: aluminum, aluminum alloys, copper, copper alloys, magnesium, and magnesium alloys.

Shielding Gas Ionization

Ionization of the shielding gases occurs when one or more electrons is caused to leave the gas atom. The force that causes the electron to leave the gas atom is called **ionization potential.** You will remember from earlier discussions that ''potential'' is another term for voltage. So it is ionization voltage—an electrical force or pressure—that causes the electron to leave the gas atom. The electron is the fundamental unit of negative electricity.

Ionization of the shielding gases is necessary to provide a preferred electrical path for the welding current which moves between the electrode and the base metal. The ions are electrically charged positive since negative electrons have been removed from the gas atoms. Since current moves from negative to positive, the ionized shielding gas is a good electrical conductor, or preferential path, for the welding current to follow.

The term "electron volts" is abbreviated eV. Ionization potential is calibrated in electron volts. As we have said, ionization potential is the voltage necessary to remove an electron from the gas atom.

It must be remembered that voltage is a force, or electrical pressure, that causes current to flow in a conductor. High frequency (HF) voltages are used with the gas tungsten arc welding (GTAW) process. HF voltages, which may reach values in excess of 3,000 volts, are used to promote shielding gas ionization. The 80 volts maximum open circuit voltage of the welding power source has little to do with shielding gas ionization. An ionization chart for some shielding gases is illustrated in Figure 156. Note that the compound gas carbon dioxide (CO_2) is included in the chart although CO_2 is not used for the gas tungsten arc welding process. Additional information about various shielding gases and shielding gas mixtures will be presented in another chapter of this text. The data is particularly correlated to other types of gas shielded welding processes.

Shielding Gas	Ionization Potential (Electron Volts)
Argon (Ar)	15.7 eV
Helium (He)	24.5 eV
Carbon Dioxide (CO_2)	14.4 eV

Figure 156. Some Shielding Gas Ionization Potentials.

The selection of the correct shielding gas for use with specific metals and metal alloys is very important to weld quality. It is not uncommon for an incorrect shielding gas to cause poor welds and rejected weldments.

The shielding gas helium is lighter than air and has low density with a relatively high ionization potential. It is normally difficult to initiate a gas tungsten arc in a helium atmosphere. In most cases, argon is used to initiate the welding arc and then, if necessary, helium is added in the percentages required for welding.

Arc Initiation

One of the basic problems with the gas tungsten arc welding process is arc initiation. There is no simple answer to the question of how to obtain repetitive, reliable arc starts with this process. Among the variety of causes for poor arc starting are the following: improper shielding gas; improper shielding gas flow rates; erratic high frequency; improper gas nozzle diameter; improper or fouled tungsten electrode; base metal that is improperly cleaned before welding.

All of the factors listed plus others not shown can cause poor arc starting. The best way to proceed is to handle one variable at a time and check it out carefully.

For example, the gas nozzle (cup) may be too large or too small for the application and the gas flow rate. In many cases the nozzle is too large in diameter, and the gas flow too high in CFH, to permit even partial ionization of the shielding gas. The remedy, of course, is to try a smaller diameter gas nozzle and a lower shielding gas flow rate. This is especially effective when the tungsten electrode is small in diameter and the welding amperage is very low in value.

Another problem that often occurs is a tungsten electrode that is too large in diameter for the amount of amperage being used. The electrical resistance may be too high and the resulting arc will be very erratic. The classic symptom will be movement of the arc around the periphery of the electrode tip. The solution is to change to a smaller diameter tungsten electrode. For extremely low amperages the tungsten electrode may have to be dressed to a tapered point.

High frequency voltage, or the lack of it, can be a deterrent to reliable arc starting with the GTAW process. Electrode non-touch arc starts are required when welding aluminum and magnesium. This helps prevent fouling and contamination of the tungsten electrode. High frequency voltage is necessary for this purpose. If the high frequency system is not functioning correctly, the result may be high frequency wander, intermittent high frequency, or even complete loss of high frequency at the electrode tip. This subject is discussed in detail in the chapter titled **High Frequency Systems.**

The best remedy for high frequency loss is a complete electrical check of the total system including all welding lead connections. In particular, a firm electrical ground at the work connection is essential for good high frequency operation at the welding arc.

Other reasons for poor arc initiation with the GTAW process include improperly cleaned base metal, low welding current values, low open circuit voltage, faulty connections in the welding circuit and aspiration of air into the shielding gas flow.

Thermal Placement in the Welding Arc

Thermal placement in the welding arc simply means where the heat of the arc is concentrated for a given electrical polarity. Remember that each half-cycle of AC power is actually flowing in one specific direction for 1/120th second so it does have polarity for that period of time. This can be related to welding current flow for any arc welding process.

The thermal placement of the arc is shown by percentages in Figure 157. This refers to the available arc energy and does not consider losses to the atmosphere. Note the polarity of the two electrodes (DCSP and DCRP) in the illustration. Each polarity may be related to one half-cycle of the AC sine wave trace. DCSP means "electrode negative" and DCRP means "electrode positive". This may be shown as DCEN and DCEP.

The type of welding current used in the welding process will have a great effect on the penetration pattern of the deposited metal as well as the top and bottom bead configuration. Typical weld deposit characteristics for DCSP, DCRP, and AC are shown in Figure 158.

The normal characteristics of the DCSP GTAW deposit are deep penetration with a relatively narrow weld bead width. Conversely, the DCRP weld deposit will have relatively shallow penetration with greater bead width dimensions. Weld deposits made with alternating

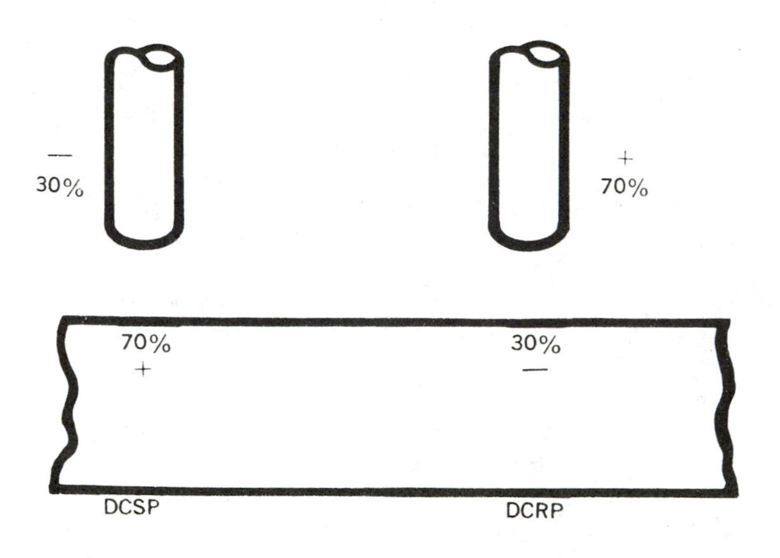

Figure 157. Thermal Placement in the Welding Arc.

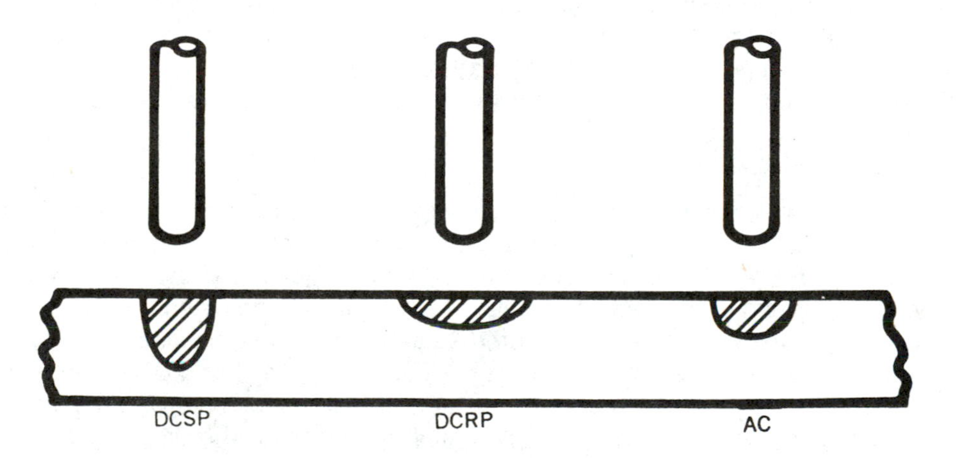

Figure 158. Typical Weld Deposit Cross Sections.

current (AC) will normally have a medium penetration depth and moderate bead width. The type of shielding gas used will also have an effect on the weld.

The reason that AC weld deposits have a modified form compared to the DC weld deposits lies in the fact that alternating current is a combination of DC reverse polarity and DC straight polarity. An AC sine wave trace and polarized electrodes for each half-cycle is shown in Figure 159.

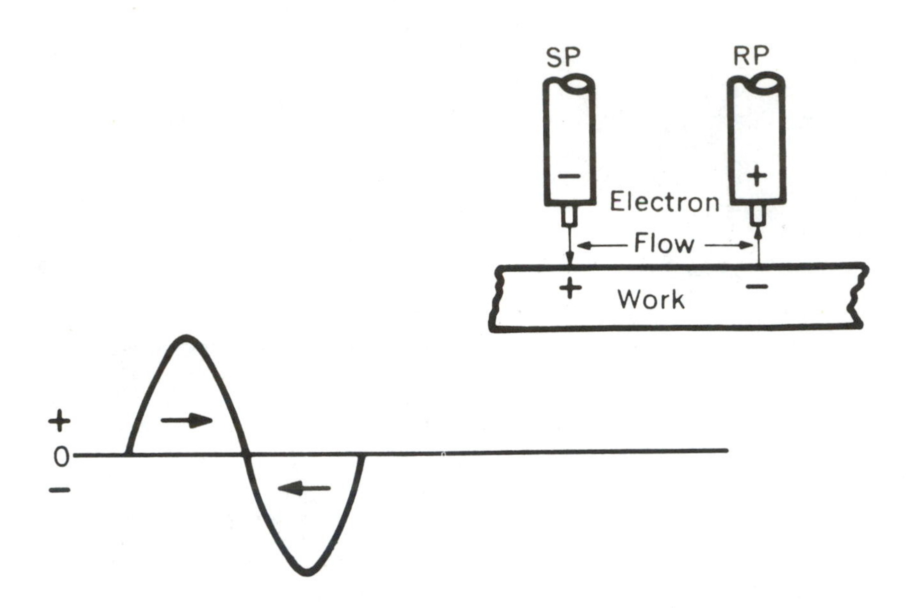

Figure 159. AC Sine Wave Trace and Polarized Electrodes.

The drawing shows the alternating current wave trace separated into two half-cycles. The zero line remains a time function. The electrode-work portion of the drawing shows the electrical characteristics of both straight and reverse polarity. As you can see, for straight polarity the electrode is negative and for reverse polarity the electrode is positive.

Tungsten Type and Diameter

Another area of concern when welding with the gas tungsten arc process is the selection of tungsten electrode type and diameter for a specific polarity and shielding gas. The applicable tungsten electrode diameters and current ranges are presented in Figure 160. This chart is designed to be used as a guide only. All settings indicated have been successfully used with a shielding gas mixture of 75% argon and 25% helium.

It should be remembered that electrode diameters for DCRP and DCSP are considerably different at the same welding current. The thermal distribution in the arc and the electron flow characteristics of each polarity are two major reasons for the differences.

For example, a 1/16'' diameter tungsten electrode easily has the capacity to carry 125 amperes when welding with DCSP (electrode negative). When welding with DCRP (electrode positive), however, a 1/4'' diameter tungsten electrode is required to carry 125 amperes. At this current level the 1/4'' tungsten electrode is at its maximum current carrying capacity. Any more welding current will cause the electrode to ''spit'' tungsten across the arc.

There is very little DCRP gas tungsten arc welding done in the welding industry because of the limiting factor of electrode heating. Usually DCSP is used for steels, stainless steels, copper alloys, low alloy steels and other relatively dense metals. Alternating current is normally used for GTAW welding of aluminum and magnesium and their alloys.

Pure Tungsten Electrodes		Current Ranges	
Electrode Dia.	ACHF-Argon	DCSP-Argon	DCSP-Helium
0.010"	up to 15	up to 15	up to 20
0.020"	10 to 30	15 to 50	20 to 60
0.040"	20 to 70	25 to 70	30 to 90
1/16"	50 to 125	50 to 135	60 to 150
3/32"	100 to 160	125 to 225	140 to 250
1/8 "	150 to 210	215 to 360	240 to 400
5/32"	190 to 280	350 to 450	390 to 500
3/16"	250 to 350	450 to 720	500 to 800
1/4 "	300 to 500	720 to 990	800 to 1100
1% and 2% Thoriated Tungsten Electrodes.			
0.010"	up to 20	up to 25	up to 30
0.020"	15 to 35	15 to 40	20 to 50
0.040"	20 to 80	25 to 80	30 to 100
1/16"	50 to 140	50 to 145	60 to 160
3/32"	130 to 250	135 to 235	150 to 260
1/8 "	225 to 350	225 to 360	250 to 400
5/32"	300 to 450	360 to 450	400 to 500
3/16"	400 to 550	450 to 720	500 to 800
1/4 "	500 to 800	720 to 990	800 to 1100

Figure 160. Tungsten Diameters, Current Ranges.

Figure 161 shows the heat distribution in the welding arc, the penetration patterns to expect from DCSP and DCRP, and the direction of electron (current) flow for each polarity. The tungsten electrode is a good electrical resistor and does not transfer heat readily. With DCRP approximately 70% of the available arc energy is in the tungsten electrode tip. The heat cannot dissipate rapidly so there is danger of overheating the electrode if excessive amperage is used.

Cleaning Action

The term "cleaning action" refers to the removal of surface oxides from materials such as aluminum and magnesium during the actual gas tungsten arc welding operation. The oxide removal occurs during the reverse polarity half-cycle when welding with the GTAW process and alternating current. Figure 162 shows the action that takes place during the cleaning half-cycle when welding aluminum with gas tungsten arc and AC. In particular, note the direction of movement of electron flow as well as the direction of gas ion movement.

As shown in Figure 162, electron flow is toward the electrode and gas ion flow is toward the work in the DCRP half-cycle of alternating current. A gas ion is a shielding gas atom that is deficient in electrons. Remember that almost all of the mass of the atom, and all of the positive electrical charge, are in the nucleus of the atom. It stands to reason, therefore, that the gas ion must have greater weight than the electron. The force exerted by the gas ion colliding with the surface of the base metal is thought to promote the physical breakup

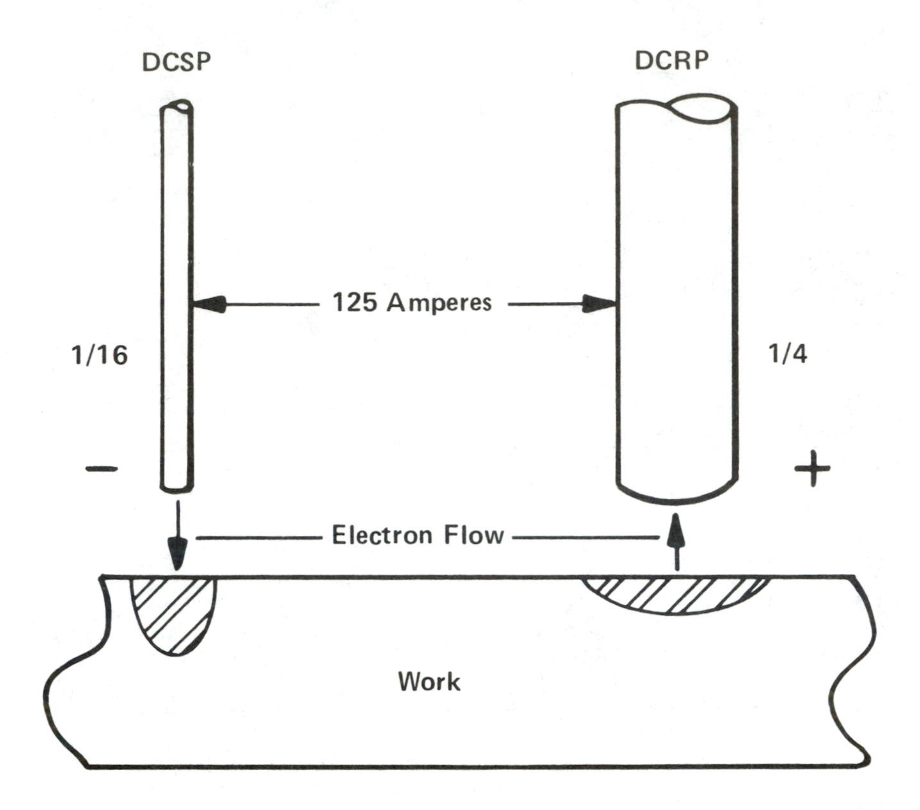

Figure 161. Tungsten Electrode Diameters and Penetration Patterns.

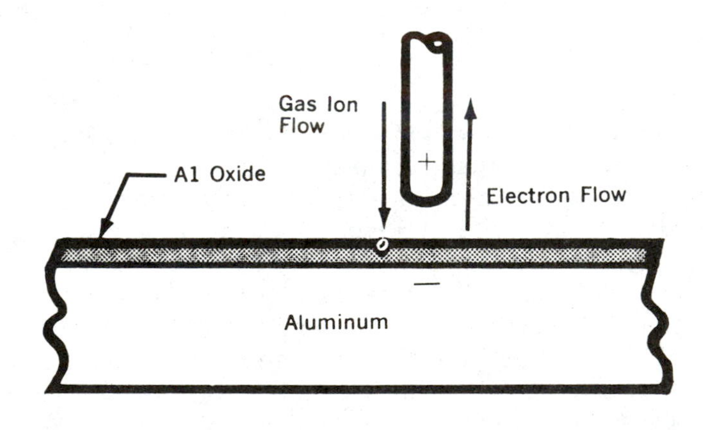

Figure 162. DCRP Cleaning Action on Aluminum.

of the surface oxide layer. The movement of electrons away from the work surface and towards the electrode actually lifts the surface oxides away from the base metal. Some portion of the oxide materials is vaporized in the heat of the welding arc. Some of the oxide residue will collect on the tungsten electrode beginning about 1/8" back from the tip of the tungsten.

Cleaning action is important when gas tungsten arc welding aluminum and magnesium because both metals have dense oxide layers that form very rapidly upon exposure to the atmosphere. The rate of oxide formation is accelerated with a rise in temperature such as would occur when welding either metal. Inclusion of the oxide, or oxide residue, in the weld deposit would decrease the strength and integrity of the weld. Such inclusions would be classed as "non-metallic inclusions".

DC Component

The subject of "DC component", what it is and what it does has been discussed at length since the introduction of the gas tungsten arc welding process. DC component is peculiar to AC welding with the GTAW process and is normally associated with the welding of aluminum and magnesium.

Actually, DC component is created during the reverse polarity half-cycle when welding with AC and the gas tungsten arc welding process. Referring to Figure 163, the electron flow characteristics are again shown. The workpiece illustrated is aluminum and the oxide layer is shown at the surface as a darker cross-hatched area. The electrodes are considered to be tungsten.

The AC sine wave trace should be familiar. The zero line is a time function with the area above the line termed positive and the area below the line termed negative. The positive half-cycle is the reverse polarity half-cycle. The negative half-cycle is the straight polarity half-cycle.

Starting with the DCSP half-cycle (negative side of the line) we will follow the welding current as it goes through two complete cycles of alternating current welding power. The negatively charged tungsten electrode has excellent electron emission characteristics so the first half-cycle (DCSP) provides almost perfect wave form shape. This is shown in Figure 159.

The second half-cycle is the positive half-cycle. It is shown as DCRP and the aluminum workpiece is the negative pole in the welding arc. Aluminum is not nearly as good an emitter of electrons as the hot tungsten electrode and, of course, it has a surface oxide layer that is very dense. The electron flow is retarded to some degree during the reverse polarity half-cycle by two factors: (1) less electron emission from the negative pole of the arc and (2) dense surface aluminum oxides. The AC sine wave trace is distorted by the lesser amount of amperage that is allowed to pass through the arc in this situation. It is commonly said that the reverse polarity half-cycle is "attenuated". To define:

Attenuate = "To weaken; to lessen the force or value of something".

If some amperage flows during the reverse polarity half-cycle the effect is termed "partial rectification in the welding arc". This is evident in the first positive half-cycle. If the oxide layer is heavy enough, and there is no current flow, the effect is full half-wave rectification in the welding arc. This shows in the second positive half-cycle illustrated.

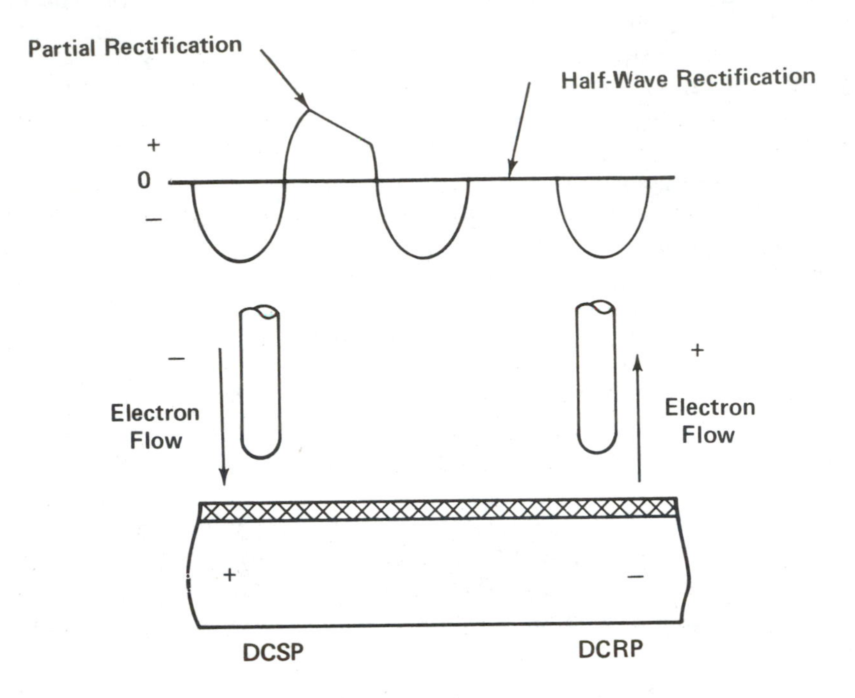

Figure 163. Formation Of DC Component.

It is at the time of partial rectification, or full half-wave rectification, that DC component is formed. The term "rectification" is the key word because to rectify alternating current power is to change it from AC to DC power. This is the function of any rectifier element.

The DC component is energy that is not used in the welding arc. Instead, it becomes a circulating current. This DC circulating current, which may be metered on a DC ammeter, flows back to the welding power source through the welding cables. If it is not filtered out of the circuit, or dissipated in some manner, it flows into the main transformer where it is dissipated as heat.

It is for this reason that AC power sources, not specifically designed for the gas tungsten arc welding process, must be de-rated for amperage when used with the process. If the AC power source is not de-rated for current output, the main transformer may overheat and be destroyed. The overheating is caused by the DC component that is dissipated as heat in the main transformer secondary coils. Data is provided in the text chapter "Power Sources For Gas Tungsten Arc Welding" telling how to de-rate AC welding power sources safely for the GTAW process.

In the NEMA Class 1 power sources designed for the gas tungsten arc welding process there is usually some form of filtering device to dissipate the DC component. The devices used include ni-chrome resistor bands that dissipate the DC component harmlessly into the atmosphere.

Another method used, which produces a balanced wave form in the amperage output circuit, is capacitor banks in the power source. The capacitors are connected in series with the welding arc. The beauty of the capacitors is that they will pass alternating current but will not pass the DC component. Still another method that has been used is to place wet cell storage batteries in series with the welding arc. While it is effective, it is awkward and cumbersome.

DC component is created in the gas tungsten welding arc. There is no welding power source that can prevent the formation of DC component. What can be done, of course, is to minimize the effects of DC component on the welding equipment and the welding arc.

CHAPTER 14

Consumable Weld Inserts

The consumable insert welding techniques are designed basically for welding root passes on joints where access is from one side only. Consumable weld inserts are used in butt joints of various types of pipe and for some structural and pressure vessel welds. The weld inserts are available in five distinct standard shapes and AWS classifications as shown in AWS A5.30 ''Specification for Consumable Inserts''.

The AWS classification has a prefix ''IN'' which indicates a consumable insert material. The balance of the designation is based on the chemical content of the insert material. The various designations are shown in several tables in AWS 5.30. For example, there are tables for mild steel inserts, chromium-molybdenum inserts, austenitic chromium nickel steel inserts, and nickel alloys including copper-nickel.

The standard consumable insert AWS Classes and shapes are explained in the following paragraphs. The specific AWS Styles are explained subsequent to the Class 5 discussion.

Class 1

The Class 1 shape is sometimes referred to as the ''Mushroom'' shape, the ''A'' shape or the ''Inverted-Tee'' shape (MIL-I-23413) consumable insert. This shape is designed for joints which have correct alignment with matching root face, bevel and internal diameter (ID); for all nuclear power piping and other rigidly controlled applications where Code or other specific requirements must be met. It is the best selection when the joint requires the ultimate predictability in underbead reinforcement. The maximum allowable mismatch of the joint is 1/32'' when using the AWS Class 1 shape consumable insert. Class 1 consumable inserts are available in Styles A, B, and C.

Class 2

This consumable insert, commonly called the ''J'' shape, has about 30% less filler metal than the Class 1 insert. It is often recommended for commercial applications where the fit-up and preparation of the joint are minimal. Naturally, it is always wise to achieve the best fit-up possible for the highest root pass welding quality.

Use care when welding subsequent passes in the joint. The second and third passes should be made in such a manner that burnthrough of the root pass is not permitted. The

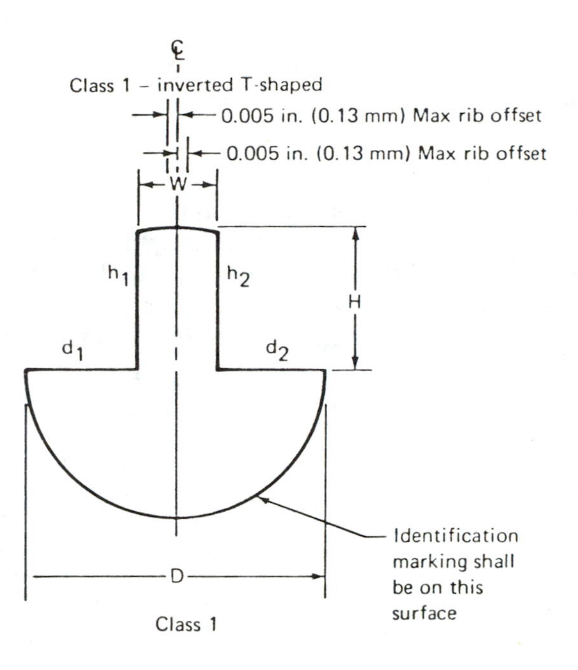

Figure 164. AWS Class 1 Consumable Insert Shape. Notes: (1) Lands(d_1, d_2) on either side of the rib shall be on the same plane within 0.0005 in. (0.13 mm); (2) Rib surfaces (h_1, h_2) shall be parallel within 0.0002 in. (0.05 mm) and square with lands (d_1, d_2) within 0.005 in. (0.13 mm); and (3) Dimensions and tolerances, see Table.

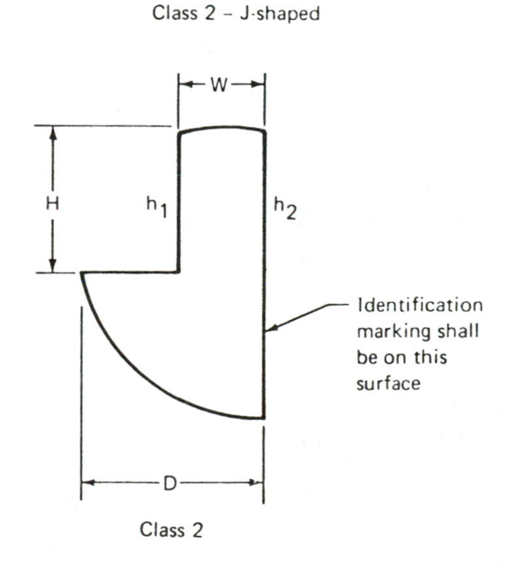

Figure 165. AWS Class 2 Consumable Insert Shape.

maximum mismatch allowable is 1/16" with the Class 2 consumable insert. The Class 2 consumable insert is available in Styles A, B, and C.

Class 3

This 1/16" × 3/16" rectangular shape is flat on both sides. Commonly called the "K" insert, it is designed for use where there is poor alignment of the pipe joint and/or field prepared joint root face and bevels. It is excellent for use with the automatic GTAW pipe welding machines presently available. The maximum mismatch allowable is 3/32" with the Class 3 consumable insert. Class 3 consumable inserts are available in Styles D and E.

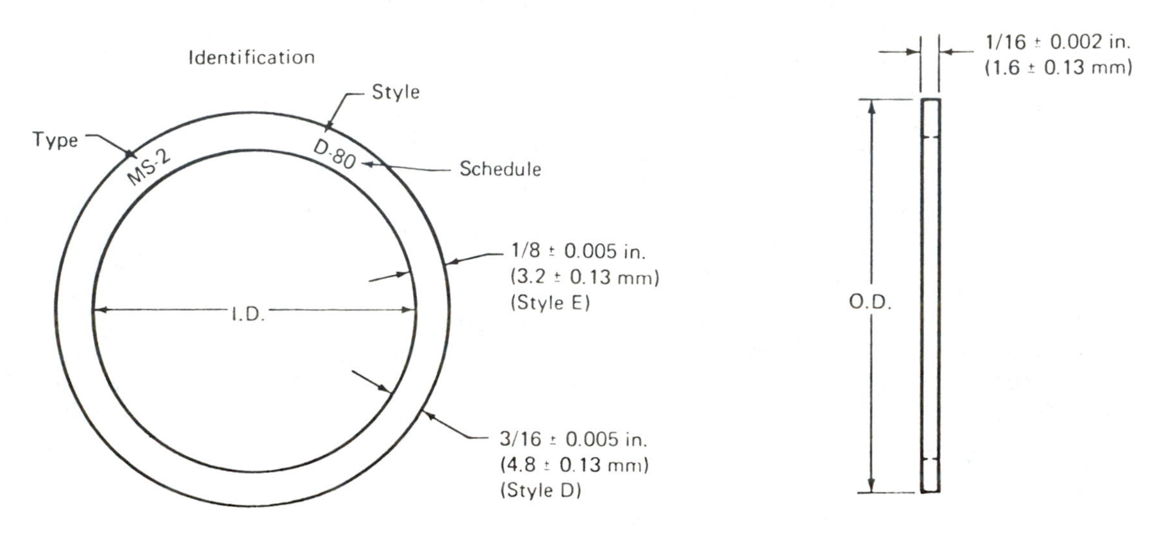

Figure 166. AWS Class 3 Consumable Insert Shape.
Class 3 - Solid ring inserts - Plan view and cross-sectional configurations.

Class 4

The AWS Class 4 consumable insert is "Y" shaped and is designed for pipe root pass welding of very high quality. Class 4 consumable inserts are available in Styles A, B, and C.

Class 5

This 1/8" × 5/32" rectangular shape is flat on both sides with slightly rounded corners. It is available in Styles A, B, and C.

The AWS Styles are explained as follows:

Style A. Coiled lengths of consumable insert.
Style B. Preformed rings with open lap joint.
Style C. Preformed rings with open butt joint.
Style D. Solid rings with 3/16" rim width.
Style E. Solid rings with 1/8" rim width.

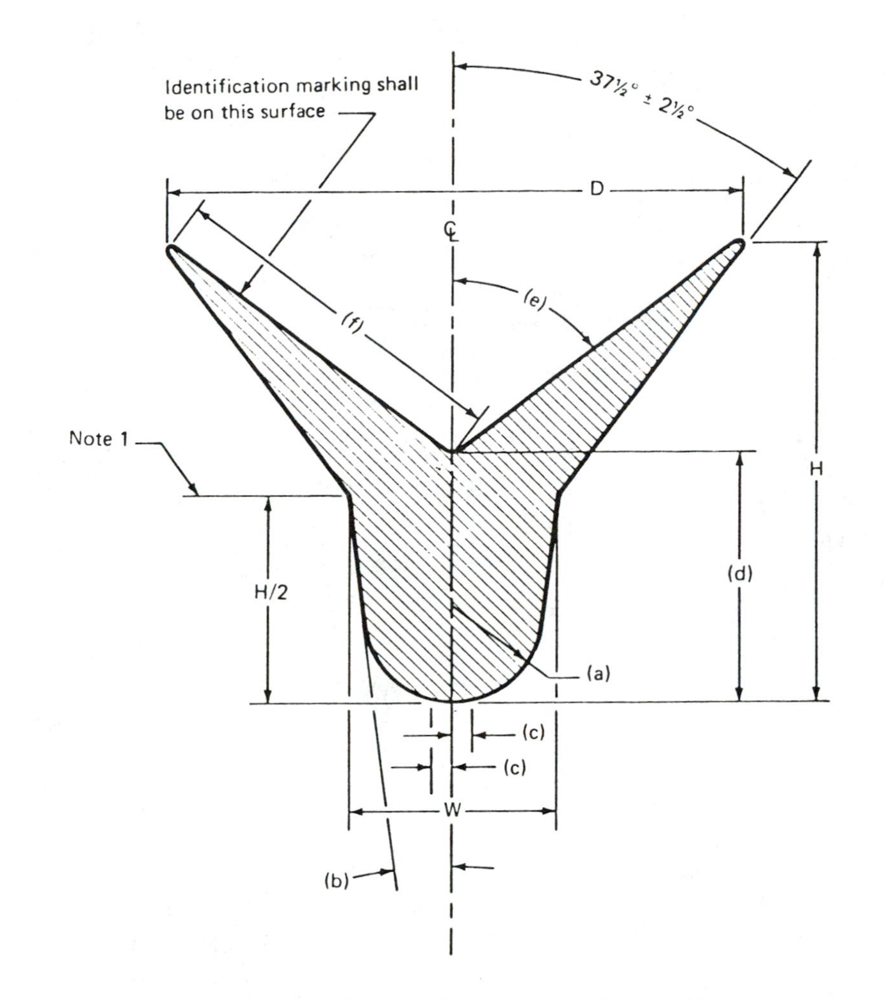

Figure 167. AWS Class 4 Consumable Insert Shape.
Notes: (1) Reference diameter for correlating with pipe I.D.; (2) Dimensions and tolerances, see Table 1 and (3) When specified, rings 1-1/2 to 2 in. (38.1 to 50.8 mm) in diameter shall be formed of 5/32 in. (4.0 mm) material.

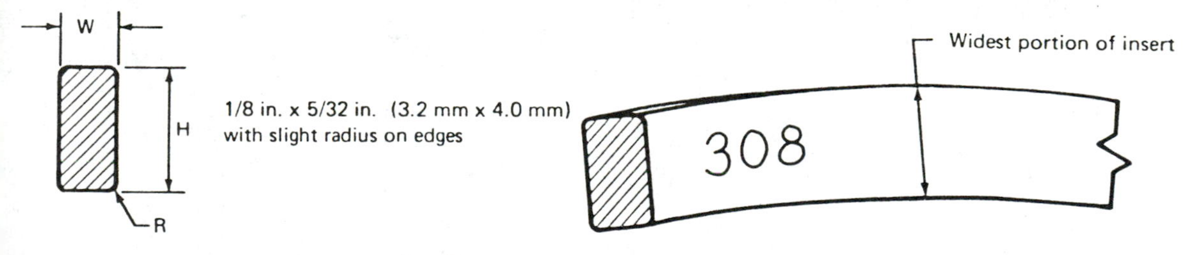

Figure 168. AWS Class 5 Consumable Insert Shape.

Table 1. Classes 1, 2, 4, and 5 inserts — cross-sectional dimensions and tolerances

Insert size		Legend (Fig. 1)	Class one[1] inverted T-shaped				Class two,[1] J-shaped				Class four,[2] Y-shaped				Class five, rectangular-shaped			
			Dimensions		Tolerances		Dimensions		Tolerances		Dimensions		Tolerances		Dimensions		Tolerances	
in.	mm		in.	mm	in.	mm	in.	mm	in.	mm	in.	mm	in.	mm	in.	mm	in.	mm
1/8	3.2	D	0.125	3.18	±0.004	±0.10	0.086	2.18	+0.011 -0.005	+0.28 -0.13	0.165	4.19	±0.010	±0.25	---	---	---	---
		W	0.047	1.19	+0.002 -0.012	+0.05 -0.30	0.047	1.19	+0.002 +0.012	+0.05 -0.30	0.078	1.98	±0.010	±0.25	0.0625	1.59	±0.010	±0.25
		H	0.055	1.40	+0.012 -0.002	+0.30 -0.05	0.055	1.40	+0.012 -0.002	+0.30 -0.05	0.140	3.56	±0.010	±0.25	0.125	3.18	±0.010	±0.25
		H/2	---	---	---	---	---	---	---	---	0.072	1.83	±0.010	±0.25	---	---	---	---
		R	---	---	---	---	---	---	---	---	---	---	---	---	0.0156	0.40	±0.005	±0.13
5/32	4.0	D	0.156	3.96	±0.005	±0.13	0.110	2.79	+0.012 -0.010	+0.30 -0.25	0.205	5.21	±0.015	±0.38	---	---	---	---
		W	0.063	1.60	+0.003	+0.08	0.063	1.60	+0.003	+0.08	0.093	2.36	±0.015	±0.38	0.125	3.18	±0.015	±0.38
		H	0.063	1.60	+0.014 -0.003	+0.36 -0.08	0.063	1.60	+0.014 -0.010	+0.36 -0.25	0.175	4.45	±0.010	±0.25	0.156	3.96	±0.015	±0.38
		H/2	---	---	---	---	---	---	---	---	0.093	2.36	±0.010	±0.25	---	---	---	---
		R	------	---	---	---	---	---	---	---	---	---	---	0.03125	0.794	±0.005	±0.13	

1. The offset between the center of the rib (W) and center of the land (D) shall not exceed 0.005 in. (0.13 mm).
2. Additional dimensions and tolerances - class 4, Y-shaped inserts (see Fig. 1):

	1/8 in. size	3.2 mm size	5/32 in. size	4.0 mm size
(a) Radius of rib	0.044 ± 0.005	1.12 ± 0.13	0.050 ± 0.005	1.27 ± 0.13
(b) Angle between side of rib and center line	1° – 2°	1° – 2°	1° – 2°	1° – 2°
(c) Rib offset	±0.010	±0.25	±0.015	±0.38
(d) Height of rib along center line	0.100 ± 0.010	2.54 ± 0.25	0.115 ± 0.010	2.92 ± 0.25
(e) Angle between top surface of inclined arm and center line	50° ± 5°	50° ± 5°	50° ± 5°	50° ± 5°
(f) Length of inclined arm	0.085 ± 0.010	2.16 ± 0.25	0.125 ± 0.010	3.18 ± 0.25

From a practical inspection standpoint, the D, W, H, and H/2 dimensions have the most to do with the usability and weldability. The (a) through (f) dimensions are furnished to complete the description of cross-sectional configuration.

Class 3, style E inserts — dimensions

Pipe dimensions					Ring diameters[2]			
Nominal diameter		Schedule number[1]	ID[2]		Ring OD for nominal pipe diameter		Ring ID for nominal pipe diameter	
in.	mm		in.	mm	in.	mm	in.	mm
1/4	6.4	10S	0.410	10.41	0.57	14.5	0.32	8.1
		40	0.364	9.25	0.52	13.2	0.27	6.8
		80	0.302	7.67	0.46	11.7	0.21	5.3
3/8	9.5	10S	0.545	13.84	0.70	17.8	0.45	11.4
		40	0.493	12.52	0.65	16.5	0.40	10.1
		80	0.423	10.74	0.58	14.7	0.33	8.4
1/2	12.7	5S	0.710	18.03	0.87	22.1	0.58	14.7
		10S	0.674	17.12	0.83	21.1	0.58	14.7
		40	0.622	15.80	0.78	19.8	0.53	13.5
		80	0.546	13.87	0.70	17.8	0.45	11.4
3/4	19.1	5S	0.920	23.37	1.08	27.4	0.83	21.1
		10S	0.884	22.45	1.04	26.4	0.79	20.1
		40	0.824	20.93	0.98	24.9	0.73	18.5
		80	0.742	18.85	0.90	22.9	0.65	16.5
1	25.0	5S	1.185	30.10	1.34	34.0	1.09	27.7
		10S	1.097	27.86	1.25	31.8	1.00	25.4
		40	1.049	26.64	1.21	30.7	0.96	24.4
		80	0.957	24.31	1.11	28.2	0.86	21.8
1-1/4	32.0	5S	1.530	38.86	1.69	42.9	1.44	36.6
		10S	1.442	36.63	1.60	40.6	1.35	34.3
		40	1.380	35.05	1.54	39.1	1.29	32.8
		80	1.278	32.46	1.43	36.3	1.18	30.0
1-1/2	38.0	5S	1.770	44.96	1.93	49.0	1.68	42.7
		10S	1.682	42.72	1.84	46.7	1.59	40.4
		40	1.610	40.89	1.77	45.0	1.52	38.6
		80	1.500	38.10	1.65	41.9	1.41	35.8
2	51	5S	2.245	57.02	2.40	61.0	2.15	54.6
		10S	2.157	54.79	2.31	58.7	2.06	52.3
		40	2.067	52.50	2.22	56.4	1.97	50.0
		80	1.939	49.25	2.10	53.3	1.85	47.0
2-1/2	64	5S	2.709	68.81	2.87	72.9	2.62	66.5
3	76	5S	3.334	84.68	3.49	88.6	3.24	82.3
3-1/2	89	5S	3.834	97.38	3.99	101.3	3.74	95.0
4	102	5S	4.334	110.08	4.49	114.0	4.24	107.7

When ordering consumable inserts it is necessary to have the AWS A5.30 Specification available. This is the governing document to assure that the consumable inserts you order are manufactured according to AWS standards. Always provide the following minimum information on the purchase order.

1. The governing document title, date, and number (A5.30-79).
2. Pipe schedule to be welded, insert class, shape, and size.
3. For stainless steel, the required ferrite number.

Joint Design and Preparation

The joint design and preparation are very important when using consumable weld inserts and the GTAW process. If possible, the pipe ends should be prepared by machining. This will provide the best high quality joint preparation. For less critical applications, flame cutting, followed by surface grinding, has been satisfactory. In some field operations, a milling tool is often used to create the selected joint configuration on the ends of the pipe to be joined. This is especially true of automatic pipe welding operations for transmission pipe lines.

Some joint designs are shown in ANSI B16.9 "Pipe Fitting Standards". These joint designs may be modified for specific wall thicknesses and applications. For carbon steels and austenitic stainless steels the recommended root face is 0.055", plus or minus 0.010". (The root face is sometimes called the "land" in a weld joint).

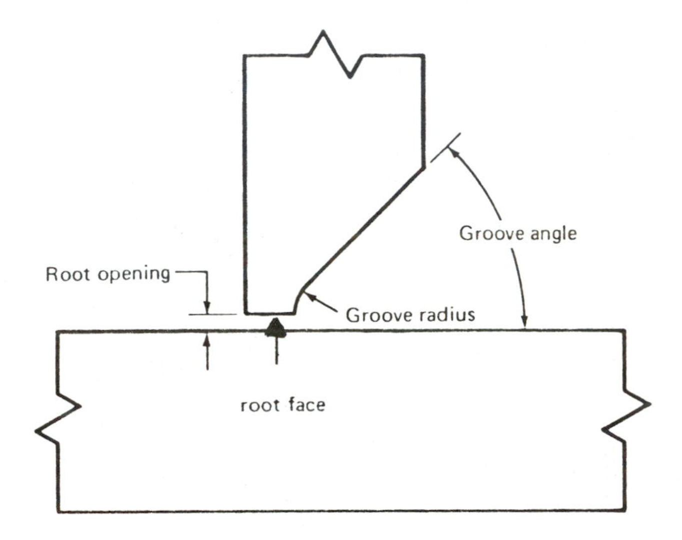

Figure 169. Root Face and Root Opening of Weld Joint.

Setting Up To Weld

Setting up the correct welding procedure requires practice welding to determine the various parameters to use. Much of this data should be determined prior to tack welding the consumable insert ring in place on the pipe.

In general, the following data should be used as a **guideline** for the actual welding of the root pass. The welding procedure must be developed for the specific job to be done. This procedure considers the typical application to be on carbon steel pipe. The pipe is in a fixed position and may be welded in all welding positions; that is, flat, horizontal, vertical, and overhead with the GTAW process.

The shielding gas is argon. The backing gas is argon. The GTAW torch is rated at 250-300 amperes, 100% duty cycle. The power source is a 300 ampere, 60% duty cycle, constant current unit designed for the GTAW process. It has remote output control capability, a built-in high frequency system, gas valve and solenoid, water valve and solenoid, gas and water post-flow timer, pulse control, and start control system. A remote foot control is normally used with this system.

The tungsten electrode may be either 1% or 2% thoria bearing tungsten. The electrode end must be prepared prior to welding. Some welding authorities like a long taper to a pointed end on the tungsten. Others prefer making a hemispherical ball on the electrode end.

The taper may be ground or created by using chemical salts and a hot tungsten electrode. The taper should not be longer than 2 times the tungsten electrode diameter, in my opinion. The end of the electrode should not be tapered to a point but should be not less than 1/2 the electrode diameter. The electrode end should be rounded to eliminate sharp corners.

The hemispherical ball may be created by using DCRP (electrode positive), minimum amperage output, minimum range on the power source, start high frequency, a remote foot control, and argon shielding gas. A copper block for arc starting is recommended.

The arc is initiated on the copper block. The electrode is melting at the tip when you see a transluscent arc column. *The ball that is formed should never be larger in diameter than the electrode diameter.* After forming the ball, be sure to return the power source to DCSP (electrode negative) before attempting to weld.

Shielding and Backing Gases

Use accepted methods for providing backing gas behind the weld to be made. The backing gas may be argon or helium for carbon steel. For austenitic stainless steel, the backing gas may be argon, helium, nitrogen or an argon-hydrogen mixture. Pressure seals are not normally required to contain the backing gas.

Some of the devices used to contain the backing gas behind the weld zone are as follows:

1. Simple paper discs fastened to the internal walls of the pipe with masking tape.
2. Removable paper cones have been used to simplify removal after the weld is completed.
3. Water soluble paper discs can also be used. Their removal after welding is simply a matter of flushing the system.
4. Metal discs with rubber or plastic gaskets are often used. The shielding gas is piped in through a fitting in the disc or gasket.
5. Metal cones are also used. Two cones are normally used with the smaller end of the cones both pointed in the direction in which they will be removed. A gas carrying hose is attached to a gas fitting in one cone.
6. Plastic discs are often used for training welders with the GTAW process on pipe. They are attached to the pipe ends with masking tape. A fitting for the gas hose is supplied on one plastic disc. A gas escape hole should be left in the other plastic disc. This provides a vent for the atmospheric gases to escape while they are being replaced with the backing gas.

Argon is, and has been, the preferred shielding and backing gas used for welding the root pass with consumable insert rings. Helium may be used although it has less density and is lighter than air. Argon is slightly heavier than air. As noted previously, nitrogen has proved to be satisfactory when used as a purge gas for austenitic stainless steels.

GTAW Tack Welding With Consumable Inserts

1. Prepare the joint edge with the selected joint design and method of preparation. Use all standard procedures for maintaining cleanliness in the weld area.
2. Place a consumable weld insert, in ring form, on a prepared pipe joint end. There should be some overlap of the ends of the insert ring.
3. Fit one end of the insert ring to the pipe joint and tack weld it in place with the GTAW process. Be sure the tack welds fuse the insert with the pipe joint edge. Make small tack welds at suitable intervals about half way around the pipe joint.
4. Using a hacksaw with not less than 24 teeth per inch, trim off the excess insert material from the ring. Leave not more than 1/32'' between the ends of the consumable insert ring.
5. Carefully tack weld the remainder of the insert ring to the pipe joint edge. The final tack weld should fuse the two ends of the insert with the pipe joint edge. The ends of the consumable insert ring should be butted together in the final tack.
6. Bring the pipe joint ends together. Align the joint carefully so that any mismatch is held to less than the maximum allowable for the insert shape being used. If tooling for aligning the pipe joint is not available, three lengths of suitably sized angle iron will normally work well if placed at equal distances around the pipe joint.
7. After aligning the pipe section, begin tack welding about midway between the original tack welds. The second pipe section is tack welded to the insert and the tack is carried across to include the original pipe section. At this point the parts should be ready for setup for final welding.

Welding the Consumable Insert Root Pass

The welding procedure that has been developed should be used in setting up the conditions for welding the pipe joint. The electrode extension beyond the gas cup is normally 3/16''-1/4''. The shielding gas flow rate will be about 10-12 CFH.

When welding pipe, the electrode is always held in such a manner as to be pointing at the centerline of the pipe. The welder should be as comfortable as possible in order to provide free movement of the GTAW torch as required for the joint.

Both amperage and speed of travel have a definite effect on the internal weld bead contour. It is wise, therefore, to know the heat input needed to make the weld properly. This may be determined by using the following equation:

$$H = \frac{\text{Volts} \times \text{Amperes} \times 60}{\text{Speed of Travel (IPM)}}$$

where

H = joules of heat per linear inch of weld
Volts = welding arc voltage
Amperes = welding amperes
60 = 60 seconds per minute
Speed of Travel = speed in IPM (inches per minute)

All that is necessary to use the equation is to plug in the settings from your welding procedure. To increase speed, it is obvious that you will have to increase amperage so that the total joules of heat input per linear inch of weld remains the same.

To initiate the welding arc, it is often preferable to use a starting block such as a copper bar rather than making a touch start on the weld joint. If at all possible, it is recommended that a high frequency non-touch arc start be used. Vertical welds on pipe are normally begun at a point just off center at the bottom of the joint. One side is welded upward and then the second side is completed.

The correct arc length is considered to be about 3/32"-1/8". This arc length will provide adequate weld puddle control and correct fusion of the consumable insert. REMEMBER: when welding heat input per linear inch of weld is too low because of low amperage or high travel speeds, there will be less of the consumable insert melted and pulled into the welded joint. This leaves excess reinforcement under the weld. A condition such as this could cause a notch effect at the root of the weld bead.

When the molten weld puddle is formed, the welder should begin a slight rocking, or weaving, motion with the torch. This will equalize the heat input across the consumable insert and fuse it with both of the pipe joint edges. When the heat input is sufficient to melt the consumable insert, the weld puddle will appear to rise slightly. This indicates that full fusion of the consumable insert and the joint edges is taking place.

At this point, the welder should begin moving along the weld joint at the correct travel speed. Complete fusion will be achieved as long as the welder can see the "rise" of the weld puddle as he progresses. At the completion of the weld the arc is moved away from the weld puddle and up the side of the joint to eliminate crater cracking.

Inspection and Repair of the Root Pass

The basic requirement for visual inspection (VT) of the root pass welds made with consumable inserts is that the weld surface have a uniform shape. Any significant depression along the finished weld bead would normally indicate that there is incomplete fusion and melting of the consumable insert.

This type of weld defect can be repaired by simply re-initiating the welding arc, creating a molten weld puddle and watching the bead until the typical rise in the molten puddle is seen. Upon completion of the weld repair the arc is moved away from the weld puddle and up the side of the joint to eliminate the weld crater and possible crater cracking in the weld.

It is often necessary to use other inspection methods for root passes made with consumable inserts and the GTAW process. For example, the dye penetrant (DPT) inspection method may be used to check for surface cracks in the weld. Radiographic inspection (RT) is often required for welding procedure qualification tests, welder certification tests, and some percentage of the finished production welding root passes.

Completing the Weld

The welding process used to complete the weldout of the joint may be the shielded metal arc welding (SMAW) process, the gas tungsten arc welding (GTAW) process or the gas metal arc welding (GMAW) process.

The SMAW process provides the greatest portability of equipment of any of the three welding processes. Sound welds can be achieved in all welding positions. Unfortunately, there is a problem with slag removal after each welding pass. The possibility of non-metallic inclusions is always present when using the SMAW process.

The gas tungsten arc welding process is always selected for welding of the highest quality. Although the GTAW process is the slowest of the arc welding processes, it does provide the weld deposit integrity required by some Codes. The actual weld application may be by either automatic or manual methods.

The GMAW process is capable of providing clean, fast weld deposits. There is no slag or flux because the process employs an externally supplied gas shield for the electrode and the weld area. The gas metal arc welding process is limited in mobility to some degree. It is much faster than either of the other two arc processes mentioned.

Any of the three manual welding processes may be successfully used for the weldout of the pipe joints. In addition, automatic welding of the joints can be done with certain automatic equipment presently available.

CHAPTER 15

Tungsten Electrodes

The development of the gas tungsten arc welding process was predicated on the use of an electrode material that would not vaporize in the heat of the welding arc. The element tungsten was selected for several reasons, some of which are discussed in this chapter.

While employed as a manufacturing research engineer, I was concerned at the excessive loss of tungsten electrodes during welding operations. Some reasons for the losses were easily explained. For example, when a welder dips the tungsten electrode into the weld puddle, or touches it with a filler rod while welding, the tungsten electrode is contaminated. It must be cleaned by removing a portion of the electrode and then pointing it in some manner for good welding characteristics.

Other losses came from improper methods of breaking the contaminated ends off tungsten electrodes. For example, holding a tungsten electrode at one end and trying to break off the tip of the other end usually results in excessive tungsten loss and a shattered electrode end. This sort of thing can be overcome by training welders to use correct methods.

The losses I had trouble understanding occurred when welders were grinding tungsten electrodes. Very often the tungsten electrodes would literally shatter in the welder's hands while he was grinding. The result was loss of the total electrode. My concern was, of course, what caused this to happen and what could be done to minimize the tungsten losses? The cost involved was considerable.

This led me to search out some facts and those results are presented here for your information.

Tungsten Manufacturing

Although most welding people never have occasion to be concerned with the manufacturing processes involved with tungsten, or tungsten electrodes for welding, it is important to have an understanding of how the material is put together. From this knowledge may come better methods of preparing and using the tungsten electrodes necessary to the gas tungsten arc welding and cutting processes.

The process of manufacturing tungsten electrodes is a form of **powder metallurgy.** Tungsten particles that are in powder form, and purified to a minimum 99.95%, are used for electrode manufacturing. To maintain purity of the base metal a continuous program of analysis and inspection is performed throughout the various manufacturing steps. Quality control is very rigid at all stages of the manufacturing process.

The high purity powdered tungsten is pressed into ingots, or "compacts", under many tons of pressure per square inch. The as-pressed ingots have very little strength and are quite fragile.

The next step is "sintering" the ingots in a hydrogen atmosphere. This is done at a temperature high enough to cause agglomeration. The packed mass formed provides adequate strength to support the weight of the ingot. To define:

Sinter = "To heat a mass of fine particles for a prolonged period of time, at a temperature below the melting point of the material, usually to cause agglomeration."

Sintering = "The bonding of adjacent surfaces of particles in a mass of metal powders, or a compact, by heating."

The material to be sintered is supported and suspended between electrical contacts and a controlled electrical current is passed through the ingot. In this manner the ingot is heated to very near its melting point (6,170° F.) and held at temperature for a specific period of time.

The "treated tungsten ingot", so-called because of the electric heating operation, is tested for crystal structure and density. Spectrographic analysis is used to assist in maintaining the quality level of the product.

The tungsten ingots are then swaged into rod form. This is a mechanical metal working process in which the ingot is forged hot. The initial working temperature is approximately 1,500° C. (2,750° F.). The actual swaging operation is done with a rotary hammer. The sintered ingot must be heated before mechanical working. Tungsten ingots are brittle at room temperature and cannot be cold worked to any extent without fracture.

The swaging operation develops a type of fibrous structure that imparts some measure of ductility and toughness into the tungsten rod. Rods that are reduced in diameter to arc welding electrodes are drawn through various sizing dies of hardened steel. A final sizing finish draw is made through an industrial diamond die.

This is the usual basic procedure followed for drawing tungsten electrode materials of different diameters. As finally drawn, the electrodes have a dense black oxide coating.

Tungsten Electrodes for Arc Welding Processes

Tungsten is employed as a non-consumable electrode for the atomic-hydrogen welding process, the gas tungsten arc welding process, the plasma welding process and the plasma cutting process. The term "non-consumable" means that the electrode is not intended to become part of the filler metal in the weld deposit. This is not to say that the tungsten doesn't get into the weld at times! Tungsten that is included in weld deposits is considered a weld defect. The hard tungsten inclusions will raise stress points in the weld due to its low ductility and characteristic hardness. Remember: the only time tungsten should be in the weld deposit is when you are welding tungsten base metal.

Methods of cleaning and finishing tungsten electrodes are explained in other sections of this chapter.

The element tungsten has the highest melting point of any of the metals. In fact, of all the elements it is second only to carbon which has a melting temperature of 6,740° F. (3,727° C.). The relative melting and boiling points for carbon and tungsten are illustrated in Figure 170.

Element*	Symbol	Melting Point Degrees C.	Melting Point Degrees F.	Boiling Point Degrees C.	Boiling Point Degrees F.
Carbon	C	3,727#	6,740#	4,830	8,370
Tungsten	W	3,410	6,170	5,930	10,706

*Metals Handbook, ASM, Eighth Ed., 1961
#Sublimes

Figure 170. Carbon and Tungsten Data.

Although carbon has a considerably higher melting point than tungsten the boiling point of carbon is substantially lower than that of tungsten. Since carbon also **sublimes** at its melting temperature, it will vaporize and not be suitable for most present day welding applications where tungsten is used. To define:

Sublime = "To pass directly from the solid state to the gaseous state without the intermediate liquid state of matter."

The metal tungsten, with a melting point of 6,170° F., has a boiling point of 10,706° F. The temperature gradient (4,536° F.) is such that it is virtually impossible to vaporize the tungsten electrode in the heat of the welding arc. This statement is based on the welding amperage being within the current carrying capacity of the specific type and diameter of tungsten electrode.

Tungsten Plus Additions

The element tungsten is a good emitter of electrons. When the tungsten electrode tip is heated, electron emission is even better. This factor assists in the initiation and stabilizing of welding arcs. Thoria oxide (ThO_2), commonly called "thoria", and zirconium oxide (ZrO_2), commonly called "zirconia", have been added in small percentages to some tungsten electrodes to improve electron emission and arc initiation.

There are three types of thoria bearing tungsten electrodes. The standard 1% and 2% thoria bearing tungsten electrodes are well known. They are used mostly for welding metals other than aluminum and magnesium. The third type is a "striped" tungsten electrode which has a small wedge of 1%-2% thoria tungsten running lengthwise in each electrode. The *total thoria addition is considered to be about 0.5%.* The zirconia tungsten electrodes have approximately 0.4% zirconia added to the tungsten matrix.

The various types of tungsten electrodes are described in the AWS A5.12 "Specification for Tungsten Arc Welding Electrodes". The tungsten electrodes are classified as follows:

AWS Class	Tungsten% minimum	Thoria%	Zirconia%	Max. Other elements
EWP	99.5	---	---	0.5
EWTh-1	98.5	0.8-1.2	---	0.5
EWTh-2	97.5	1.7-2.2	---	0.5
EWTh-3	98.95	0.35-0.55	---	0.5
EWZr	99.2	---	0.15-0.40	0.5

Figure 171. AWS A5.12 Chemical Requirements of Tungsten Electrodes.

The addition of thoria and zirconia was originally intended to promote better arc starting characteristics in the tungsten electrodes. In the process it was found that the addition of these elements provided increased current carrying capacity for the tungsten electrodes. Investigation proved that the addition of up to about 0.6% thoria to the tungsten electrode matrix would increase the current carrying capability of the electrodes. Additional amounts of thoria will not increase the current carrying capability of the tungsten electrodes although it will improve the arc starting characteristics.

Tungsten Size and Finishes

Tungsten electrodes for arc welding are available in sizes from 0.020'' diameter to 1/4'' diameter. Tungsten electrodes are normally furnished to the welding public in boxes of ten electrodes per package.

Clean finished tungsten electrodes are normally bright silver in color. They have had all the surface contaminants, oxides, etc., removed. Clean finished electrodes may or may not be concentric to the electrode centerline.

Centerless ground tungsten electrodes, as the name implies, are ground to a bright, shiny finish. They are concentric to the electrode centerline. Centerless ground tungsten electrodes, being perfectly round, make the best possible electrical contact between the torch collet and the electrode. This means there is minimum electrical resistance and minimum resistance heating in the torch head. It is natural that centerless ground tungsten electrodes are used where minimum electrical resistance losses at the collet-electrode contact are desired.

Pointing Tungsten Electrodes

Tungsten electrodes are sold with square cut ends. In many GTAW applications it is necessary, or desirable, to have some type of end preparation on the electrode. The end preparation may be natural or it may be accomplished through mechanical, electrical, or chemical methods.

The form of electrode preparation will depend on the type of tungsten used. For example, pure tungsten always has a molten ball at the electrode tip when welding. When cool, the electrode end forms a shiny, smooth ball. This is also true of zirconia bearing tungsten electrodes.

Thoria bearing tungsten electrodes do not melt under the heat of the welding arc. This type of electrode may be tapered through grinding or by chemical means. Thoria bearing tungsten electrode ends may also be prepared by electrical methods. This is done with DCRP (electrode positive) power and special operating techniques.

Mechanical End Preparation

Most mechanical tungsten electrode ''pointing'', or end preparation, is done by grinding the electrode to some taper. Many welding authorities are of the opinion that tungsten electrodes should not be ground on an abrasive wheel. Other equally competent welding authorities are of the opinion that grinding tungsten electrodes is perfectly all right. Probably both sides have some merit in their arguments.

The requirement for grinding tungsten electrodes suggests the electrode is too large in diameter for the welding application. Usually the only time a tungsten electrode should require grinding is when the welding application is on extremely thin material. An approximate range of material thicknesses where grinding of tungsten electrodes would be acceptable is from 0.001" to 0.050".

If grinding tungsten electrodes is necessary it should be done within certain criteria. The proper technique for grinding tungsten electrodes is shown at the left side of Figure 172. The incorrect method of grinding electrodes is shown at the right side of the drawing.

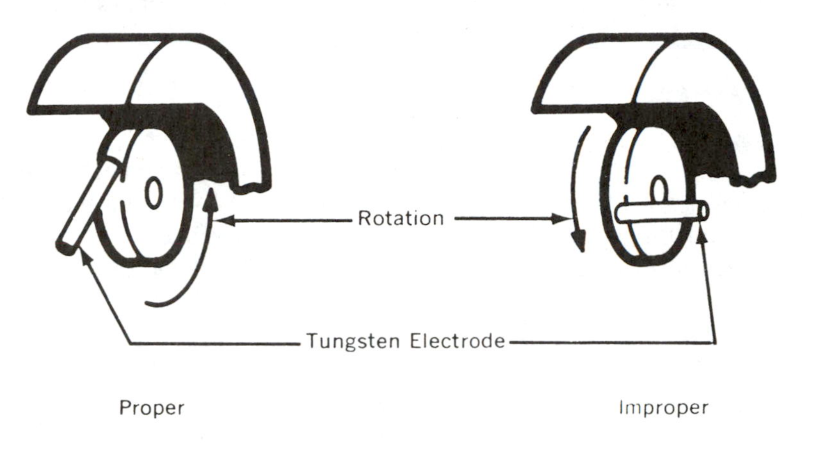

Figure 172. Tungsten Electrode Grinding Methods.

If tungsten electrodes are ground the work should be done on a special fine-grit extra hard abrasive wheel. The abrasive wheel should be used for no other material except tungsten electrodes! Other materials will contaminate the grinding wheel, and the tungsten electrodes, with foreign particles that can become defects in the weld deposit.

What effect does improperly grinding tungsten electrodes have on actual welding operation costs? There is no single answer to the question but the following discussion will supply at least two answers.

Many times when a welder is grinding a tungsten electrode it will splinter into bits and pieces. The electrode is suddenly just a mass of fragments. Remember that tungsten is a hard and brittle material with very little ductility. It is certainly harder than any grinding wheel you may purchase. The abrasive wheel is relatively soft material, compared to tungsten, and therefore the tungsten particles are not really ground off. They are, instead, literally chipped away particle by particle. You may remember that the tungsten electrode started out as very fine high purity powdered particles.

The action that takes place when grinding a tungsten electrode may be compared to an Indian chipping a stone arrow point. The particles of tungsten are removed by a percussive action, or series of blows, made by the grinding wheel. The shock of the impact on the tungsten electrode, which has very little ductility, can cause the electrode to splinter.

Any grinding operation is bound to leave machining marks on the material being ground and tungsten is no exception. When the electrode is improperly ground the machining marks will be concentric to the centerline of the electrode. The danger is in the ridges of electrode material that melt and pass across the arc column to the weld puddle. Such tungsten spatter from the electrode will show up on radiographic film as a grouping of fine white spots. The extremely dense tungsten materials will normally be located at the point where an arc was started with a freshly ground tungsten electrode.

Any tungsten inclusions in the weld would be classed as defects which must be removed. It is entirely possible that small portions of the grinding wheel would become lodged in the ground portion of the tungsten electrode between the machining grooves. Under the heat of the welding arc these non-metallic particles would transfer across the arc and become defects in the weld deposit.

Electrical End Preparation

Electrical end preparation is a relatively simple method that produces a hemispherically shaped end on thoria bearing tungsten electrodes. This method is not required for pure and zirconia bearing tungsten electrodes. The ball formed on the end of the thoria bearing electrodes should never be larger than the diameter of the electrode.

Electrical end preparation is accomplished using DCRP (electrode positive). The welding power source is set for the lowest amperage range, minimum amperage setting on the rheostat, "start" high frequency, and a remote foot control is used. A small piece of copper plate or bar stock should be used as the base metal when initiating the welding arc. The high frequency voltage will provide an ionized gas path for the welding current to follow. UNDER NO CIRCUMSTANCES SHOULD A TUNGSTEN ARC BE INITIATED ON CARBON OR GRAPHITE BLOCKS. The carbon will vaporize and immediately contaminate the tungsten electrode tip. If used on production work, the carbon or graphite vapors would also contaminate the workpiece.

Use the remote foot control to attempt arc initiation. The welding arc may not start immediately if the electrode is too large in diameter for the low amperage setting. If the welding

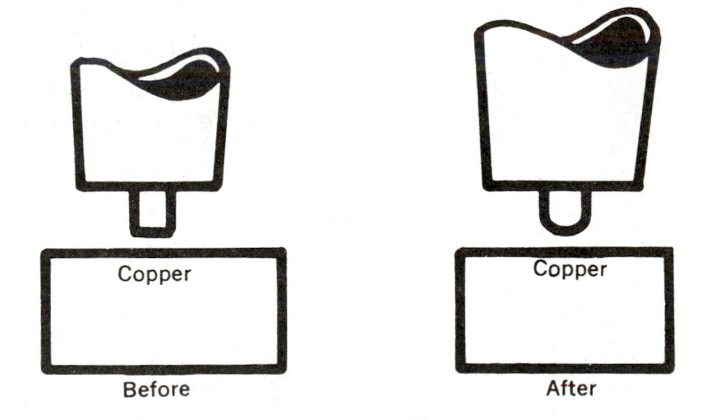

Figure 173. Pointing Tungsten Electrodes With DCRP.

arc doesn't start when the remote foot control is fully depressed, stop the operation and increase the rheostat setting for more amperage output. Attempt arc initiation again.
As soon as the welding arc is established, the electrode end will show a bright orange color. This will change to a brilliant white as more amperage is applied through the remote foot control. Increasing the amperage even more will turn the arc column to an incandescent white. At this point the electrode tip is melting. For small electrodes 1/16'' diameter and less, power should be removed immediately before the molten tungsten electrode tip becomes too large. The molten ball should never be larger than the diameter of the electrode. The tungsten electrodes before and after electrical end preparation are shown in Figure 173.

The resulting hemispherical ball formation on the electrode tip is excellent because the shape allows the electrical current to find its own level on the electrode. It is a fact that current, like water, will seek its own level on an electrode diameter. This will, in turn, provide welding arc stabililty.

Chemical End Preparation

Chemical salts are presently available for tapering tungsten electrode ends. The technique is relatively simple. The tungsten electrode is held between thumb and forefinger in such a manner that it can be rotated. A holding chuck may be used for this operation. The tungsten end is heated to a dull red color and placed in the container of chemical salts. The tungsten is then rotated rapidly which increases the temperature of the tungsten end. In this way the taper is formed on the tungsten electrode. At this writing there have been no reports of any weld contamination from the chemical salts used.

Tungsten Selection

The question often arises as to which tungsten electrode type to use for a specific application. There are a number of variables that should be considered as criteria when selecting tungsten electrodes. For example, the type of metal to be welded, the type of welding current used, the cost of various tungsten electrodes, the size of the weld bead desired, and the depth of penetration required would all be important.

Pure tungsten electrodes are unalloyed and carry the lowest cost of any of the tungsten electrodes. They also have the lowest current carrying capacity of any of the electrodes. Pure tungsten melts immediately at the electrode tip when the arc is initiated. The molten end forms a hemispherical ball. When the arc is extinguished, and the molten ball solidifies, it will be bright and shiny if the shielding gas has been correct. Pure tungsten is normally selected for use with aluminum, aluminum alloys, magnesium, and magnesium alloys.

The thoria bearing tungsten electrodes will carry more amperage for a given electrode diameter than pure tungsten. Thoria bearing tungsten electrodes do not melt under the heat of the welding arc. These electrodes must have some form of end preparation for maximum effectiveness. Thoria bearing tungsten electrodes are preferred for all metals except aluminum and magnesium.

Zirconia bearing electrodes have higher current carrying capacities than pure tungsten electrodes. They melt under the heat of the welding arc and form the same hemispherical ball that pure tungsten does. These electrodes are used principally for aluminum and magnesium.

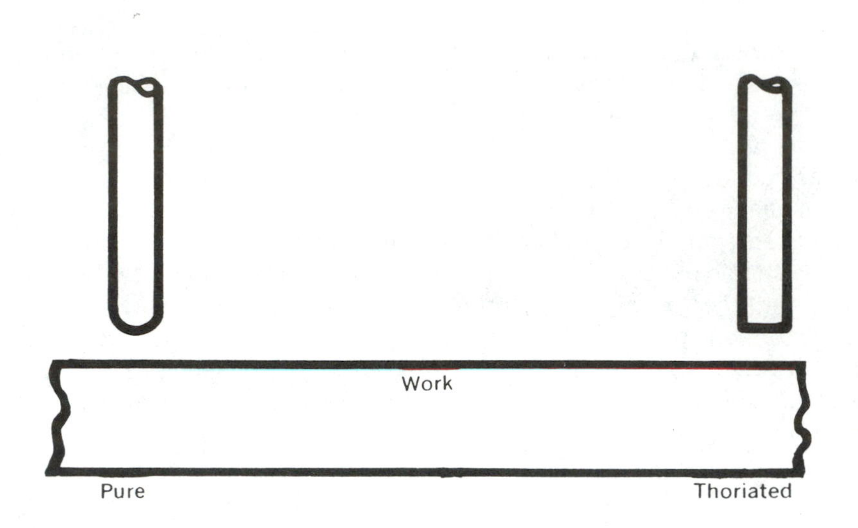

Figure 174. Tungsten Electrode Types and End Shapes.

Summary

Incorrect end preparation of a tungsten electrode by grinding or any other method may cause defects in the weld deposit when the electrode is used for welding. The correct general shape and dimensions for grinding or chemical pointing tungsten electrodes are illustrated in Figure 175.

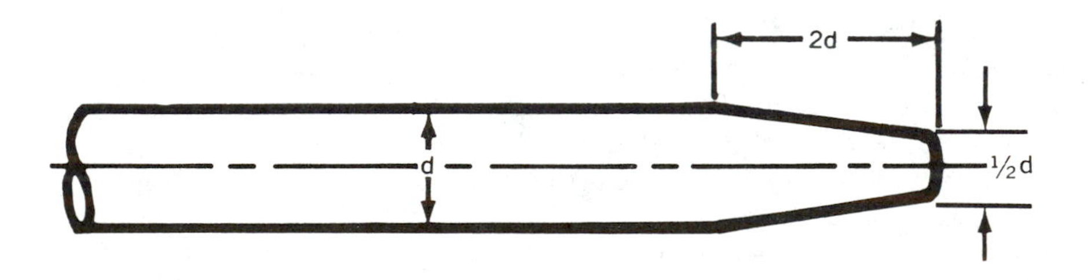

Figure 175. Typical Tapered Dimensions For Tungsten Electrodes.

Note the taper is shown as 2d where "d" is the diameter of the electrode. The electrode end is shown as 1/2d.

Making tungsten electrode end preparations to a needle point is to be discouraged. It is much better to use a smaller diameter electrode. The needle point will inevitably melt back or migrate across the arc column. As an inclusion in the weld deposit, the tungsten particles will decrease the ductility of the weldment and create a localized stress point. In either case the results are not desirable.

CHAPTER 16

High Frequency Systems

The term high frequency, as it refers to welding processes and power sources, normally refers to electrical pulses in the frequency range of 50,000 to 3,000,000 cycles per second (hertz). Actually, high frequency covers a major portion of the total frequency spectrum.

There are a number of reasons for using high frequency energy for welding applications and with certain types of welding power sources. Some things to be careful of when using high frequency energy are discussed in this chapter. To be very frank, high frequency energy is a necessary evil in the welding industry. If a method could be developed that performed the functions of high frequency without the problems of high frequency it would be outstanding! In truth, high frequency causes almost as many problems as it solves for welding.

Arc Stabilization

Arc stabilization is probably the most important function of high frequency energy. When welding with alternating current (AC) there is an arc outage each 1/120th of a second (each half-cycle). It occurs each time the AC sine wave trace passes through the zero line. The actual time of arc outage will depend somewhat on the maximum open circuit voltage of the AC power source and the re-ignition characteristics of the unit. High frequency, as a total part of the welding power source circuitry, provides the stable arc re-initiation effect necessary to maintain a steady, stable arc condition.

The typical AC sine wave trace shown in Figure 176 illustrates the two arc outage points per cycle. It is at this time that high frequency energy really does its work. There is zero voltage at the moment of arc outage—and voltage is needed to sustain the welding arc. The high frequency energy supports re-ignition of the welding arc by supplying up to appoximately 3,000 volts in the arc area.

Gas Ionization

The ionization of shielding gases occurs when an electron is caused to leave the gas atom. The force that causes the electrons to move and leave an atom is called ionization potential (voltage). As you have learned, ionization potential is measured in electron volts (eV).

When a negative electron leaves a neutral gas atom, the remaining sub-atomic portion of the atom (an ion) is electrically charged positive. The ionized shielding gas column is a preferred path for welding current to follow.

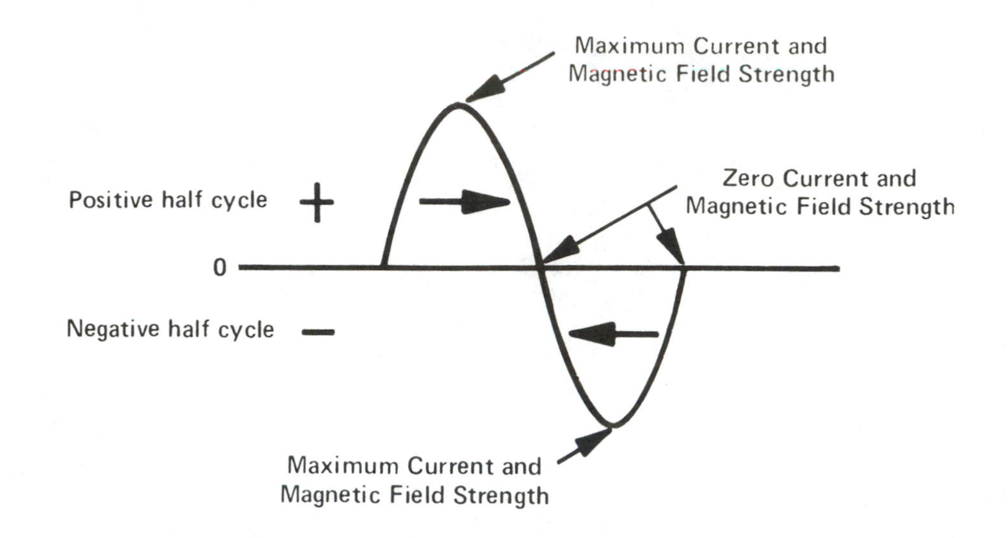

Figure 176. AC Sine Wave With Arc Outage Points.

The 80 volts maximum open circuit voltage of a welding power source is not sufficient power to have an effect on the ionization of shielding gases. There are literally millions of millions of shielding gas atoms passing through the intended arc zone every second. The open circuit voltage would only ionize a very few gas atoms at any given time.

High frequency power provides extremely high voltage (about 3,000 volts) at the electrode tip. The ionization voltages (potentials) of the two common shielding gases used for the GTAW process, argon and helium, are relatively low. The ionization potential of helium is 24.5 electron volts and of argon, 15.7 electron volts.

The high frequency voltage can, and does, create a minimum ionized electrical path through the shielding gas for the welding current to follow. High voltage energy is considered relatively safe for the welder to use in making non-touch starts with tungsten electrodes. Non-touch arc starts are preferable when welding materials that could contaminate the electrode such as aluminum and magnesium.

Arc Initiation

High frequency voltage, by ionizing a minimum gas path between the tungsten electrode and the work surface, helps bridge the physical distance across the arc column. This makes arc initiation possible without touching the electrode to the base metal. High frequency energy promotes electron emission from the tungsten electrode for more stable arc initiation.

It is important that the welding power source have sufficiently high maximum open circuit voltage to enable welding current to flow across the arc column. Of course, when the welding arc is established, shielding gas ionization is accomplished by the heat of the arc. This is called thermal ionization.

High Frequency Oscillators

Spark gap oscillators have proven to be the most reliable and effective units for producing high frequency power at reasonable costs. Some high frequency units are built-in the power sources while others are totally separate devices. A typical separate high frequency spark gap oscillator is shown in Figure 177.

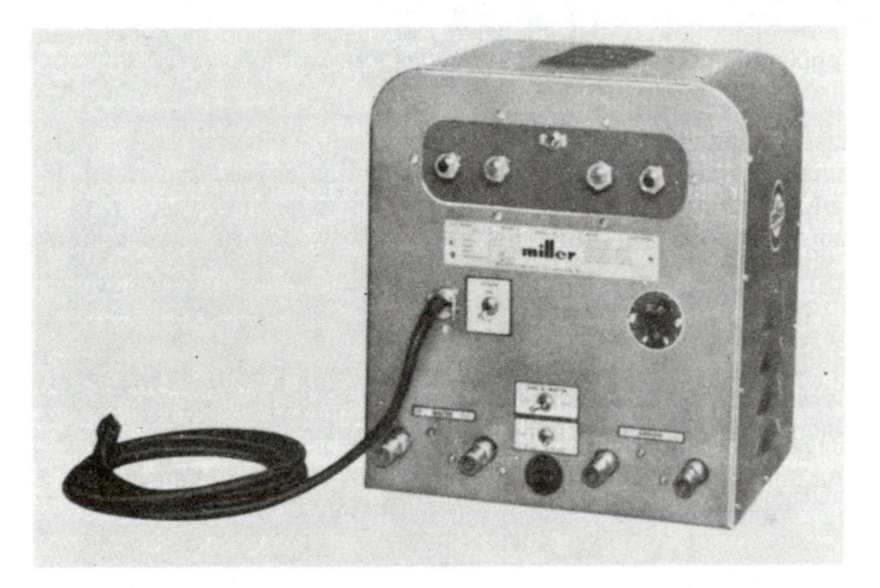

Figure 177. Separate High Frequency System.

Although relatively small in physical size, the separate HF units may be placed very near the actual gas tungsten arc welding operation. This helps minimize the losses of high frequency energy between the high frequency generator and the welding arc.

Separate high frequency systems, such as that shown in Figure 177, may be used with constant current power sources originally designed for use with the SMAW process. The HF units have excellent mobility and may be placed anywhere within the welding circuit. In addition to providing welding power terminals, the units usually provide both gas and water solenoids and valves.

Most welding power sources designed and built for use with the gas tungsten arc welding process (GTAW) have built-in high frequency systems and circuitry. The spark gap points are usually easily accessible from the front of the power source. This permits adjustment of the tungsten-faced points without removing the power source cover.

Spark gap oscillator high frequency systems are practical because they are relatively inexpensive to manufacture in the frequency ranges necessary for welding processes. Rugged construction enables the spark gap type HF unit to perform under normal shop operating conditions without the shock mounting that would probably be required for vacuum tube oscillators. This does not mean that spark gap type oscillators should be subjected to rough treatment. Common sense handling of any welding equipment will prolong its useful life.

High frequency energy travels at the surface of electrical conductors via the "skin effect". High frequency power for welding applications is usually rated in milli-amperes at several thousand volts. The skin effect of high frequency power transfer is illustrated in Figure 178.

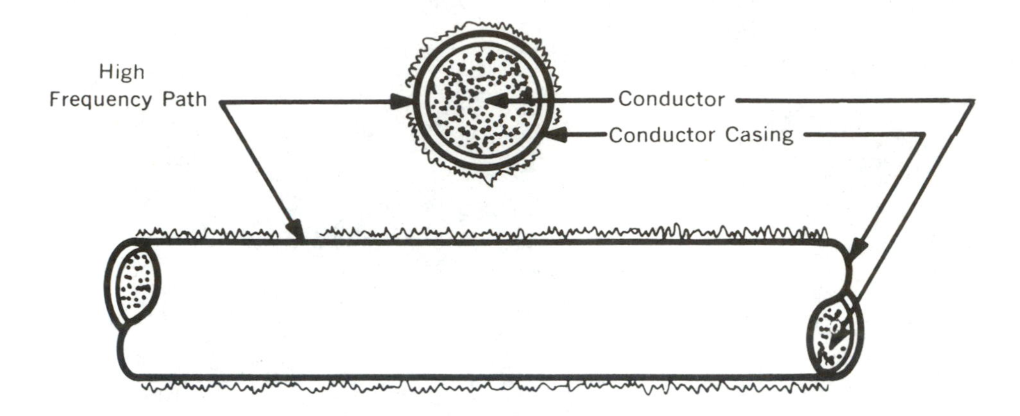

Figure 178. High Frequency Power Transfer for Welding.

High frequency power is normally super-imposed on the welding current conductor through air core coupling coils. The high frequency power is induced into the welding power circuit just before the welding current reaches the output terminals of the welding power source. For separate HF units, such as is shown in Figure 177, the HF power is normally super-imposed much closer to the actual welding arc.

High Frequency Circuits

The circuitry of the high frequency system is relatively simple yet very unique when compared to a standard welding power source transformer. The circuit diagram shown in Figure 179

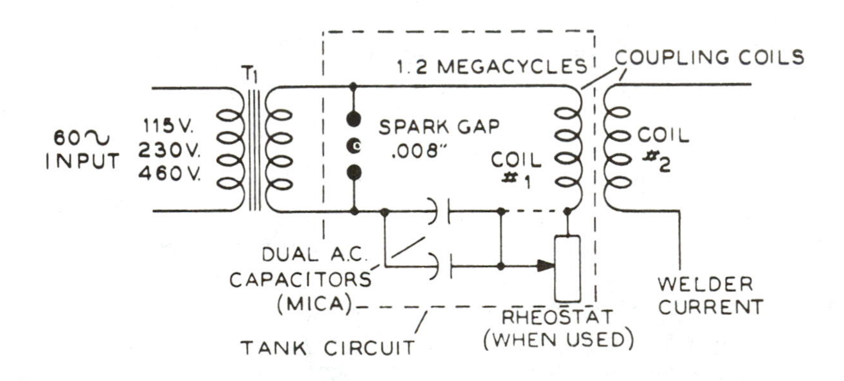

Figure 179. High Frequency Circuit Diagram.

represents a typical high frequency system. Although minor modifications may be incorporated in some models by certain manufacturers, the basic concept will remain the same.

Primary voltage for separate high frequency systems may be any commonly supplied voltage such as 208, 230, 380, 460, or 575 volts. In most cases, the primary voltage to be used with the HF unit must be specified at the time of ordering.

The primary voltage for high frequency systems built-in the welding power source are usually 115 volt. This voltage is normally taken from an isolated control circuit in the power source.

The power flow sequence through the high frequency system is a simple, straightforward operation. The input voltage is impressed at 60 hertz. (Most commercially generated electrical power is 60 hertz). The primary power energizes the primary coil of a high leakage, **step-up transformer** to bring the secondary voltage to approximately 3,000 volts no load. This voltage is apparent at the secondary coil of the high frequency transformer.

It is important to remember that frequency stays the same on both the primary and secondary coils of any transformer. That is the main reason the HF transformer is very carefully insulated and isolated from ready access. 3,000 volts secondary would be lethal at 60 cycles per second frequency.

The secondary power is applied to the two dry mica capacitors shown at the bottom of the circuit diagram. The capacitors accept the electrical charge until there is enough voltage to overcome the resistance of the air space at the spark gap points. When this occurs, the capacitors discharge their stored electrical energy. An arc is created at the spark gap points and the stored energy is transferred to the other side of the same mica capacitors. The high frequency power flows **from** one side of the mica capacitors, **through** the spark gap points, and **to** the other side of the same mica capacitors.

When the capacitors have again charged, and the voltage is sufficient to overcome the electrical resistance at the spark gap points, the stored energy flows in reverse to its previous direction. An arc is established at the spark gap points and the stored energy is returned to its original starting place. In essence, the high frequency power is applied to the mica capacitors and **oscillated** back and forth through the spark gap points.

The high frequency energy generated in this manner is impressed on the air core coupling coils which transfer the HF energy to the welding circuit conductor. In this way relatively safe high voltage can be brought to the electrode tip where it is required to perform its functions in the gas tungsten arc welding process.

The high frequency circuit diagram in Figure 179 shows the oscillatory system is within dotted lines. The total electrical circuit within the dotted lines is called the "tank circuit". It is within this circuit area that the high frequency current oscillates from one side of the mica capacitors to the other.

The air core coupling coils are shown in the circuit diagram as coil #1 and coil #2. Coil #1 is the high frequency coil and is normally very light gage wire or strip metal, heavily insulated. This is logical when you consider that this part of the circuit carries very low current values although the voltage is very high. Coil #2 is the welding current carrying coil. The high frequency energy is induced onto the welding power coil and carried to the electrode through the welding cables.

Key Data About HF Power

1. IT IS THE VALUE OF THE CAPACITORS AND THE INDUCTANCE OF THE COUPLING COILS THAT DETERMINES FREQUENCY.
2. HF CIRCUIT VOLTAGE IS CONTROLLED BY THE SPARK GAP POINT SETTING AND THE VALUE OF THE CAPACITORS.
3. THE INTENSITY RHEOSTAT CONTROLS THE AVAILABLE CURRENT IN THE HF CIRCUIT.
4. THE AVAILABLE CURRENT IN THE HIGH FREQUENCY CIRCUIT IS DETERMINED BY THE SIZE OF THE CAPACITORS AND THE VOLTAGE APPLIED.

A major manufacturer of high frequency equipment sets the spark gap points of all HF systems at 0.008''. This is done at the factory. The spark gap distance may be increased to approximately 0.012'' although this is normally only done in cases of excessive losses of high frequency power in the secondary welding circuit.

REMEMBER: As the spark gap point distance is increased, the voltage at the capacitors increases in value. Excessive voltage will cause the capacitors to fail in service prematurely.

Fundamental Problems With High Frequency Energy

High frequency energy is necessary for some gas tungten arc welding applications and very useful for others. In addition to knowing when and where to apply HF to good advantage, it is also advantageous to know some of the problems that may occur with high frequency. The problems discussed here are some of the more common ones that occur.

Two of the most common general problems that plague the owners of high frequency stabilized welding equipment are: (1) lack of high frequency energy at the welding arc and, (2) radio-TV interference from high frequency radiation.

Absence of HF at the Arc

The *symptom* would be complete lack of high frequency energy at the electrode tip. The logical place to check first would be the high frequency spark gap points of the HF oscillator. If there is no high frequency spark at the points, the problem is isolated in the HF circuit and system.

The first part of the HF circuit that should be checked are the two dry mica capacitors. The proper procedure for testing the capacitors is to isolate them from the HF circuit one at a time. It is not necessary to physically remove them from the high frequency panel for the test. First, **using electrically insulated tools,** ''bleed off'' the stored voltage by short circuiting across the two capacitor terminals. Then disconnect the circuit conductors from the terminals of *one* capacitor.

If available, an ohm meter may be used for testing the capacitor. Place each of the electrical probes on a separate terminal of the isolated capacitor. If there is no continuity the capacitor is functioning correctly. If there is continuity, the capacitor is short circuited and is defective.

If there is no ohm meter available the capacitors can still be tested. Remove the electrical conductors from one capacitor as previously directed. Don't forget to ''bleed off'' the capacitor-stored voltage before attempting to disconnect the terminals. (If you do, you'll only do it once!) Turn on the power source and the high frequency unit. If there is high frequency power at the spark gap points, the disconnected capacitor is defective.

If there is still no high frequency spark, shut everything off. Discharge the active capacitor as before and disconnect it from the HF circuit. Re-connect the first capacitor. Turn on the power source and the high frequency unit. You should see HF energy at the spark gap points.

If the trouble is in the dry mica capacitors, usually only one of them will be defective. The solution, if another capacitor is not readily available, will still let the welder operate with the power source. First, remove the defective capacitor from the HF circuit and HF panel. Connect the operating capacitor in the HF circuit. Re-set the spark gap points at 0.004'' using a machinists feeler gage. The power source and HF system will operate satisfactorily in these conditions.

The second dry mica capacitor should be installed as soon as possible and the spark gap points re-set to 0.008''. Figure 180 illustrates a typical high frequency spark gap point and holder installation. The spark gap points are faced with a very thin layer of tungsten for better electron emission. The tungsten-faced points are a consumable item and should be replaced periodically as needed.

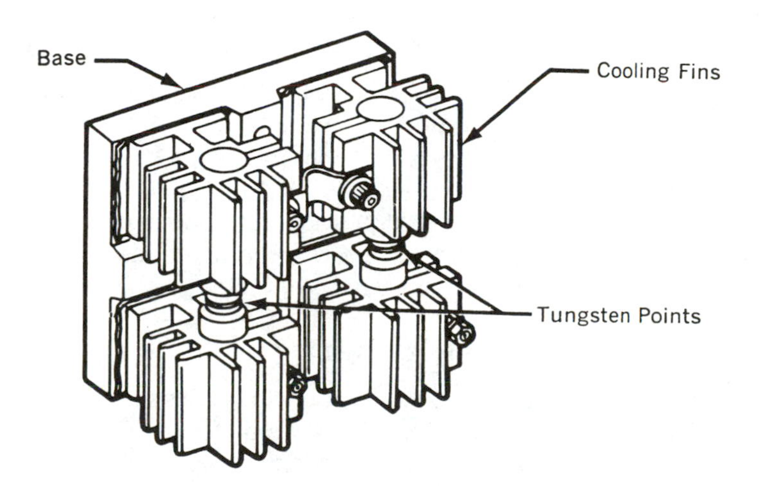

Figure 180. High Frequency Spark Gap Points and Holders.

Going back to the initial check of the high frequency spark gap points, suppose there is high frequency visible at the points but still no HF energy at the electrode tip. In this situation, the high frequency energy is probably going to earth ground somewhere in the secondary welding circuit. The HF panel circuitry is operating correctly. A careful check of the circuit from the welding power source terminals to the electrode holder and work, or ''ground'', connection will usually show where the HF losses are occurring.

CAUTION: DO NOT ATTACH A STANDARD VOLTMETER TO THE WELDING POWER SOURCE OUTPUT TERMINALS WHEN HIGH FREQUENCY IS IN OPERATION. THE EXTREMELY HIGH VOLTAGE WILL RUIN THE VOLTMETER.

Be sure to disconnect the primary input power circuit when work is to be done on any part of a welding power source, including the high frequency panel. The easiest method is to disconnect the primary power switch at the primary input fused disconnect switch box.

Radio Interference and Radiation

Radio and TV interference from high frequency stabilized welding installations is always a difficult problem to deal with. To minimize such interference it must be established exactly where the radiation is coming from and how it is being transferred to the affected area.

Every high frequency stabilized welding power source built by U. S. manufacturers must have an FCC certification procedure as part of the Installation, Operation and Maintenance manual shipped with the power source. The certification procedure discusses in detail the correct methods of installation for the power source and HF system to be in compliance with FCC regulations.

There are four common ways that high frequency radiation may escape from the welding area. Brief discussions of each of these problem areas may provide ideas of where to look and what to look for if trouble occurs.

Direct Radiation From the Power Source

High frequency energy has no preferred orientation of direction for radiation. It can, and will, go through any opening and in any direction. The radiation level is very pronounced in the immediate area if the power source doors and access panels are left open. The same is true if the power source is not electrically grounded properly.

It is necessary that all openings in any high frequency stabilized welding power source be kept closed. When proper installation and operating procedures are followed according to the manufacturers directions there should be no problem with this type of radiation.

Direct Radiation From the Welding Cables

Direct HF radiation from welding and work grounding cables is very pronounced in the immediate area of the cables. The intensity decreases quite rapidly with distance from the affected area. By keeping the welding cables as short as possible this type of interference may be minimized. It is best to keep the electrode and work leads as close together as possible.

The use of welding cables having a foil wrapping over the current carrying conductor, which is covered by some type of insulating material, is common in the welding industry. The purpose of the foil shield is to deter grease and oil from contaminating the conductor material. Hypolon is the insulation most used and recommended for welding cables that carry high frequency power.

Welding power cables that carry high frequency power should not be suspended overhead since they would have the effect of being radiating antenna. If extensive runs of welding cable are necessary to the welding operation it is suggested that fastening the cables to dry wood boards, using plastic cable clamps, will lessen the amounts of high frequency energy lost to earth ground.

An observed shop practice shows that high frequency radiation intensity may be altered considerably by changing the relative position of the work connection and the electrode lead in the work area. This is also true for magnetic arc blow in certain instances. The best method is to make the distance between the welding arc and the work lead connection as short as possible.

Direct Feed-Back to Primary Power Lines

High frequency radiation may leak to the primary power lines by direct electrical coupling inside the power source. The primary power line then serves as a radiating antenna. By proper installation of primary power to the power source (through solid conduit) direct electrical coupling may be avoided.

Most manufacturers of HF stabilized welding power sources specifically state in the installation instructions that "no rubber covered primary electrical conductors should be used with this type of equipment". High frequency energy will cause rapid deterioration of rubber.

Pick-Up and Re-Radiation From Power Lines

While high frequency radiation intensity will decrease rapidly with distance, the field strength in the immediate area of the welding leads will be very high. Unshielded wiring and ungrounded metal objects within this strong magnetic field may pick up the radiation directly from the welding circuit, conduct the high frequency energy for some distance, and produce a strong electrical interference in a totally unrelated area.

This type of high frequency interference can be troublesome and hard to locate. It can be minimized by carefully following the installation directions in the instruction manual provided with the welding power source.

Summary

It is not possible to write a simple formula to follow when dealing with high frequency problems. Each interference case seems to be unique. Following the installation procedures as outlined by the power source manufacturer and the Federal Communications Commission (FCC) will certainly decrease the possibility of high frequency interference above the allowable minimums.

Good electrical grounding, short welding cables, properly insulated welding cables, tight electrical connections, properly shielded primary wiring and plain common sense are the best assets to have when dealing with high frequency problems.

CHAPTER 17

Power Sources for Gas Tungsten Arc Welding

Most welding power sources used with the gas tungsten arc welding (GTAW) process are transformer or transformer-rectifier type devices. The general category is "constant current" although certain electronic controlled (solid state) power sources may be modified for either constant current or constant voltage operation. The output may be AC, AC/DC, or DC only. The type of welding current used is normally based on the type of metal to be welded.

Selection of welding power for aluminum and magnesium is based on the refractory surface oxides inherent on both metals. Alternating current is normally used with the GTAW process because of the "cleaning half-cycle" of AC. As was previously explained, the positive half-cycle of AC power is the cleaning half-cycle.

All other metals are normally gas tungsten arc welded with direct current straight polarity (electrode negative). Direct current reverse polarity (electrode positive) is not used commercially for the gas tungsten arc welding process because of the wide weld puddle formed and minimum penetration patterns.

The constant current power sources used for the GTAW process may also be used for the shielded metal arc welding (SMAW) process. Some power sources are specifically designed for the GTAW process. The output control may be a simple mechanical system, a more advanced electrical control method with remote control capability, or an electronic solid state control.

The primary power needs and main transformer design will depend on the output characteristics of the power source. Power sources with AC or AC/DC welding output will always operate from single phase primary power. The reason is that AC welding power must have a direct connection between the transformer secondary coils and the output terminals of the power source. It would be literally impossible to maintain stability and phase control in the AC welding current if three phase power were used.

Welding power sources with DC output only for the GTAW process may operate from either single phase, or three phase, primary power. To achieve the best DC welding power output, three phase transformer-rectifier power sources are recommended. Single phase based rectified DC welding power provides an output characteristic that is less stable than three phase based power.

The welding power sources considered in this chapter are all single operator units. Much of the basic data would also be applicable to multi-operator welding power sources. In most cases, this type of welding power source was originally designed for the shielded metal arc process. To use it for the gas tungsten arc welding process usually means touch starting of the arc.

Probably the most common time cycle frequency of electrical power in use is 60 hertz (60 cycles per second). There is, however, considerable 50 hertz power in Europe and Great Britain. Either frequency is suitable for arc welding although the power source design, especially the cooling system, is modified to some degree. The data presented here is applicable to either frequency. This text is generally based on the use of 60 hertz power.

An AC sine wave trace is illustrated at the left in Figure 181. The AC power is changed to DC by a rectifier. The resulting rectified DC is shown at the right side of the drawing.

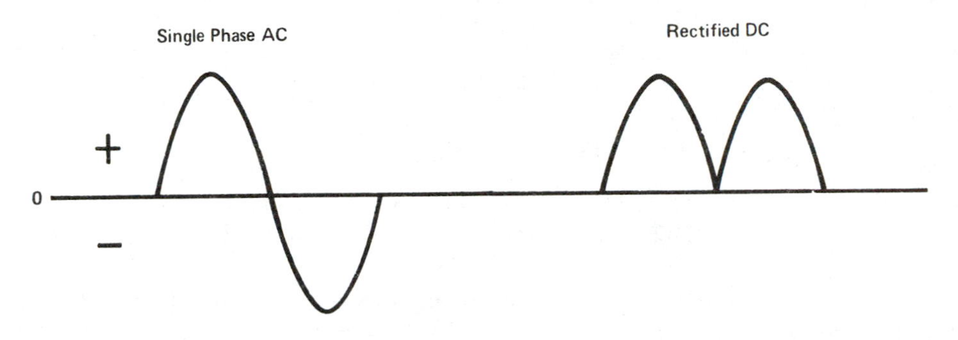

Figure 181. AC Sine Wave and Rectified Single Phase

The illustration shows that AC has both positive and negative values. The rectified DC, however, has either a positive value or a negative value. Since direct current flow is unidirectional; that is, it flows in one direction only, the direction of current flow can only be changed by either of two methods: (1) switching the welding cables at the power source output terminals or, (2) by re-positioning the polarity switch on the power source (if one is provided).

Three phase AC power is shown at the left side of Figure 182. Rectified three phase power is illustrated at the right side of the drawing. This is similar to the single phase AC drawing in Figure 181 with one major exception. There are three separate AC sine wave traces shown but within the same time span (1/60th of a second) as the single phase AC sine wave trace. The three AC phases are exactly 120 electrical degrees apart; that is, relative to each other.

Rectified three phase DC welding power exhibits very smooth arc characteristics. The rectified three phase power also has substantially higher average power levels than rectified single phase welding power. The higher average power level is to be expected since one phase of the three phase power is always ascending to maximum electrical strength.

The NEMA Class 1 constant current power sources designed especially for the gas tungsten arc welding process are normally equipped with certain auxiliary control devices not found on standard constant current power sources. Some of the controls are listed here:

- High frequency control system
- Gas and water solenoids
- Power factor correction
- Remote control circuit
- Gas and water valves
- Post-flow timer for gas and water
- Line voltage compensation circuit
- Primary or solid state contactor

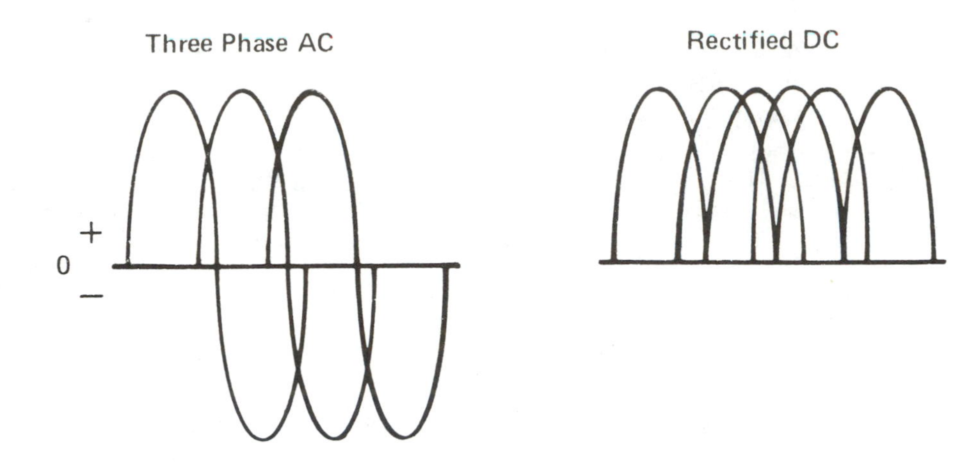

Figure 182. Three Phase AC and Rectified DC.

Other optional equipment is available for almost any application desired. For example, volt and ammeter sets, spot welding timers, AC wave balance control, gas and water pre-flow circuit, pulsar circuit, and full programmer drawer and panel for automatic operations. The use of micro-processor integrated circuits expands the versatility of solid state controlled power sources for fully automatic applications of the gas tungsten arc welding process. Fully controlled amperage upslope and downslope with programmable welding conditions is also available from most high technology power source manufacturers.

Transformer type AC power sources that are not designed for the gas tungsten arc welding process must be de-rated for welding output amperage when used with the GTAW process. The amperage de-rating procedure is necessary to protect the power source from the effects of DC component heating.

DC component is unique since it only occurs when welding with AC power and the GTAW process. It may be caused by either partial, or full wave, power rectification in the welding arc when welding materials with refractory oxides such as aluminum. Certainly a contributing factor to the formation of DC component is the difference in electron emission of the tungsten electrode and various base metals.

Tungsten electrodes have excellent electron emission characteristics which actually improve when the electrode is heated. Conversely, aluminum has comparatively poor electron emission qualities, part of which is due to the refractory oxide on the metal surface. Aluminum oxide forms rapidly at any temperature. The formation of aluminum oxide is accelerated when the base metal is heated.

There is no particular problem in de-rating an AC welding power source for amperage. Two methods may be used: (1) the AC welding power source may be de-rated for AC-GTAW current while retaining the same duty cycle it normally has, or (2), it may be de-rated to a current value at 100% duty cycle for processes other than gas tungsten arc welding. A second step is necessary to de-rate the power source for AC-GTAW at 100% duty cycle.

The current de-rating procedures are only necessary when gas tungsten arc welding with alternating current. There is no DC component when welding with direct current welding power and the GTAW process. De-rating power sources for AC-GTAW at 100% duty cycle is a bit more involved but it does guarantee that the welding power source will not be overloaded when welding. It is the safest method to use.

A cautionary note:

DE-RATING A CONSTANT CURRENT AC WELDING POWER SOURCE BY 30% FROM ITS RATED AMPERAGE WILL PROVIDE A SAFE WELDING AMPERAGE VALUE FOR AC GAS TUNGSTEN ARC WELDING BUT AT THE SAME DUTY CYCLE AT WHICH THE POWER SOURCE WAS ORIGINALLY RATED.

The chart shown in Figure 183 lists the factors involved for de-rating almost all types of constant current welding power sources to 100% duty cycle. The chart applies to all AC or DC welding power sources with the following design exception: If the AC power source has a tapped secondary coil for output control, it will remain at its original duty cycle (normally 20%) at all taps or plug-in settings. The 20% duty cycle refers to NEMA Class 3 AC power sources only. This type of power source is usually rated at about 180-225 amperes and is designed for limited use with the SMAW process. It is not designed for the GTAW process.

Present Duty Cycle	Rated Amps Times:	Duty Cycle	For AC-GTAW: Rated Amps Times:
60%	75% =	100%	70%
50%	70% =	”	”
40%	55% =	”	”
30%	50% =	”	”
20%	45% =	”	”

Figure 183. Duty Cycle Factor Chart.

To use the chart, you must know the present duty cycle of the power source you wish to de-rate for current output. Multiply the rated amperes of the power source times the engineering factor listed in column 2. The answer will be the maximum amperes to use at the new duty cycle shown in column 3.

For example, assume the power source is rated at 300 amperes, 32 load volts, 60% duty cycle. The engineering factor is 75%. The calculation would be done as follows:

300 amperes × 0.75 = 225 amperes @ 100% duty cycle.

This de-rated amperage would be suitable for all power source applications EXCEPT AC-GTAW. For gas tungsten arc welding with AC, the de-rated amperes value (225 amps) is further de-rated by 30%.

225 amperes × 0.70 = 157.5 amperes, 100% duty cycle, AC-GTAW.

This provides an absolutely safe AC welding current for use with the gas tungsten arc welding process.

The reason is simple. Every power source has conductors. As welding current flows in the power source circuits, the conductors increase in temperature due to resistance heating. The circuitry is designed to accommodate a certain temperature rise without damage. Exceeding the designed maximum temperatures in the conductors will cause damage to the insulation of the circuitry.

By reducing the amount of amperage in the welding power circuit, the resistance heating temperature is also decreased. When the energy from DC component is dissipated in the AC power source secondary coils, it does not damage the insulation. The welder can weld at any amperage up to the maximum de-rated current with complete safety.

Overheating in the main transformer coils can cause two severe problems. They are: (1) breakdown of the insulation on the coil conductor wire and, (2) decrease in the electrical efficiency of the transformer due to high electrical resistance in the heated coils. The disastrous effect of excess DC component is shown in Figure 184. The coil illustrated is from a standard AC welding power source that was not de-rated for current when used with the GTAW process.

Figure 184. Burned Coil Caused by DC Component.

The typical output volt-ampere characteristics of constant current power sources used for the GTAW process are illustrated in Figure 185. The open circuit voltage is normally about 80 volts. The maximum short circuit current output is sharply limited due to the design characteristics of the power source.

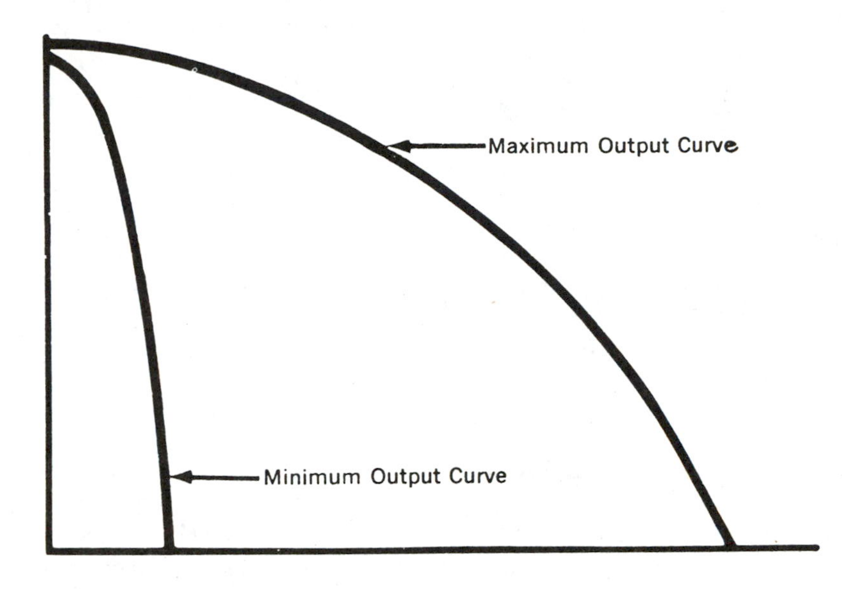

Figure 185. Typical CC Power Source Volt-Ampere Curve.

Summary

The design of constant current power sources hasn't changed substantially in the past ten years or so. What has changed are the methods of output control of the power source. Of course, the greatest change has to do with the "solid state" electronic controls. In essence, electronic controls have made mini-computers of several different types of welding equipment, including power sources for the GTAW process.

CHAPTER 18

Gas Metal Arc Welding

The gas metal arc welding (GMAW) process was first developed in 1948. It was a natural development based somewhat on the idea of the gas tungsten arc welding (GTAW) process. The principle difference in the two welding processes is that the GTAW process uses non-consumable tungsten electrodes. The GMAW process uses continuous solid wire consumable electrodes. There are also some differences in the shielding gases which will be discussed in a later section of the text.

The gas metal arc welding process requires a constant voltage (constant potential) DC welding power source, some method of controlling and feeding the electrode wire to the arc, and some type of hose/cable assembly and gun through which the electrode wire reaches the welding arc. Electrical contact is made at the barrel end of the gun, through a copper contact tube, to electrically energize the welding electrode. Some type of shielding gas is used with this process. The shielding gas may be inert or chemically active to the base metal.

The speed of metal deposition will vary depending on the method of metal transfer used, the shielding gas, the electrode type and diameter, the welding position, and the base metal classification. In almost every instance the gas metal arc welding process is faster than the shielded metal arc welding process. If you were to categorize the GMAW techniques by the speed of metal deposit per hour of running time, they would probably line up as follows. The fastest deposition is at the top of the list.

1. Spray Transfer

This method of metal transfer employs a bare, solid electrode wire in an argon-rich gas shield. (An argon-rich gas is considered to have a minimum 90% argon in the gas mixture.) The metal transfer is extremely fast. Arc voltage and arc amperage are relatively high for the electrode diameter. The molten weld puddle is very fluid and this technique is normally used only for flat grooves or fillets and horizontal fillet welds.

2. Buried Arc Transfer

The buried arc transfer technique was developed for high speed welding of mild steel plate materials. The shielding gas used is CO_2 only. In most applications, the weld joint is square butt up to 5/8'' thickness.

3. Pulsed Current Transfer

This method of metal transfer is a form of spray transfer but at a much slower rate than true spray transfer. Pulsed current transfer may be used in all positions of welding. Because of the discrete separation of the molten metal droplets, larger diameter electrodes may be used with some base metals than would be used with the short circuit technique.

4. Globular Transfer

Globular transfer is not normally considered a good method of metal transfer for most welding applications. There is a large amount of metal spatter that must be removed from the base metal surface surrounding the weld bead. The excessive spatter from the weld causes considerable loss of electrode efficiency.

5. Short Circuit Transfer

The short circuit transfer method of metal transfer is the slowest of all the GMAW welding techniques because of the low amperage and voltage at which the technique functions. The nature of the metal transfer method requires intermittent arc outages at nominal rates of 150-170 short circuits per second. The short circuit method of metal transfer is usable in all welding positions.

Process Names

The gas metal arc welding process has had a variety of names since its beginning. For example, the term "MIG" is sometimes used to identify the process. This acronym is taken from the term Metal Inert Gas. The term "MIG" originated in the early days of the process when the only shielding gases used were helium and argon, both of which are inert to the products of the weld zone. With the introduction of shielding gases having active elements, such as oxygen and carbon dioxide (CO_2), the term MIG is not technically applicable to the process. The American Welding Society (AWS) has officially adopted the name Gas Metal Arc Welding as the name of the welding process.

Some manufacturers of GMAW equipment have used registered trade names to describe their particular brands of welding equipment. This is fine for company recognition but it should not be used for technical publications such as welding procedure specifications. Always use the terms as defined by the AWS. This will provide the widest area of understanding.

The Process Patent

The gas metal arc process was patented in 1950. Part of the **process patent** states that the process is to be used with direct current, reverse polarity (DCRP). This is true regardless of the base metal type. Steel, low alloy steel, stainless steel and aluminum are all welded with DCRP (electrode positive). The only exception has been with the short circuit method of metal transfer on very thin gage mild steel. Then it is appropriate to use DCSP (electrode negative) with 0.030" diameter electrodes and the correct shielding gas. Mild steel in the thickness range of 0.015" to 0.035" has been welded with this technique.

Alternating current has been tried with the GMAW process but without success at this writing. The extremely high open circuit voltages necessary for arc stabilization, and re-ignition, are not safe for commercial welding applications. The problem is the arc outage each 1/120th of a second with AC. This causes loss of ionization of the shielding gas. High frequency is to no avail here because it would get to the drive motor on the wire feeder system. The very high voltage of the HF system would destroy the electric motor.

In my opinion, there could be some significant use of AC constant voltage power sources with electronic control. The solid state controls could produce square wave AC in which the arc outage would be micro-seconds. The possibility of eliminating magnetic arc blow with this type of welding power source makes it worth looking into.

Direct current, straight polarity (DCSP) is not normally used with the gas metal arc welding process, except for thin metals, as we said before. The penetration patterns with DCSP are very shallow, the weld bead usually high crowned. There is the possibility of "arc wander" caused by the instability of the cathode spot on the electrode tip. The weld transfer would be globular instead of the true spray normally expected with DCRP and argon shielding gas.

Specially coated electrode wires have been manufactured for welding with DCSP and the gas metal arc welding process. The rare earth coating (usually Cesium) tends to decrease the melt rate of the electrode that would be expected with DCSP. Applications of DCSP and the GMAW process are presently limited to relatively high speed welding of light gage sheet metal and auto body shop jobs where minimum penetration is required.

The gas metal arc welding process is designed for use with all the commonly welded metals. The joint design and the specific welding technique will normally determine if out-of-position welding can be done. With most base metals, only pulsed current transfer or short circuit transfer can be accomplished out-of-position.

Current Density

Current density is calculated by dividing the electrode cross sectional area, in square inches, into the welding amperage used. The current density is a key factor in the gas metal arc welding process. In all applications there is the use of relatively high amperage values with relatively small diameter electrodes. This combination will, of course, provide high current density at the electrode tip. The result is good penetration in a properly prepared weld joint.

The shielded metal arc welding process (SMAW) has a relatively low current density arc. The electrode diameter is relatively large, and the welding amperage relatively low, for the process. This limits the penetration into the base metal with the SMAW process.

For example, the largest solid electrode used with the GMAW process is 1/8" diameter. The welding amperage required to weld with this continuous consumable electrode may range 500-650 amperes. You would expect, and get, high deposition rates and good penetration characteristics with a qualified welding procedure.

With the SMAW process, an E60XX 1/8" mild steel electrode will weld at about 100-110 amperes. The penetration will be approximately 1/8" maximum into the base metal with the SMAW process and the deepest penetrating electrode (E6010) used for mild steel welding. The deposition rates will be low and the current density will be very low.

Basic Equipment Required

The fundamental equipment necessary to operate with the gas metal arc welding process includes a welding power source, a wire feeder-control system, the necessary interconnecting gas hose and cable assembly, and a gun or torch. A source of cooling water is necessary if the gun is water cooled. The line diagram shown in Figure 186 indicates the basic equipment required for welding with gas metal arc.

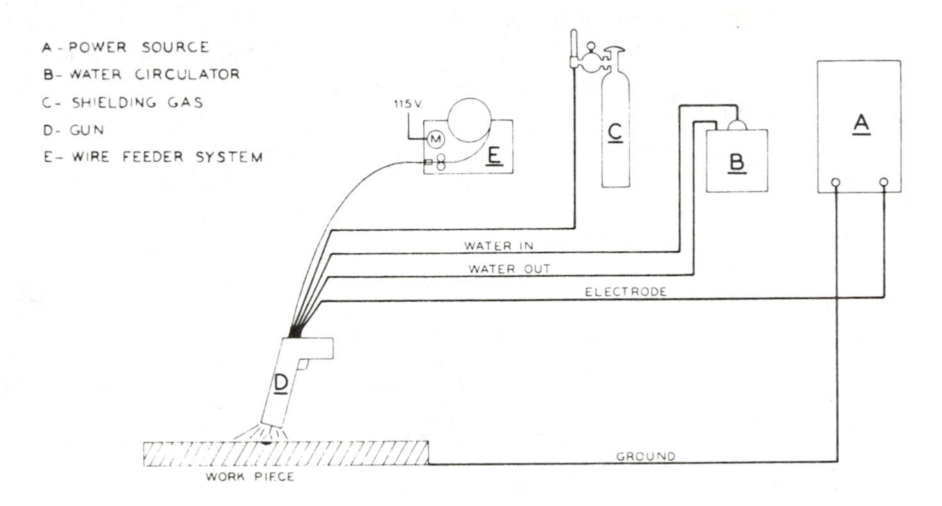

Figure 186. Basic Gas Metal Arc Welding Equipment.

There has been much discussion over the name for the electrode wire dispenser at the arc. Is it a "gun" or is it a "torch" After asking many different people, including AWS staff personnel, about this I have concluded that nobody particularly wants to label the device. I think this is because as soon as a commitment is made to a name, others will probably seize the opportunity to disagree with, and perhaps disparage, the person who "stuck his neck out".

The following logic is what I use to determine the name of the device used for dispensing electrode wire at the arc. If the unit is shaped like a pistol with a trigger device and a hand grip, I call it a "gun". If, on the other hand, it has the shape of an oxy-acetylene welding torch with tip I call it a "torch". Torches are normally air-cooled. Guns are often water cooled.

As the requirements of the welding industry have grown to meet exacting manufacturing codes and standards, so has the development of welding equipment progressed. Many pieces of basic welding equipment such as power sources and wire feeder-controls have totally different control systems than were even known about ten years ago.

Welding Power Source Development

When the gas metal arc welding process was developed there was a limited selection of welding power sources with DC output available. The DC power sources then in use were designed for the shielded metal arc welding (SMAW) process. The volt-ampere output curve characteristics had a sharply negative effect. Maximum open circuit voltage was normally limited to about 100 volts DC. The maximum short circuit current was about 150%-175% of the amperage rating of the power source.

With the increased demand on the gas metal arc welding process to perform more and varied applications came the need for welding power sources that could extend the versatility of the process equipment. Due in part to these demands the constant voltage (constant potential) type power sources were developed.

The terms "constant potential" and "constant voltage" are synonymous when used with welding power sources. The constant potential type welding power sources have a relatively low maximum open circuit voltage. The maximum short circuit current, however, may be several times the rated amperage of the power source. Typical volt-ampere output curve charateristics for both constant current and constant potential power sources are illustrated in Figure 187.

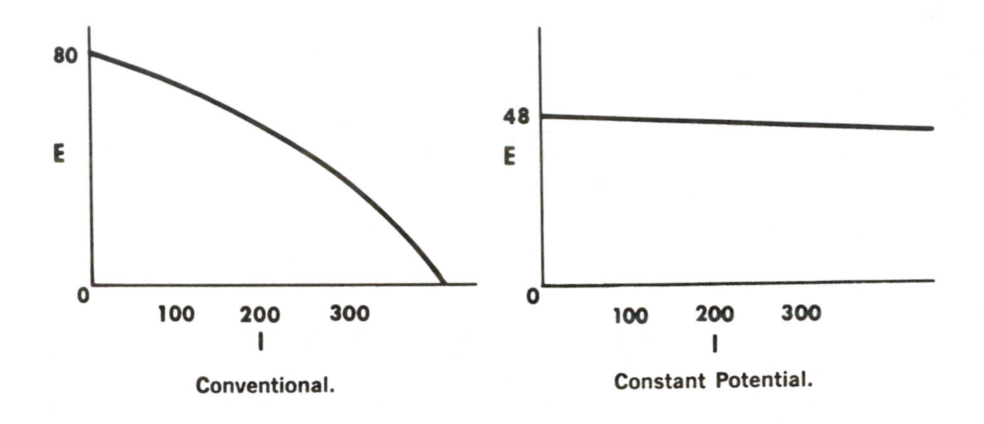

Figure 187. Comparative Volt-Ampere Curves.

Additional data concerning the development of the various types of welding power sources used for gas metal arc welding may be found in a subsequent chapter entitled POWER SOURCES FOR THE GMAW AND FCAW PROCESSES.

Types of Metal Transfer

There are several types of metal transfer employed with the gas metal arc welding process. Although many times they are referred to as "welding processes", the various types of metal transfer are actually welding methods or techniques within the scope of the total process.

For example, there are actually only two basic types of metal transfer with the gas metal arc welding process. They may be classified as the **gas shielded open arc method** and the **gas shielded short circuit method.**

The gas shielded open arc method of metal transfer considers that molten metal is separated from the welding electrode, moved across a physical air space within the arc column, and deposited as weld filler metal in the joint to be welded. Terms such as "spray transfer", "buried arc transfer", "globular transfer", and "pulsed current transfer" are applied to the various gas shielded open arc methods of metal transfer.

The short circuit method of metal transfer deposits the molten weld metal by direct contact of the welding electrode with the base metal. There is no metal transfer across the arc with the short circuit transfer technique.

There are several manufacturers trade names used to describe the equipment used for short circuit transfer welding. No matter what name is used for the process equipment, the welding technique and metal transfer characteristics remain the same. Some of the trade names used include Short-Arc, Dip-Matic, Micro-Wire, Dip-Transfer, and Mini-Arc.

Spray Transfer

Spray transfer is accomplished by the movement of a stream of tiny droplets of molten weld metal from the electrode, across the welding arc column, to the base metal. The arc has a characteristic humming or buzzing sound if the welding condition is properly set.

The spray transfer method is accomplished with relatively high load voltages and high amperages. Typical load voltage would range from about 23 to 33 volts. The amperage is determined by the electrode type and diameter. The weld puddle is very fluid and carries a fairly high volume of weld metal. The contact tube end is **recessed** about 1/8" back from the end of the gas nozzle. It is for these reasons that spray type transfer is not normally used for out-of-position welding. Spray transfer is usually designated for flat groove and fillet welds and horizontal fillet welds.

The shielding gases used for spray transfer require an argon-rich atmosphere. This means a minimum 90% argon content in any gas mixture. Other gases are often added to argon to achieve specific metal transfer characteristics.

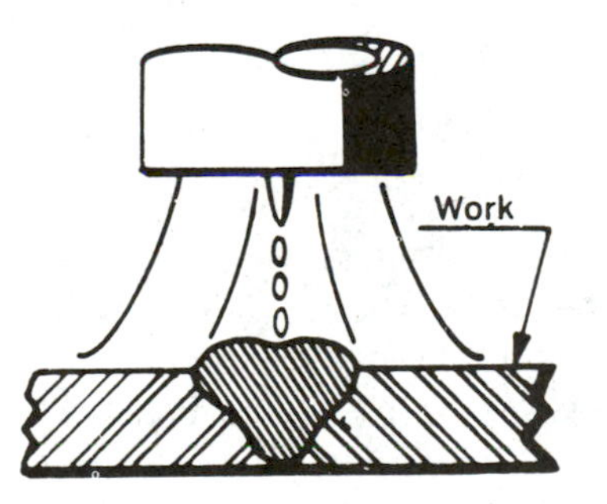

Figure 188. Typical GMAW Spray Transfer.

A typical GMAW spray transfer is illustrated in Figure 188. In particular, notice the electrode tip and the acicular manner in which it is shaped. This is typical of the spray transfer method of welding. The electrode melt rate is so fast that molten metal literally runs off the electrode in an almost continuous stream.

Buried Arc Transfer

The buried arc technique is truly an open arc method of welding low carbon mild steel. All metal transfer occurs below the surface of the base metal. A cavity is created by the arc force. The actual opening in the base metal surface is relatively small in diameter. The solid electrode wire is fed into this opening and the arc is sub-surface. The contact tube end is normally flush with the gas nozzle end. All spatter is retained in the weld cavity area as weld filler metal. There is very little reflected heat from the weld so the welding "torch" or "gun" is not subject to overheating as readily as with spray transfer welding.

The shielding gas used is carbon dioxide (CO_2). CO_2 is one of the factors that helps keep the buried arc process cost-efficient. The spatter inherent with open arc CO_2 shielded welding is contained in the unique cavity created by the arc force. The gas nozzle of the gun or torch is held very close to the surface of the base metal, usually not more than 1/8" away. The weld puddle is very fluid and the penetration patterns are deep. It is an excellent welding process for automatic welding of heavier thicknesses of mild steel.

The basic concepts of buried arc welding mild steel with CO_2 shielding gas are relatively simple. High amperages and high load voltages are used. Electrode diameters usually range through 0.045", 1/16", and 3/32". For example, the following welding conditions are typical:

Electrode	Weld Current	Weld Voltage
0.045" diameter	350-450 amperes	33-36 load volts
1/16" diameter	450-650 amperes	34-39 load volts
3/32" diameter	500-800 amperes	35-40 load volts

Figure 189. Typical Buried Arc Transfer Conditions.

The ranges presented in Figure 189 are for use as guides only. The minimum mild steel plate thickness recommended for the buried arc welding technique is 1/4" due to the deep penetration of the arc.

The buried arc welding techniques were designed for use with CO_2 gas shielding and low carbon mild steel. The intent was to provide fast, high quality welding of steel plate thicknesses with minimum joint edge preparation. Plate thickness will determine which diameter of electrode wire that should be used. The welding speed of travel can be quite high with this welding technique especially if automatic equipment is employed. The main problem to watch is that travel speed doesn't outrun the solidification pattern of the weld metal deposit.

There are several interesting points about the buried arc welding technique shown in Figure 190. The relatively high speed of welding can certainly be a cost savings to the user. The fact that all metal transfer occurs below the surface of the base metal is important.

A cavity is created below the base metal surface by arc force. Most of the arc heat, and all of the filler metal spatter, is retained within the sub-surface crater. This has two specific

Figure 190. Buried Arc Transfer.

advantages: (1) the welding gun will operate at cooler temperatures because there is not the tremendous amount of reflected heat on the gun barrel and, (2) the electrode deposition efficiency is higher because the spatter is trapped within the arc crater and is used as filler metal.

Buried arc transfer is a high speed, deep penetration welding technique. Constant voltage power sources with a slightly negative output volt-ampere curve are preferred for this type of welding because they provide very fast response to changing arc conditions. The constant voltage power sources used for buried arc transfer should have minimum ratings of 600 amperes, 40 load volts, 100% duty cycle. Preferably, the power source would have remote control capability.

Pulsed Current Transfer

The pulsed current transfer method of gas metal arc welding is a form of spray transfer. It is not as fast as true spray transfer because of the discrete transfer of metal droplets from the electrode. The number of current pulses per second are usually based on the input primary current frequency. *Electronic pulsing may range from 50-250 pulses per second.*

For example, 60 hertz power as the primary frequency provides either 60 or 120 pulses per second on some present power source models. There is one molten metal droplet transferred with each pulse. Almost any wire feeder-control system may be used with this welding technique although the constant potential type power source used is unique.

In one type of welding power source used for pulsed current welding there are two levels of current supplied. Both levels of current come from the same power source although from two separate transformers.

The main transformer is a three phase system which supplies the "background" current. The objective of this amperage setting is to maintain the welding arc at all times without melting the electrode. It also keeps a steady flow of heat into the weld joint and maintains shielding gas ionization. Background current is adjustable and may be changed to suit the electrode type and diameter used.

The second current level is the "pulsed" current. This is supplied from a single phase transformer which is connected to the main transformer through a set of reactors. The pulse amperage is adjustable and should be set just above the spray threshold amperage for the specific type and diameter of electrode used. (It is important to know that every GMAW electrode type and diameter has an amperage level above which it will automatically go into spray transfer in an argon-rich atmosphere).

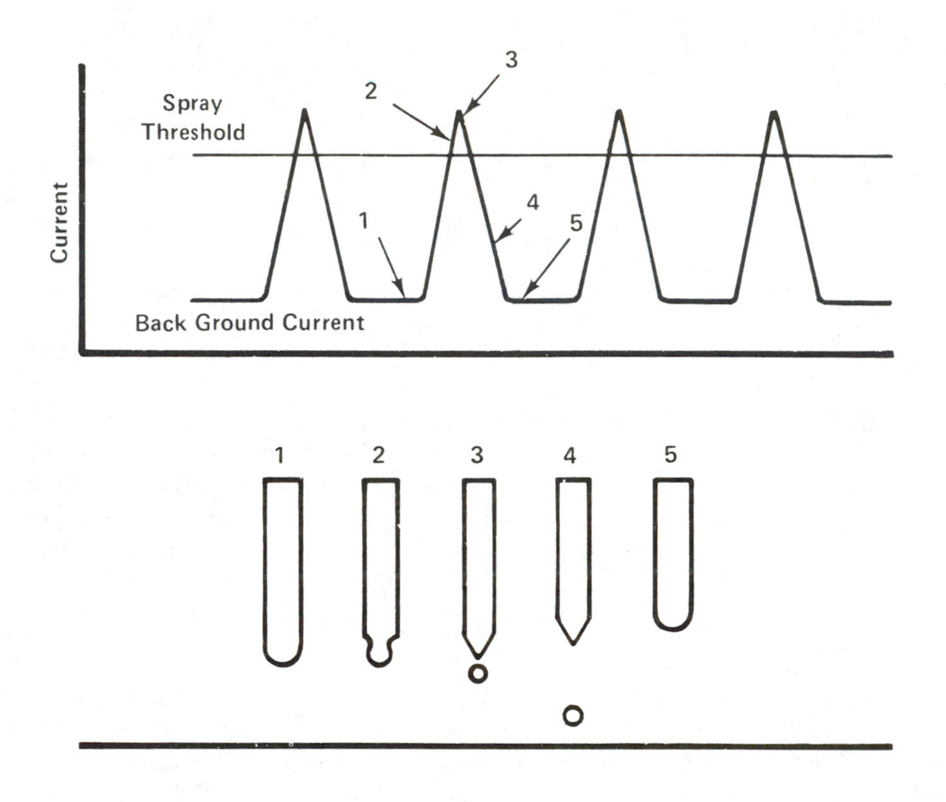

Figure 191. Pulsed Current Metal Transfer.

The pulsed current transfer technique provides the capability to perform a modified type of spray transfer under controlled heat input conditions. With this welding method, a form of spray transfer may be accomplished in all welding positions.

Pulsed current spray transfer is unique in that only one droplet of metal transfers across the arc with each pulse of current. If the unit is set for 60 pulses per second there will be 60 droplets of metal moving across the arc every second. This is, of course, much slower than true spray transfer which has a continuous stream of molten metal droplets moving across the arc. Electronically pulsed current transfer provides faster pulsing at 50-250 pulses per second.

Globular Transfer

As the name implies, globular transfer of weld metal is in the form of large, irregularly shaped globules of molten metal transferring from the electrode across the welding arc column. They land in the molten weld puddle with a splash and literally evacuate some of the metal as spatter. This technique of gas metal arc welding is typical of open arc mild steel welding with CO_2 shielding gas.

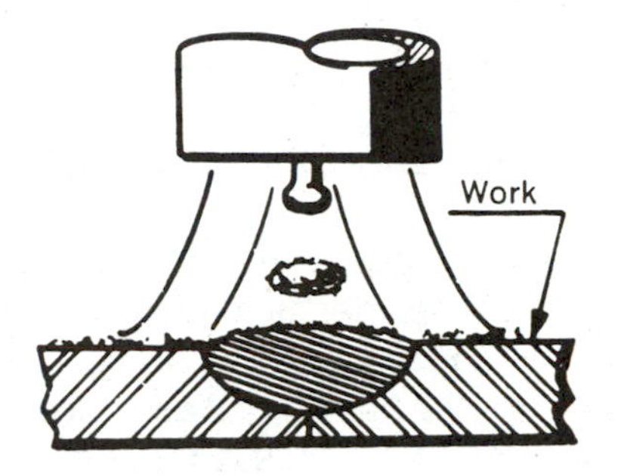

Figure 192. Globular Transfer.

Globular transfer may occur because the welding condition is not correctly set. The most frequent causes of unintentional globular transfer are improper settings for amperage and/or voltage. Figure 192 illustrates the molten ball bulging at the electrode tip which is typical of globular transfer welding.

The use of globular transfer is not normally a desirable condition because of the excessive spatter losses that occur. GMAW electrode efficiency can decrease from about 95% to as low as 70% with excessive spatter losses from globular transfer.

Excessive spatter from any type of welding is expensive! The costs consider the time of welding that is lost, the volume of electrode filler metal that is lost, the cleanup and removal of spatter from the surface of the base metal, and the possible repair of surfaces damaged by the molten spatter.

Short Circuit Transfer

The short circuit method of metal transfer is unique in that no molten metal transfers across the welding arc column. Weld metal transfer takes place only when the electrode makes physical contact with the base metal. This is where the name "short circuit transfer" comes from. The end of the contact tube usually extends about 1/8" past the end of the gas nozzle.

The actual number of short circuits per second depends on the electrode type and diameter, the shielding gas used, the open circuit voltage, the load voltage used and the wire feed speed. With all other factors held constant, the number of short circuits per second will be directly influenced by the wire feed speed. The faster the electrode is fed, the greater the number of short circuits per second. Excess wire feed speed will, however, cause the electrode to stub into the weld deposit.

Figure 193 shows that short circuit transfer is a series of changing voltage and amperage values which are controlled by the pre-set welding conditions.

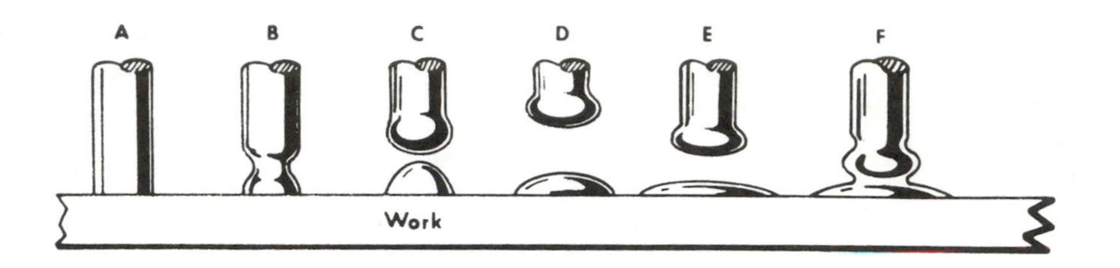

Figure 193. Short Circuit Transfer.

The illustration in Figure 193 shows one short circuit transfer cycle. Consider that this is happening approximately 150-170 times per second with 0.035'' dia. mild steel electrode and a good welding condition. The sequence of action typically follows this pattern:

The electrode makes contact with the base metal as shown at point ''A''. The resistance of the electrode wire to welding current flow causes it to heat and ''neck down'' as shown at point ''B''. Several factors contribute to the reduction in cross section of the electrode at point ''B''. These include the loss of columnar strength in the electrode and the development of a very strong magnetic field around the electrode. The loss of columnar strength is due to the resistance heating of the electrode when it is in contact with the base metal. The strong magnetic field is caused by the rapid increase in welding current value when the electrode short circuits to the base metal.

The action begins to pick up at point ''C''. The electrode has separated from the base metal and an arc is established. Remember that the **wire feed speed is constant** and does not change at any time in the short circuit weld cycle. The electrode melt rate is faster than the wire feed speed at point ''C''.

Point ''D'' in the diagram is a point of equilibrium where the wire feed speed and the electrode melt rate are exactly equal. The arc length, and load voltage, are both at maximum. At point ''E'' the electrode tip is shown closer to the work surface. The melt rate of the electrode has slowed still more and now it is slower than the wire feed speed. A ball of molten metal has formed at the tip. At point ''F'' the electrode again short circuits with the base metal, transferring the weld metal by direct contact, and the welding cycle begins again. Depending on the electrode type and diameter, the number of short circuits per second can range from about 50 to 250.

The short circuit method of metal transfer is essentially a low voltage, low amperage technique. It is capable of being used in all welding positions because the rate of metal transfer is slow. The welding power sources used for this welding technique are constant potential type units with some form of slope control either built-in as a fixed reactance or adjustable through a range of slope settings. Some power sources have an inductor (stabilizer) control that may be adjustable or built-in as a fixed inductance device. This subject will be discussed in detail in the section of the text entitled **Welding Power Sources For Gas Metal Arc Welding.**

The slope settings are such that the maximum short circuit current of the power source is limited to a relatively low value. This precludes the possibility of explosively high current levels at the electrode tip which would cause electrode spatter in the welding arc. The use of slope control reactors also slows down the rate of response of the constant potential type power source to changing arc conditions.

Typical load voltages used with the short circuit method of metal transfer will range approximately 15-20 load volts. **IMPORTANT:** If load voltage exceeds 20 volts, there is no longer a short circuit welding condition. Above 20 volts is globular transfer. This will be easily recognized by the dramatic increase in spatter levels.

Although slope control has not been discussed, the following data are pertinent to this section of the text. Slope settings presented here are based on the use of a 14 turn slope reactor. (You will recall that an electrical turn is one wrap of the conductor wire around the periphery of a coil). The exact slope setting will be determined by the type of metal being welded and the diameter of the electrode wire. The shielding gas for mild steel could be either argon 75%-CO_2 25% or plain welding grade CO_2.

For example, slope settings for mild steel will normally be 6-8 turns based on a 14 turn reactor coil. For 0.030'' dia. electrode 6 turns would be very adequate. The small volume of molten metal permits the faster response of the power source without spatter.

For 0.035'' and 0.045'' dia. mild steel electrode, probably 8 turns of slope would be best. This will slow the rate of response of the power source enough to delete spatter problems.

Slope settings for austenitic stainless steels (300 series) welded with the short circuit transfer method will be 10-12 turns. An ideal setting for 0.035'' dia. electrode wire is 10 turns of slope using the Tri-Mix gas containing 90% He-7.5% Ar-2.5% CO_2. This shielding gas was designed especially for short circuit transfer with stainless steel. The bead will be relatively flat in profile in groove joints. A point to remember is that stainless steel electrodes with high silicon content will require different welding conditions than similar stainless steel classifications without the high silicon content.

Automatic Amperage Control

Amperage is thought to be controlled by wire feed speed when gas metal arc welding with constant potential type welding power sources. The faster the wire feed speed, the higher the amperage output of the power source. At the same time that amperage is increasing, the load voltage is decreasing on the specific volt-ampere curve. All of the foregoing data is true but it isn't the whole story. Wire feed speed is only one factor in setting the welding condition with the gas metal arc welding process.

It is the shape of the static volt-ampere curve, as determined by open circuit voltage and the ''slope'' setting of the power source, that actually dictates the amount of amperage that is supplied at the welding arc. The wire feed speed simply determines the balance of load voltage and welding amperage at the welding arc.

Amperage control is a function of the output characteristics of the constant potential type welding power source. Figure 194 illustrates the **automatic amperage control** inherent with the constant voltage (constant potential) type power source. Please keep in mind that the power source is going to maintain a constant arc voltage while welding as long as none of the conditions are changed.

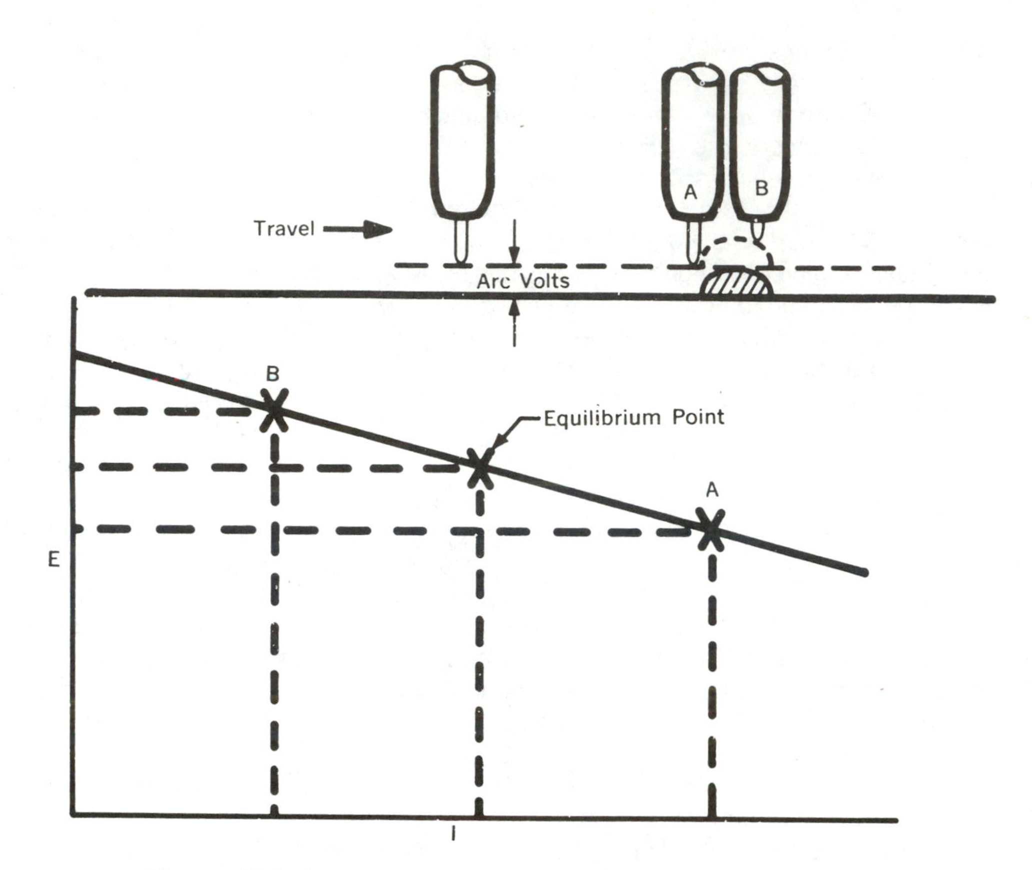

Figure 194. Automatic Amperage Control With GMAW.

When the wire feed speed is increased in inches per minute, the power source increases the amount of amperage available at the arc to melt the electrode. This will continue until the wire feed speed is increased to the point where the power source cannot supply enough amperage, with sufficient load voltage, to melt the electrode wire. When this happens, the electrode wire "stubs" into the base metal. By "stubbing", we mean that the electrode wire will make solid physical contact with the base metal and will not sustain a welding arc. The electrode wire just turns red from resistance heating and piles up on the base metal surface.

The welding setup shown in Figure 194 considers the gas nozzle at a fixed distance from the base metal surface. The actual arc length between the end of the electrode and the base metal surface is determined by the open circuit voltage, the power source slope setting, and the wire feed speed setting. The welding condition is deemed correct for the base metal, the electrode type and diameter, etc.

The volt-ampere curve, with the "X" at the point of equilibrium, reflects the welding condition set in the upper portion of the drawing. (The point of equilibrium is where the wire feed speed and the electrode melt rate are exactly equal). The diagram shows the weld progressing from left to right. The problem appears to be a tack weld in the weld joint. For this example we will assume the tack weld deposit does not melt significantly.

At the instant the welding arc reaches the tack weld "A" the arc length momentarily shortens. Since arc voltage is a function of arc length the arc voltage decreases momentarily. The volt-ampere curve in Figure 194 shows the decreased voltage but a considerable increase in amperage at point "A".

The increased amperage provides a **faster melt rate** for the electrode wire; in fact, faster than the electrode is being fed to the weld joint. Remember that wire feed speed remains constant! The faster melt rate causes an increase in arc length, with a corresponding slide up the volt-ampere curve, until the point of equilibrium is once again reached. At the point of equilibrium, of course, the melt rate of the electrode is exactly equal to the wire feed speed.

The back side of the tack weld is indicated as point "B" in the illustration. At this point the arc length increases momentarily. The increase in arc length also increases arc voltage while decreasing amperage. This is shown on the volt-ampere curve at point "B". The melt rate of the electrode is less at lower amperage than the wire feed speed. The result is that the electrode drives down toward the base metal thus decreasing arc length and arc voltage. The slide down the volt-ampere curve increases amperage until the equilibrium point is once again established.

It is apparent that there is built-in amperage adjustment in the constant voltage type welding power sources while welding. The terms "constant potential" and "constant voltage" mean that the welding power source is designed to maintain a relatively constant arc voltage. This automatically controls the amperage level of the power source.

CHAPTER 19

Shielding Gases and Electrodes for GMAW

There are a variety of shielding gases and shielding gas mixtures in common use with the gas metal arc welding process. Some of the gases and gas mixtures have a broad range of applications while others are restricted in their use. It is very important to understand the effects that each type of shielding gas has on the welding arc. For example, certain gases and gas mixtures promote true spray transfer with specific materials such as aluminum. Other types of shielding gases and gas mixtures with active elements would not permit aluminum welding at all.

Two of the shielding gases explained here have been discussed concerning their use with the gas tungsten arc welding (GTAW) process. This section will look at their use with the gas metal arc welding (GMAW) process.

Argon (Ar)

Argon is a chemically inert gas that will not combine with the products of the weld zone. It has an ionization potential of 15.7 electron volts (eV). Please recall that "ionization potential" is the energy necessary to remove an electron from the gas atom thereby making it an ion. The ionized, or electrically charged, gas column provides a better electrical conductor for the welding current to follow from the electrode to the base metal surface.

Argon has low thermal conductivity. The arc column is constricted with the result that high current densities are present. Current densities are measured by dividing the welding current amperage by the cross sectional area of the arc column in square inches. The constricted arc column permits more of the arc energy to go into the base metal as heat. The result is a relatively narrow weld bead width and good penetration.

Argon causes a more concentrated arc column than any of the other shielding gases used with the gas metal arc welding process. This is one reason that argon has a reputation as a "cleaning gas" for removing surface oxides. It is actually the arc column concentration, and therefore the heat energy concentration, that causes refractory oxides to loosen and disperse. The fact that DCRP (electrode positive), the cleaning half-cycle of polarity, is used with the GMAW process certainly helps in removing the tenacious oxides.

Argon has a purity of 99.995% minimum by specification. In fact, argon sold for welding purposes usually has better than 99.997% purity. Argon is used as a shielding gas for all of the commonly welded metals. It is also the preferred primary gas used in most shielding gas mixtures. Argon is the only inert gas that provides true spray transfer with the gas metal arc welding process. It is about 1.1 times heavier than air.

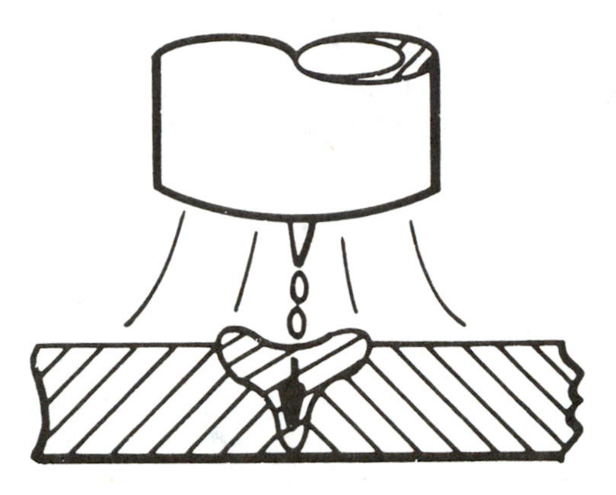

Figure 195. Argon Shielded GMAW Weld in Aluminum.

The gas metal arc welding process brings heat plus molten metal to the weld joint. When using spray type transfer it would appear that there is an instantaneous molten weld puddle. Many welders see this and begin to travel along the weld joint. BUT—There is no instantaneous penetration of the base metal. The molten metal laying on the base metal surface is mostly molten electrode material. It transfers across the arc in an axial, or straight line, flow of small droplets. On striking the base metal surface, the molten filler metal flows out to the periphery of the weld puddle. The weld puddle edges are where the heat transfer into the base metal exceeds the heat input to the weld puddle. This weld deposit/base metal interface is the beginning of the heat-affected zone (HAZ) in the base metal.

A typical penetration pattern on aluminum base metal (Figure 195) would show the heat "soaking" into the base metal across the entire weld puddle. Very quickly the quenching action of the base metal mass will slow the melting of the base metal under the weld deposit.

The constricted argon-shielded arc column brings the arc heat into the center of the molten weld puddle. The heat cannot transfer radially at the surface because the molten weld metal is already hot. Remember, heat will transfer much more readily through cold metal than it will through hot metal. The heat from the arc column goes to the cold metal deep down in the base metal under the weld. This is how deep central penetration occurs with argon shielding gas and the gas metal arc welding process.

Helium (He)

Helium is also an inert gas and is lighter in weight than air. The ionization potential of helium is 24.5 electron volts (eV). It has excellent thermal conductivity. The helium arc column will expand under heat producing thermal ionization of the gas and reducing current density. The bell-like shape of the helium arc column impinges over the entire molten weld puddle in most applications.

With the gas metal arc welding process, heat plus molten metal is brought to the welding area. The molten metal spreads to form the weld puddle very quickly but with very little penetration. The heat input to the helium shielded weld puddle is essentially equal over the entire surface area of the arc column impingement.

Penetration into the base metal by the molten weld deposit occurs by conduction of the heat from the weld puddle. Since heat input is as great at the periphery of the arc column as it is at the center, heat transfer and weld penetration is deeper at the weld puddle edges than it is with argon and the GMAW process.

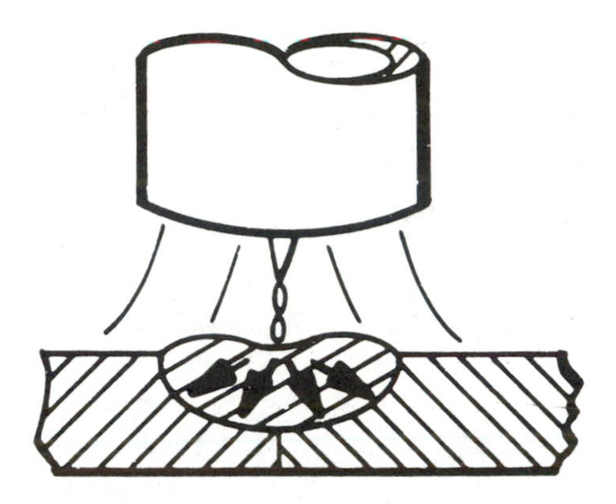

Figure 196. Helium Shielded GMAW Weld in Aluminum.

Figure 196 shows a typical cross section of a helium shielded gas metal arc weld in aluminum. Perhaps you have noticed there is a difference in the cross sectional shapes of welds made with the GTAW and GMAW processes. The illustrations are correct. The reasons for the differences are inherent in the two welding processes as described in the text.

Remember: The GTAW process brings heat only to the weld area. The GMAW process brings heat plus molten metal to the weld joint.

Argon-Helium (Ar-He)

Argon-helium mixtures are usually used to obtain the best and most favorable characteristics of both gases. The addition of helium may be in the percentages of 20%-90%. In many applications the consumer mixes the gases on-site to suit his specific welding requirements.

The normal procedure in smaller shops is to purchase a cylinder of each type of shielding gas (argon and helium). They are then mixed through a gas proportioner to the specific percentages required by the user.

Argon-helium mixtures are normally used where heavier sections of non-ferrous metals are to be joined. This would include aluminum, magnesium, and copper and their alloys. The heavier the metal thickness, the greater the percentage of helium would normally be in the gas mixture.

Argon and helium can be mixed in a single container. They do not, however, become a compound gas mixture. Each gas retains its individual atomic structure. Gas mixtures contain helium or argon are always **gas mixtures** because of the monatomic nature of the two gases.

Carbon Dioxide (CO_2)

Carbon dioxide is a compound gas which means that it has more than one element in its chemistry. Where argon and helium have only one gas atom to each molecule, carbon dioxide has two gas atoms to the molecule. Argon and helium are classed as monatomic gases and carbon dioxide is classed as a diatomic gas. To define:

Monatomic = "having one atom in the molecule".
Diatomic = "having two atoms in the molecule".

The primary constituents of carbon dioxide are atomic oxygen and carbon monoxide. CO_2 is a shielding gas but it is not an inert shielding gas such as argon and helium.

In fact, CO_2 has a characteristic different than argon or helium. This is the ability of CO_2 gas to dissociate in the heat of the welding arc and re-combine in the atmosphere. It is this factor that permits more heat energy to be absorbed in the carbon dioxide gas passing through the arc area. It also uses the free oxygen in the arc area to superheat the weld metal transferring from the electrode to the weld joint. As illustrated in Figure 197, CO_2 has a slightly wider arc column than argon but less width than helium.

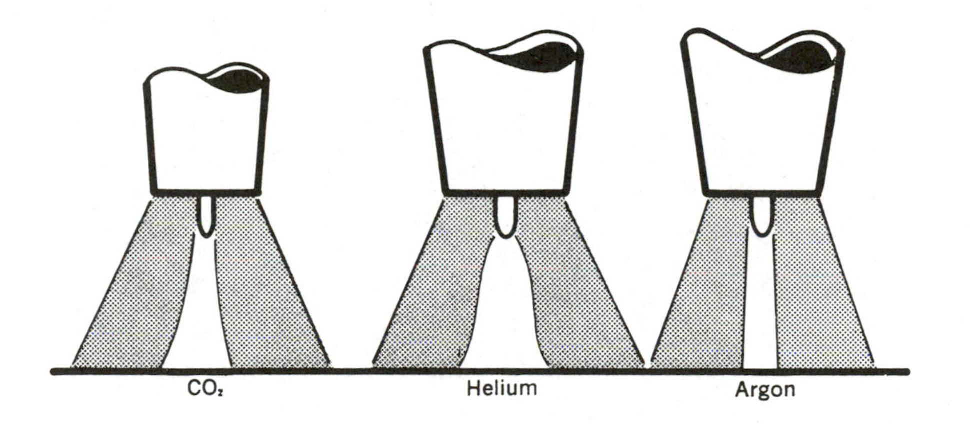

Figure 197. Comparative Arc Columns With CO_2 Helium, and Argon.

Care must be taken when gas metal arc welding with open arc techniques and CO_2 shielding gas. The globular transfer that occurs can cause plastic cold shuts, gas entrapment in the weld deposit, and intrabead cracking. There is also the probability of lack of fusion at the sidewalls of the weld joint. Weld defects of this type can be caused by cooling of the molten metal globule transferring across the arc.

Carbon dioxide (CO_2) has a tendency to cause a spattering, unstable arc when used with the GMAW open arc transfer of weld filler metals. Open arc transfer means the molten weld filler metal is physically moved across the arc column space from the electrode to the base metal. Excessive weld spatter is common with this transfer mode due to the globule of filler metal literally splashing metal out of the weld puddle. The weld spatter may be contained to some degree by maintaining a very close arc while welding. In fact, the best method is to use the Buried Arc Transfer technique previously discussed.

Welding grade carbon dioxide is normally used only for low carbon and mild steel welding applications with the GMAW process. This is because the carbon dioxide (CO_2) gas breaks down under the heat of the welding arc into approximately 33% carbon monoxide by volume, 33% free atomic oxygen by volume, and the balance remains CO_2 since it is outside the arc heat zone. Carbon dioxide is also used in a variety of shielding gas mixtures used with the GMAW and FCAW processes.

Argon-Oxygen (Ar-O_2)

Pure argon is an excellent shielding gas for the gas metal arc welding process because it permits the use of spray type metal transfer with all the commonly welded metals. When depositing flat or horizontal welds on steels or stainless steels, however, the main problem is the "fast freeze" characteristics of argon shielded welds.

This does not allow time for the weld deposit to "wet out" to the toes of the weld. Undercut will invariably form at the edges, or "toes", of the weld bead. The illustration at the left of Figure 198 shows a typical high crowned weld bead, with the usual undercut condition, of an argon shielded weld on stainless steel.

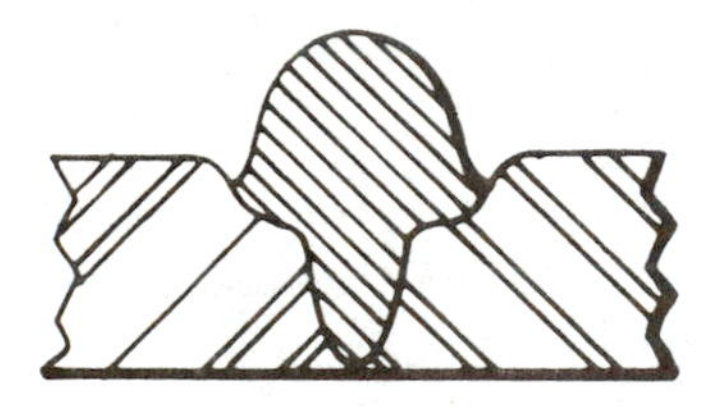

Argon Gas

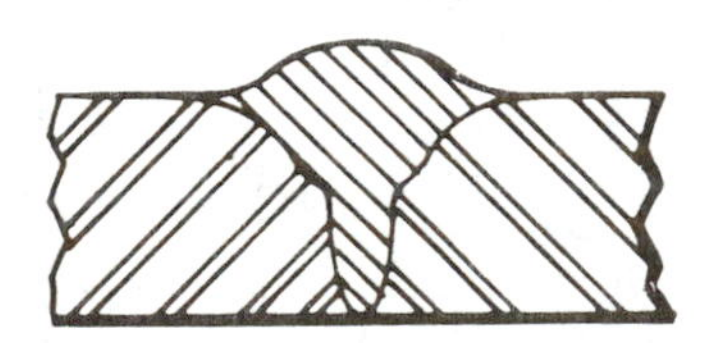

Argon-Oxygen Gas

Figure 198. Comparative Welds With Argon and Argon-Oxygen.

The right side of the illustration shows a weld with oxygen added to the argon shielding gas. The tendency to undercut may be minimized or prevented by the addition of 1%-5% oxygen to the argon shielding gas.

This mixture is commonly available in cylinder mixtures of argon plus 1%, 2%, 3% and 5% oxygen content. The oxygen increases the temperature of the molten filler metal transferring across the arc column to the weld joint.

The increased filler metal temperature retards the cooling rate of the weld puddle by a few milli-seconds. This lets the molten filler metal wet out to the toes of the weld bead which virtually eliminates the possibility of undercut. The additional time in the molten state also acts to control the bead profile by flattening the weld.

The natural question is, "Where do I use 2% oxygen and where do I use 5% oxygen?" The following data are provided as a guide. Keep in mind the reason for adding the oxygen which is to increase the temperature of the molten metal. A small volume of molten metal needs only a small amount of oxygen whereas a large volume of molten metal requires more oxygen.

1. Stainless steel thickness of 1/16"-1/8" = argon + 1% oxygen.
2. Stainless steel thickness of 1/8"-3/16" = argon + 2% oxygen.
3. Stainless steel thickness of 3/16"-1/4" = argon + 3% oxygen.
4. Stainless steel thickness of over 1/4" = argon + 5% oxygen.

Argon-Carbon Dioxide (CO_2)

Welding grade carbon dioxide (CO_2) does not provide the stable arc characteristic required for open arc welding of mild steel and low alloy steel. The globular transfer of molten filler metal will cause excess weld spatter which can scar the surface of the base metal. There is also the substantial decrease in filler metal efficiency due to the weld spatter. With low alloy steels, CO_2 shielding gas will cause high oxidation losses of certain alloying elements.

The welding industry has developed a series of shielding gas mixtures having various percentages of argon and carbon dioxide for welding carbon steel and low alloy steel. Many of the mixtures were developed by consumers to meet specific requirements in base metals and weld joints.

Some welding authorities believe the argon-CO_2 mixture should not exceed 25% CO_2. Others are of the opinion that mixtures with up to 80% CO_2 are practical. The individual user must determine the requirements for his specific application.

Cost is one of the primary reasons for using as much CO_2 as possible in a gas mixture. The price of argon is usually about 10 times the cost of welding grade carbon dioxide in most areas. By purchasing each type of gas separately, and mixing it on-site through a gas proportioner, considerable savings can be realized. This will also permit the user to vary the gas mixtures to suit the specific applications as they may change.

Argon-CO_2 gas mixtures are normally sold at the same cylinder price as pure argon. The reason is the additional cost of mixing the two gases to exact percentages. This requires extra cylinder handling with the added cost of capital equipment for mixing the gases. When the shielding gases are purchased separately by the user and mixed on-site the overall gas cost is more proportionately distributed according to the amount of each gas used. In some cases, substantial cost savings can be realized.

The argon-CO_2 shielding gas mixtures are employed for gas metal arc welding carbon steels, low alloy steels and, in some few cases, for some thin gage stainless steel applications. Various mixture percentages will produce metal transfer characteristics that range from modified spray transfer to globular transfer. The argon 75%-CO_2 25% mixture was designed especially for the short circuit transfer method of metal transfer with low alloy steel and mild steel.

Helium-Argon-Carbon Dioxide (He-Ar-CO_2)

The helium-argon-carbon dioxide shielding gas mixture was developed specifically for welding 300 series austenitic stainless steels with the short circuit method of metal transfer. It is normally mixed and sold as a pre-mixed cylinder gas having a content of 90% helium, 7.5% argon, and 2.5% carbon dioxide. The important factor is the high helium content.

Helium tends to spread out the welding arc column because of its excellent ability to transfer heat. This provides greater heat input over a wider surface area of the weld joint. The molten weld puddle stays liquid a few milli-seconds longer, permitting the relatively sluggish stainless steel filler metal to wet out to the toes of the weld.

The combination of gases imparts a unique characteristic to the weld metal. With the tri-gas mixture, it is possible to make a weld with very little buildup of the top bead profile. The result is excellent where the high weld crown might cause the re-entry angle to be too steep.

In particular, the welding of austenitic stainless steel pipe with the short circuit method of metal transfer has been accomplished with relative ease. Even the relatively small weld puddle stays molten long enough to flatten the crown and fill in the toes of the weld bead.

The small addition of carbon dioxide has not shown carbide precipitation at the grain boundaries of the weld metal. Apparently the time of reaction is so short, and the heat level in the weld maintained at such a low value, that the material passes through the critical range (1400°-800° F.) with little or no carbide precipitation occurring.

Other Gases

Other gases have been used in combination with argon and/or helium. For example, a trace of chlorine was added to argon in one experimental mixture for welding aluminum. It was thought the chlorine would supply a fluxing action that would tend to eliminate porosity from the weld deposit.

The basis for this opinion comes from the fact that chlorine is used as a purifying and fluxing agent during the manufacturing process of aluminum. Massive purges of chlorine are injected directly into the molten aluminum (approximately 1325° F. and silvery-red in color) while it is in a holding furnace and just before it is cast into ingots. At this temperature the chlorine removes the impurities from the aluminum. They float to the top of the molten metal and are removed as dross.

The use of argon-chlorine mixtures for gas metal arc welding was not commercially successful because of the inherent danger from the breakdown of the chlorine gas under the high temperature welding arc. The arc heat, which is in excess of 10,000° F., will cause the chlorine to decompose into a gas called phosgene. Phosgene is a poisonous gas that is a real health hazard.

Nitrogen has been used in Europe as a shielding gas, either in pure form or mixed with argon, for welding copper and copper alloys. The arc is very harsh with excessive spatter when pure nitrogen is used. A mixture of approximately 70% argon and 30% nitrogen has been used successfully for welding copper and copper alloys. The addition of argon stabilizes the welding arc and creates less turbulence in the weld puddle. Nitrogen was used because of the extremely high cost of helium in Europe.

Shielding Gases and Applications

This section of the text will show some of the shielding gases, and shielding gas mixtures, that are used for welding various applications with the gas metal arc welding process. Each shielding gas, and shielding gas mixture, is associated with specific types of base metals and metal transfers.

Several factors are involved with the proper selection of a shielding gas or shielding gas mixture. The actual criteria for gas selection may include any one, or all, of the following:

1. Base metal classification and thickness.
2. Electrode classification and diameter.
3. Method of metal transfer.

In all of the recommendations listed, it is considered that the shielding gas, or gas mixture, is the best for the specific application. This is not to say that the shielding gas, or gas mixture, cannot be used successfully for other metals or applications.

Brief descriptions of metal transfer methods, as mentioned in the data, are detailed below for your information. They are to be used as guidelines in your work.

***Spray Transfer.** Molten metal droplets are always smaller than the solid electrode wire diameter. Creates a very fluid weld puddle. Normally only welded in flat groove and fillet or horizontal fillet positions.

***Modified Spray Transfer.** Molten metal droplets are larger than true spray droplets but never larger than solid electrode wire diameter. Normally welded in flat or horizontal positions.

***Modified Globular Transfer.** Molten metal globules are slightly smaller than regular globular transfer. Globules will be less than two times the solid electrode wire diameter.

***Globular Transfer.** Molten metal globules will be at least two times the solid electrode wire diameter. Not normally desirable as a metal transfer method because of excessive spatter.

***Short Circuit Transfer.** All metal transfer is by direct contact of the molten end of the electrode with the base metal. No molten metal transfers across the welding arc.

Each shielding gas and shielding gas mixture is accompanied by the type of welding atmosphere it produces, the type of metal transfer normally obtained, and the type of base metals with which it is best used. Electrode diameters are provided where applicable.

Argon. Inert atmosphere. *True *spray transfer.* Necessary for non-ferrous metals such as aluminum (Al), magnesium (Mg), copper (Cu) and their alloys. For Al and Mg up to 1/4" thickness; Cu up to 1/8" thickness.

Helium. Inert atmosphere. **Globular transfer.* Used mostly as an element in shielding gas mixtures for non-ferrous metals such as Al, Mg, and Cu. Not normally used as a single gas with the GMAW process.

Carbon Dioxide (CO_2). Oxidizing atmosphere. **Short circuit transfer or buried arc transfer.* Excellent for low carbon and mild steel. Not normally used for open arc GMAW because of excessive spatter levels while welding. Excellent for most carbon steel and low alloy steel flux cored electrodes.

Argon 75% - Helium 25%. Inert atmosphere. **Modified spray transfer.* For non-ferrous metals. Al and Mg: 1/4"-1/2" thickness; Cu: 1/8"-1/4" thickness.

Argon 50% - Helium 50%. Inert atmosphere. **Modified globular transfer.* For non-ferrous metals. Al and Mg: 1/2"-3/4" thickness; Cu: 1/4"-3/4" thickness.

Argon 25% - Helium 75%. Inert atmosphere. **Modified globular/globular transfer.* Non-ferrous metals. Al and Mg: over 1" thickness; Cu: 1/2" and over.

Argon 98% - Oxygen 2%. Oxidizing atmosphere. *True *spray transfer.* For carbon steels, low alloy steels, 300 series austenitic stainless steels. Electrode diameters up to and including 0.045" diameter solid wire.

Argon 95% - Oxygen 5%. Oxidizing atmosphere. *True *spray transfer.* For carbon steels, low alloy steels and 300 series austenitic stainless steels. Electrode diameters 1/16" diameter and greater.

Argon 90% - Carbon Dioxide (CO_2) 10%. Oxidizing atmosphere. **Modified spray transfer.* For carbon steels and low alloy steels, all thicknesses. May be used for certain low alloy flux cored electrode applications.

Argon 88% - CO_2 9% - Oxygen 3%. Oxidizing atmosphere. **Modified spray/modified globular transfer.* For carbon steels and low alloy steels, all thicknesses. May be used for certain low alloy flux cored electrodes.

Argon 85% - CO_2 15%. Oxidizing atmosphere. **Modified globular transfer.* For carbon steels and low alloy steels, all thicknesses. May be used for certain carbon steel and low alloy steel flux cored electrodes. Deeper penetration than gas mixture selections with less CO_2.

Argon 75% - CO_2 25%. Oxidizing atmosphere. **Modified globular/globular open arc transfer or short circuit transfer.* Oxidizing atmosphere. Developed for short circuit transfer with carbon steels and low alloy steels.

Helium 90% - Argon 7.5% - CO_2 2.5.%. Oxidizing atmosphere. **Short circuit transfer.* Developed for short circuit transfer with 300 series austenitic stainless steels.

In many applications, the shielding gas is blamed for excess spatter in the welding conditions. As a matter of fact, the shielding gas has very little to do with the spatter levels in a welding condition. Improper voltage and amperage conditions are the two factors most often found to cause excess weld spatter. There should be very little, if any, spatter in a properly set up welding condition using the GMAW process.

GMAW Solid Core Electrodes

The electrode materials used with the gas metal arc welding process are solid core, continuous length, and consumable in the welding arc. They have no flux coating or inner core of flux. All of the alloying and deoxidizing elements necessary to the weld are part of the total chemistry of the electrode material. Some type of externally supplied shielding gas is used to cover the weld area and exclude the atmosphere.

The electrode materials used with the GMAW process are classified by the American Welding Society (AWS) in various Specifications. These Specifications are in a continuing state of development and upgrading by AWS Committees assigned the responsibility. Committee members are from the welding industry throughout the world.

Every welding shop, large or small, should have the applicable electrode Specifications relating to the types of electrodes used in the shop. All welding electrode materials should be ordered by citing both the AWS Specification and the AWS electrode classification on the purchase order. It is also a good practice to routinely request, in writing, that "Certificates of Conformance" for the electrodes be furnished by the supplier. They should come with the electrode shipment at no cost to the purchaser.

AWS Electrode Specifications for GMAW

There are several AWS Specifications applicable to the gas metal arc welding (GMAW) process. Other AWS Specifications are available for all of the electrodes and filler metals presently used in welding processes. Copies of the electrode Specifications may be purchased from the American Welding Society. Their address is:

AMERICAN WELDING SOCIETY
P. O. Box 351040
Miami, FL 33135

The following AWS electrode Specifications are specifically used with the gas metal arc welding process.

A5.7 "Specification for Copper and Copper Alloy Bare Welding Rods and Electrodes".

A5.9 "Specification for Corrosion Resisting Chromium and Chromium-Nickel Steel Bare and Composite Metal Cored and Stranded Arc Welding Electrodes and Welding Rods".

A5.10 "Specification for Aluminum and Aluminum Alloy Bare Welding Rods and Electrodes".

A5.14 "Specification for Nickel and Nickel Alloy Bare Welding Rods and Electrodes".

A5.16 "Specification for Titanium and Titanium Alloy Bare Welding Rods and Electrodes".

A5.18 "Specification for Carbon Steel Filler Metals for Gas Shielded Arc Welding".

A5.28 "Specification for Low Alloy Steel Filler Metals for Gas Shielded Arc Welding".

There are two sections to each of the AWS Specifications. The first part is the actual Specification which provides all the legal data such as chemical compositions, electrode classifications, tests required, etc. This is where the designers and welding engineers select the correct filler metals for the weld joints. This is where the purchasing agent checks to be sure the correct classification is ordered.

When the electrode materials are received, they should be checked for conformance with the purchase order and the AWS Specification. Then they should be stored in a cool, dry clean place until they are used.

The second section of the Specification is actually not a part of the Specification at all. It is the **Appendix** to the Specification. This part tells how the electrode materials are classified by the AWS. It also provides some safety and use information.

Probably the most important part of the Appendix, however, is the section titled "Description and Intended Use". Here is where each electrode classification is explained and it's specific application defined. It is a most useful section with excellent data for engineers, supervisors and welders. Here also may be found information on special electrode storage requirements as required.

Alloys and Deoxidizers

It is very seldom that chemically pure metals are used by the welding industry. While pure aluminum is used in some areas of manufacture it is not considered a structural material. There are only a few pounds of pure iron in existence and it is considered a laboratory curiosity. The point is, there are always trace elements of impurities in "pure" metals.

All welding electrodes are alloys of two or more elements. For example, low carbon steel is an alloy of carbon and steel with fractional percentages of other elements, some of which are impurities. In most cases, the impurities are sulphur and phosphorous. Some of the elements added to electrode core wire serve as alloying agents while others function as deoxidizers and scavengers of unwanted elements that could weaken the weld. In some instances the added element performs both the deoxidizing and alloying functions.

Some alloying elements work as deoxidizers with one metal, alloys with another, and yet can be the prime metal in still other alloys. Figure 199 shows how aluminum may be used as a deoxidizer in steel, as a base metal, and as an alloying element in aluminum bronze.

To further illustrate that many elements are used as alloys, please note the following list of elements used in various aluminum alloys:

Copper, silicon, manganese, iron, zinc, chromium, nickel, lead, titanium, magnesium, and bismuth.

When considering all the carbon steels and low alloy steels available it is difficult to list all the deoxidizing and alloying elements that might be used in their manufacture. Most of the same elements are used in the manufacture of electrode materials. The type of alloying elements and their percentages in the electrode wire will be determined by the type of steel to be welded and the shielding gas used.

The reason for considering the shielding gas is that some shielding gases cause severe oxidation losses of alloying elements and deoxidizers across the welding arc. Carbon dioxide (CO_2) is an excellent example of such a gas when used for welding carbon steel. Remember: CO_2 breaks down under the heat of the welding arc into about 33% by volume carbon monoxide, 33% by volume free atomic oxygen, and the balance will remain CO_2 because it is outside the influence of the arc heat.

It is well known that both silicon (Si) and manganese (Mn) are subject to high oxidation losses in the CO_2 gas shielded welding arc. The percentage of alloy transfer may be improved by the addition of more silicon or more manganese to the electrode chemistry. The addition of a larger percentage of either of the elements (Si or Mn) to the electrode material will decrease the percentage losses of both elements across the welding arc.

The type of electrode used, solid core wire or flux cored electrode, will have a definite affect on the transfer efficiencies of both deoxidizer and alloying elements. Transfer efficiencies of Si and Mn with different shielding gases are shown for some steel electrodes:

1. Solid electrode wire in argon-oxygen shielding gas mixtures is most efficient. (Normally spray type transfer).
2. Solid electrode wire in carbon dioxide (CO_2) shielding gas is next best.
3. Flux cored electrodes with CO_2 gas shielding is least efficient.

Figure 199. Aluminum, A Versatile Metal.

Deposition Rates

Deposition rates will depend on several variables in the specific welding condition. The type and diameter of electrode wire, type of welding power source used, the shielding gas used, the method of metal transfer, and the position of the weld joint will all have an effect on the deposition rate.

Data is provided in charts in the appendix of this text for solid core electrode wire calculation and deposition rates. Use the charts as guides for your welding applications.

Flux Cored Electrodes

The flux cored arc welding (FCAW) process requires basically the same type of equipment as the gas metal arc welding (GMAW) process. The essential difference in the two welding processes is the electrode used. The flux cored arc welding process uses a tubular electrode with granular flux inside the tube.

The use of granular flux in the tubular electrode permits a wide variety of weld deposit metallurgical characteristics. Almost any element can be added to the flux of even low carbon steel flux cored electrode. This permits the manufacture of both self-shielded and gas shielded flux cored electrodes.

Self-shielded flux cored electrodes have certain gasifying elements added to their flux. These elements decompose under the heat of the welding arc to form basically CO_2 gas for shielding the weld. A slag also forms on the weld surface to protect the weld deposit as it cools. There is considerable smoke from the flux which should be removed from the welding area with some form of smoke extractor.

Gas shielded flux cored electrodes require an externally supplied shielding gas while welding. The gas used depends on the type of electrode material. For example, most carbon steel and low alloy steel electrodes can be welded with carbon dioxide (CO_2) shielding gas. Flux cored electrodes for welding chromium and chromium-nickel alloys may use carbon dioxide or a mixture of argon 98% - oxygen 2% as described in AWS A5.22 "Specification for Flux Cored Corrosion-Resisting Chromium and Chromium-Nickel Steel Electrodes", Table 1.

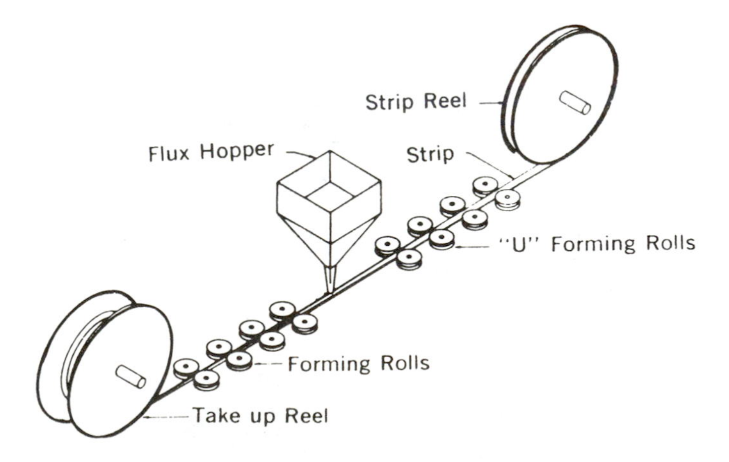

Figure 200. Fabricating Flux Cored Electrodes.

The AWS Specifications for flux cored electrodes are as follows:

A5.20 "Specification for Carbon Steel Electrodes for Flux Cored Arc Welding".
A5.22 "Specification for Flux Cored Corrosion-Resisting Chromium and Chromium-Nickel Steel Electrodes".
A5.29 "Specification for Low Alloy Steel Electrodes for Flux Cored Arc Welding".

Flux cored electrodes are normally fabricated in 3/16" diameters and reduced to the final electrode diameter in succeeding drawing operations. Flux cored electrodes are presently available in diameters of 0.035", 0.045", 1/16", 5/64", 3/32", 7/64" and 1/8". Figure 200 shows a typical flux cored electrode manufacturing sequence.

The AWS Specifications note that some flux cored electrodes are designed for single pass welding only while others are designed for multiple pass welding. The single pass electrodes require significant dilution of electrode metal with base metal to achieve their minimum required strength levels. Using these flux cored electrodes for multiple pass welding would result in less physical and mechanical strength levels in the weld deposit.

CHAPTER 20

Flux Cored Arc Welding and Air Carbon-Arc Cutting Processes

The flux cored arc welding (FCAW) process is similar to the gas metal arc welding (GMAW) process. The two major differences are the types of continuous, consumable electrodes used and the types of shielding gases used.

FCAW electrodes are tubular with granular flux contained in the tube. The carbon steel flux cored electrodes normally operate with welding grade CO_2 although mixtures of argon-CO_2 are often preferred.

Stainless steel flux cored electrodes may be used with either CO_2 or argon 98% - 2% oxygen. Other shielding gas mixtures may be employed at the consumers discretion.

Some flux cored electrodes are self-shielding and require no externally supplied shielding gas. This type of flux cored electrode also requires somewhat different welding techniques than gas shielded flux cored electrodes.

The air carbon-arc (AAC) process is used for cutting and gouging metals. The process uses carbon electrodes to melt the base metal and compressed air to literally blow the molten metal away. The metal is not burned as it is with flame cutting methods used for carbon steel. For this reason the AAC process can be used with almost any of the commonly welded metals.

Flux Cored Arc Welding

Welding with the flux cored arc welding (FCAW) process requires knowledge of torch handling techniques similar to those used with the GMAW process. The power source used is a constant voltage/constant potential type unit and, by the design of the power source circuitry, provides an output volt-ampere curve suitable for the FCAW process. The wire feed speed setting determines where on the volt-ampere output curve the amperage setting will be. Wire feed speed also sets the load voltage at the same time. A typical volt-ampere curve with the load voltage and amperage noted is shown in Figure 201.

The torch lead or lag angle and the torch work angle are set in similar fashion to those used with the GMAW process. The lag angle is often called the "drag" angle or "backhand" technique. The lead angle is often called the "push" angle or "forehand" technique. The terms "backhand" and "forehand" are used here because they are most descriptive of the process techniques.

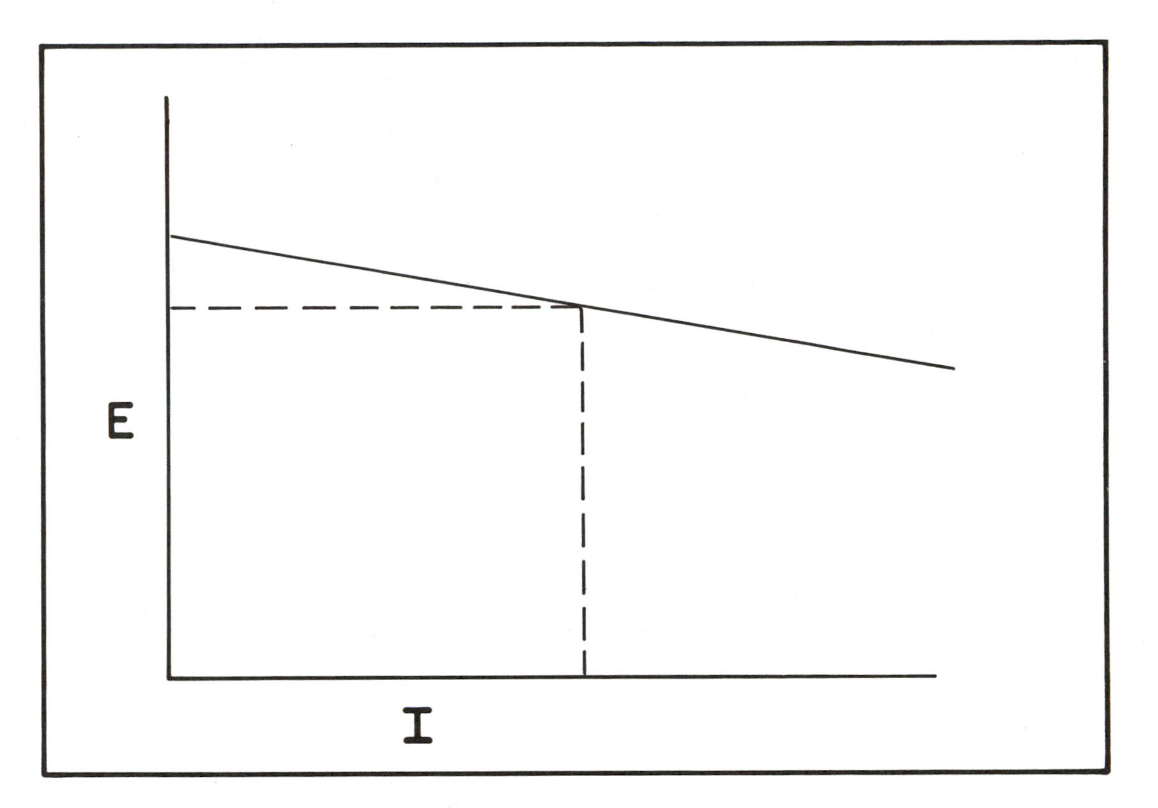

Figure 201. Volt-Ampere Curve With Welding Volts and Amps.

The **backhand technique** shows the torch head and electrode tip pointing *away from the direction of travel* and toward the weld puddle. This technique provides the deepest penetration into the base metal. It is somewhat slower than the forehand technique.

For gas shielded FCAW the backhand torch angle is about 10°-15° from the vertical plane. For the self-shielding FCAW electrodes, the backhand torch angle can be 15°-35°. The backhand torch angles presented are normally used for flat groove and fillet welds as well as horizontal fillet welds. The forehand technique is often used by skilled welders where penetration requirements are not stringent. Forehand welding permits faster travel speeds with slightly less penetration into the base metal.

Any vertical welding done with the FCAW process would be done vertical up with a slight **forehand technique.** An angle of about 5°-10° with the electrode pointed upward in the direction of travel would be suitable. Trying to weld vertical up with a backhand technique would permit molten slag to run into the weld and could lead to non-metallic slag inclusions.

The **work angle** is the angle of the torch head and electrode with respect to the base metal surface(s). For example, a flat position groove weld would normally have the electrode positioned so that it points straight down into the weld at a 90° angle to the work surface. This is illustrated in Figure 202.

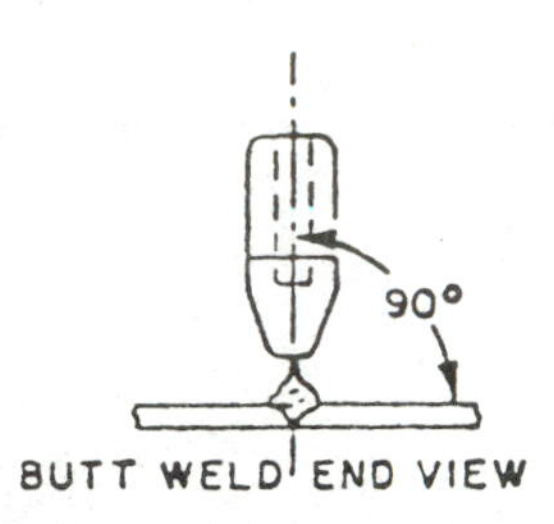

Figure 202. Torch and Electrode Work Angle Of 90°.

A horizontal fillet weld would be made normally with the torch and electrode angle at about 45° from the two joint surfaces. This is shown in Figure 203.

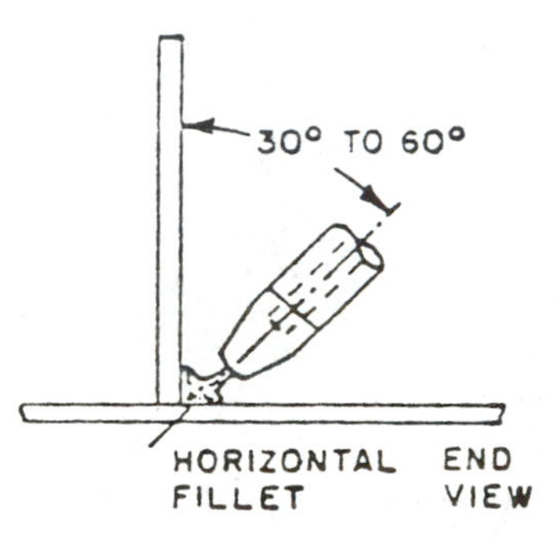

Figure 203. Torch and Electrode Work Angle Of 45°.

Remember that the terms "backhand", "lag" angle, and "drag" angle are synonymous. It is also true that the terms "forehand", "lead" angle, and "push" angle are synonymous. The term "work angle" has been left as it is for some reason, probably because it makes sense.

FCAW Process Equipment

The welding process equipment required for the FCAW process is similar to the equipment used with the GMAW process. The recommended types of welding power sources are the **constant potential/constant voltage** units. For most FCAW applications, the power source ratings range from 450-650 amperes with maximum open circuit voltage at about 65 volts. The power sources are rated at 100% duty cycle and normally have remote voltage control capability.

Wire feeder-controls for the FCAW process have either two or four drive rolls for advancing the electrode. Many users prefer the four drive roll systems because they will function as electrode straighteners while providing positive drive pressure to the flux cored electrode. Straight electrode material imposes much less resistance on the movement of the electrode through the monocoil liner of the casing assembly. In some applications where 0.045'' and 1/16'' diameter flux cored electrodes are used, the two drive roll systems are considered adequate.

The drive roll groove shape should be such that the flux cored electrode is not crushed or flattened by excess pressure. For smaller diameter electrodes (5/64'' diameter and smaller) knurled vee groove drive rolls may be used satisfactorily. For larger diameters of flux cored electrodes, ''U'' shaped grooves are often employed.

In all wire feeder-control applications, the consumable nature of several system components must be considered. For example, the **inlet guide tube** and the **outlet guide tube** can develop grooves from electrode movement through their ID's. The **drive rolls** can, and do, wear until they no longer are usable. To assure first quality welding, **preventative maintenance** must make sure worn FCAW system components are replaced before they cause weld defects.

There is a hose and cable assembly connected to the wire feeder-control unit. The flux cored electrode is fed through the outlet guide into the **monocoil liner of the hose and cable assembly.** The liner directs and carries the electrode to the **gun,** or **torch,** at the output end of the hose and cable assembly. The electrode moves through the torch body and head to the **copper contact tube.** The copper contact tube is where the flux cored electrode is energized with welding amperage. From the contact tube the electrode goes to the welding arc.

The monocoil liner of the hose and cable assembly probably causes more welding problems with the FCAW and GMAW processes than any other equipment item. The liner picks up dirt and drawing compound from the electrode passing through. Eventually the foreign elements cause wire feeding problems by restricting the electrode movement in the liner.

The solution is to remove the monocoil liner after every 60-100 pounds of electrode used or after every 8 hours of welding use. Replace it with a clean liner and put the dirty liner in a solvent solution for about 6-8 hours of soaking. Then remove the soaked liner and blow it out thoroughly with compressed air or dry nitrogen. NEVER USE OXYGEN! Coil and store the liner until needed. This bit of preventative maintenance can save considerable money in reducing equipment ''down time''and weld defects.

Flux Cored Electrodes

The significant factor of flux cored arc welding (FCAW) that makes it different from other welding processes is the granular flux internal to the tubular electrode. This is where any alloying additions, deoxidizing agents, and gasifying elements are added rather than in the solid metal of the tube material.

The **gas shielded FCAW electrodes** require some type of externally supplied shielding gas to protect the weld puddle and electrode tip from outside atmosphere. The most used shielding gas for FCAW is welding grade carbon dioxide (CO_2). Shielding gas requirements are detailed in the various AWS Specifications for flux cored electrodes. The flux of the electrode provides a slag covering for the deposited weld metal. The slag covering is usually friable and easily removed from most weld deposits. To define:

Friable = ''easily crumbled and removed''.

The **self-shielded FCAW electrodes** do not require an externally supplied shielding gas. They develop shielding gas, basically CO_2, from decomposition of the flux in the arc. This makes the gun used with self shielding electrodes simpler in design since it does not have to carry shielding gas and diffuse it through a gas nozzle. In addition to supplying the gasifying elements, the flux provides a slag that protects the deposited weld metal while it cools. Self-shielding flux cored electrodes are available for carbon steels and some 300 series austenitic stainless steels.

AWS Flux Cored Electrode Specifications

There are three specific AWS Specifications for flux cored electrodes. They are listed here for your information:

A5.20 "Specification for Carbon Steel Electrodes for Flux Cored Arc Welding"
A5.22 "Specification for Flux Cored Corrosion-Resisting Chromium and Chromium-Nickel Steel Electrodes"
A5.29 "Specification for Low Alloy Steel Electrodes for Flux Cored Arc Welding"

Each of the flux cored electrode AWS Specifications contains data that is important to the users of the various types of electrodes. Some data of significance to most users includes the chemical composition requirements, the shielding gases with which the electrodes are tested by the Specification, the methods of electrode classification, and mechanical and physical properties of the weld deposits.

The A5.20 Specification is for low carbon and mild steels. It is interesting to note that Table 1, "Chemical Composition Requirements", shows that the following electrode classifications all have the same basic chemical composition: EXXT-1; EXXT-4; EXXT-5; EXXT-6; EXXT-7; EXXT-8; EXXT-11; and EXXT-G.

There is a statement in Table 1, A5.20, that the 1.8% maximum Al is for self-shielded flux cored electrodes only. Another statement says the amount of carbon shall be determined though not by whom.

The following list of elements is the total chemical composition called out in Table 1 of the A5.20 Specification: Carbon; Phosphorous; Sulfur; Vanadium; Silicon; Nickel; Chromium; Molybdenum; Manganese; Aluminum.

Considering that this is specified as the chemical composition for carbon steel flux cored electrodes it is obvious that the element iron and its percentages is missing.

The following flux cored electrode classifications are for single pass welding and have no chemical requirements of any kind according to the A5.20 Specification: EXXT-GS; EXXT-2; EXXT-3; EXXT-10.

FCAW Electrode Classifications

Carbon Steel

This discussion concerns the methods used in classifying low carbon mild steel flux cored electrodes. The classification information is found in the Appendix of AWS Specification A5.20 and is used here with permission of the American Welding Society. As in all AWS filler metal specifications, the appendix contains a great deal of information about the electrodes classified in that particular Specification.

The two flux cored electrode classifications listed below are for essentially the same type of flux cored electrode. There is one specific difference in the numbering system, however, which will be explained as we come to it in the discussion.

*E70T-1
**E71T-1

The explanation of the classifications is as follows:

E = Designates an electrode for arc welding.
7X = Indicates minimum tensile strength, deposited weld metal (72KSI).
*0 = For flat and horizontal welding positions only.
T = Indicates a tubular, or flux cored, electrode.
-1 = Indicates usability and performance capabilities.

This explains the E70T-1 electrode classification. The E71T-1 electrode classification is the same except for the number "1" instead of "0". The numbering difference is explained as follows:

**1 = For welding all positions (F, H, V, and O).

In the AWS Specification A5.20 appendix section titled "Description and Intended Use", the various classifications of carbon steel flux cored electrode are discussed. It is strongly recommended that every welder and user of these electrodes read and understand this information. Some of the electrodes are self-shielding while others require some form of externally supplied gas shielding. This information is illustrated by the "T-X" indication as follows:

Gas shielded classifications = T-1; T-2; T-5.
Self-shielded classifications = T-3; T-4; T-6; T-7; T-8; T-10; T-11.

The two classifications "T-G" and "T-GS" are not covered under any of the presently defined classifications. The "T-G" classification is for multi-pass welds. The "T-GS" classification is for single pass welds.

Stainless Steel

The AWS Specification A5.22 concerns chromium-nickel filler metals for alloys commonly called stainless steels. The method of flux cored electrode classification is considerably different from that used for carbon steels.

Classification of stainless steel flux cored electrodes is based on the *shielding gas used and the chemical analysis of deposited weld metal.* This is further limited by defining the shielding gases recognized in the Specification as either carbon dioxide (CO_2) or argon-oxygen gas mixtures. Table 1 of AWS Specification A5.22 shows the electrode classification system and shielding gas numbering. For example:

EXXXT-1 = CO_2 shielding gas
EXXXT-2 = Ar + 2% O_2
EXXXT-3 = None
EXXXT-G = Not specified

Table 2 of the Specification shows the chemical composition requirements for both the gas shielded and self-shielded stainless steel electrodes. It is interesting to note that some of the self-shielded flux cored electrodes have different percentages of element content requirements than the gas shielded electrodes of the same designation. This is considered and explained in the appendix section of the Specification.

There is a very good although brief discussion of ferrite content in the appendix of the A5.22 Specification. It is strongly recommended reading for those using stainless steel flux cored electrodes.

Low Alloy Steel

The AWS A5.29 Specification discusses the low alloy steel flux cored electrodes. These electrodes are classified similar to the carbon steel flux cored electrodes. Some of the low alloy steel electrodes are of higher tensile strength and so the numbering system does go into three digits. It is imperative that the entire classification be used when specifying any of these classifications of flux cored electrodes.

The following electrode classification is used as an example:

E120T1-K5

E = Designates an electrode for arc welding.
12X = Shows minimum tensile strength of deposited weld metal (120KSI).
*0 = For flat and horizontal positions only.
T = Indicates a tubular, or flux cored, electrode.
1 = Shows usability and performance capabilities.
-K5 = Shows chemical composition of deposited weld metal.
*1 = For all welding positions as applicable (F, V, H, O).

There are four basic types of flux cored electrodes listed in this Specification. They are listed here with some of their unique characteristics:

T1 = Gas shielded, can be all position (smaller diameters).
T4 = Self-shielded, flat and horizontal positions.
T5 = Gas shielded, globular transfer, lime-fluoride flux.
T8 = Self-shielded, DCEN, desulfurizing of weld deposit.

Most flux cored electrodes designed for use with low alloy steels are either T1 or T5 types. This is clearly illustrated in Table 1 of the A5.29 Specification.

FCAW Process Welding Tips

This is a discussion of some factors that affect the way flux cored arc welding is done. It is intended to help the welder and user stay out of trouble with the FCAW process.

The **electrode extension** is always measured from the end of the copper contact tube where the electrode picks up electrical energy for welding. Electrode extension is sometimes called ''electrode stickout''. The term ''electrode extension'' will be used in this text.

The proper electrode extension for gas shielded flux cored electrodes is usually in the range of 3/4'' minimum to 1 1/2'' maximum. The actual electrode extension will depend on a variety of things including electrode type and diameter and the position of welding.

The reason for having a specified amount of electrode extension is to preheat the flux within the tubular electrode. This permits the flux to decompose quickly under the heat of the welding arc and perform its function.

Manufacturers of self-shielded flux cored electrodes recommend from about 1'' to about 3 1/2'' of electrode extension for their particular products. Always read the manufacturers instructions and follow them for best results. The extra electrode extension length preheats the gasifying elements in the self-shielded electrodes so that shielding gas will form immediately as the elements reach the welding arc. This is very important for the self-shielded flux cored electrodes.

Excessive electrode extension can cause severe weld spatter, erratic penetration and fusion patterns, and an unstable welding arc. Gross porosity with some piping porosity would be one result of this condition. Uneven weld bead width and height may be expected when excessive electrode extension exists with flux cored electrodes.

If the electrode extension is insufficient, several unpleasant things will occur. The most significant will be a severe and rapid buildup of spatter in the gas nozzle of the torch or gun. This will, in turn, cause poor gas flow patterns around the molten weld puddle.

The result would be visible as ''worm tracks'' on the surface of the weld. In every case that I have seen, the narrow depressions referred to as ''worm tracks'' are eliminated by increasing the electrode extension to the correct length. This conclusion is based on there being a good welding condition otherwise.

When the electrode extension is insufficient look for lack of side wall fusion in the weld joint. There will probably be inadequate penetration also. Keeping the electrode extension within the recommended ranges eliminates a lot of welding defects and problems.

Watch that the hose and cable assembly from the wire feeder-control to the torch is not put into a severe bending stress. This will cause certain problems in wire feeding due to the restriction placed on the electrode by the bend in the liner. At the most severe the problem would manifest itself as a burnback of the electrode into the contact tube. This will immediately disprove the contention that copper and steel cannot be fusion welded together!

Shielding gas flow rates for all gas shielded flux cored arc welding should be set at a sensible level. A sensible level means that adequate shielding gas is provided without excessive flow rates. Flow rates will depend on the type of shielding gas used, the atmospheric conditions while welding, and the type of base metal welded.

Shielding gas flow rates should be sufficient to protect the molten weld puddle and the electrode tip from the outside atmosphere. Written welding procedures should always show the shielding gas flow rate used in setting up the procedure. It is important that the gas flow rate be maintained within 10% of the value on the welding procedure.

It is important that proper joint designs be used for any of the gas shielded welding processes. With the FCAW process, this includes the self-shielded electrodes. Many designers and engineers forget that the welder has to be able to get the welding gun close enough in the joint to make the weld.

The Air Carbon-Arc (AAC) Process

The air carbon-arc (AAC) cutting and gouging process is relatively simple in its method of operation. The primary requirements are a source of electrical power, a source of compressed air, a modified electrode holder that also directs the compressed air toward the arc, and carbon electrodes.

The air carbon-arc process requires that an electric arc be established between the carbon electrode tip and the surface of the base metal. The purpose of the electric arc is to melt the base metal surface in a specific area. Compressed air is directed from the electrode holder at the molten metal to literally blow it away from the puddle.

The molten metal is not burned as it is with the oxygen-flame cutting process on carbon steel. For this reason, the AAC process may be used with all metals regardless of their oxidation characteristics.

When used for grooving out metals, the AAC process leaves a groove that, if properly done, will be relatively smooth in appearance. The air carbon-arc process can be used for a variety of metal removal requirements including weld groove preparation, defective weld metal removal, casting riser removal and pad washing of foundry castings.

A specially designed insulated electrode holder is used for manual AAC work to channel the compressed air to the arc and to hold the carbon electrode. The AAC electrode holder is often referred to as a "torch" by welding people, probably because it cuts and gouges metals. Equipment is available for either manual or automatic operation of the AAC process.

Safety Considerations

Safety rules as set forth in the AWS/ANSI Standard Z49.1 "Safety In Welding And Cutting" are applicable to the air carbon-arc cutting and gouging process. In addition, special care must be taken against fire hazards due to the volume of molten metal being removed. Portable fire extinguishers and fire retardant screens will help contain the sparks and spatter within a safe area.

The person operating the AAC equipment should always wear hardened safety glasses with side shields. The shaded lens in the welding helmet should be at least a shade 12 when using carbon electrodes up to 3/8" diameter. For larger diameter carbon electrodes the number 14 shaded lens should be used. This will help combat the volume of bright light from the arc and reflected glare from the surrounding area. Protective non-flammable clothing is recommended for the AAC operator.

All hoses, welding cables, flammable materials, compressed gas cylinders and tools must be removed from the path of the molten metal stream. Severe injury to personnel, and/or damage to buildings and tools can occur from improperly operated air carbon-arc equipment.

Ventilation must be in accordance with the minimum air flow requirements of the AWS/ANSI Standard Z49.1. There is a possibility of toxic fumes from the copper coated carbon electrodes. In severely restricted or confined areas fresh air respirators should be worn by AAC operators and others working in the area. Exhaust systems must be provided for removal of smoke and toxic fumes.

Another possible hazard is the noise level developed by use of the air carbon-arc process. Some form of certified ear protection must be worn when using, or working near, the AAC process. Some degree of deafness can result from exposure to the noise level if personnel are not properly protected.

No portion of the welders body should be in water when using any electric arc welding process. This certainly applies to the AAC process where the carbon electrode is always energized. Serious injury or death could result from accidental grounding in the electrical circuit.

Finally, beware of accidentally making any contact between the base metal and the electrode. The carbon electrode is always electrically energized when the power source is turned on. The resulting arc could cause severe eye damage to anyone "flashed" by the arc.

It is always best to remove the carbon electrode before putting the AAC electrode holder down. Do NOT throw the AAC electrode holder down after use. It could be seriously damaged.

Basic AAC Equipment Required

Some of the basic equipment required for the air carbon-arc process is illustrated in Figure 204. This is a typical setup for manual AAC operation which includes an electric welding power source, a specially insulated electrode holder (shown), carbon electrodes, a compressed air supply and 3/8" ID air hose.

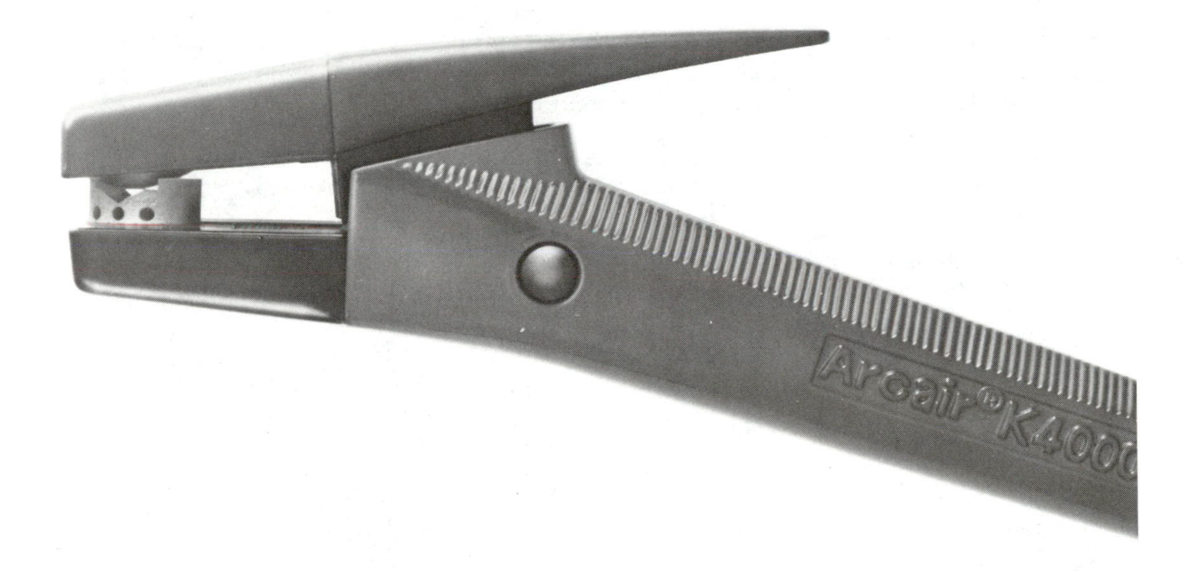

Figure 204. Basic Equipment for the Manual AAC Process.

The protective screens, personal protective clothing and equipment have already been mentioned under the "Safety" section. They are certainly part of the necessary equipment and materials required for operation with the AAC process.

Welding Power Sources

Welding power sources used with the air carbon-arc process may be either the constant current (CC) or constant voltage/constant potential (CV/CP) types. The minimum requirements for amperage, at 100% duty cycle, are based on the carbon electrode type and diameter. For the safety of both the welder and the power source used we recommend the following guide:

"The amperage rating of the welding power source at 100% duty cycle must not be less than the maximum current carrying capacity of the largest carbon electrode used".

The minimum open circuit voltage that is suitable will depend on the largest diameter carbon electrode that will be used with the power source. The arc voltages used with the AAC process range from about 27 volts for smaller diameter carbon electrodes to about 55 volts for larger diameter electrodes used with automatic units.

The minimum open circuit voltage required for using all diameters of carbon electrodes is about 74 volts with constant current type power sources. The minimum open circuit voltage required for using all diameters of carbon electrodes with constant voltage/constant potential power sources is about 65 volts.

Either alternating current (AC) or direct current (DC) may be used with the AAC process. It is well to remember that all AC and AC/DC welding power sources operate from single phase primary power. The DC welding power based on single phase primary power has lower average power levels than DC based on three phase primary power. For this reason DC based on single phase primary power is not recommended for use with the AAC process.

Welding Power Sources For The Air Carbon-Arc Process

Power Source Type	Output Current	Comments
Constant Current, 3 phase: Transformer-rectifier; Motor-generator set; Resistor-grid unit.	DC	Recommended for all carbon electrode types and diameters.
Constant Voltage, 3 phase: Transformer-rectifier; Motor-generator set; Resistor-grid unit.	DC	Recommended for carbon electrodes 1/4" diameter and larger.
Constant Current, single phase: Transformer.	AC	Use only AC carbon electrodes 3/16"-1/2" diameter.
Constant Current, single phase: Transformer-rectifier.	AC/DC	Use only AC output with AC carbon electrodes. Do not use DC output.

Figure 205. Welding Power Sources for the AAC Process

AAC Electrode Holders

There are three basic classifications of electrode holders for the air carbon-arc process. They are designed for manual operation, semi-automatic operation, or fully automatic operation. The classification of electrode holder selected will depend on the thickness of the base metal to be worked, the type of base metal, the diameter of the carbon electrode, and the position in which the work will be accomplished.

Manual AAC electrode holders are the most commonly used in the welding industry. They have a rotatable platform, or "head", which holds the electrode and has air passages built-in to direct high pressure compressed air at the arc area. Each of the various models available has a minimum-maximum carbon electrode diameter which they are designed to hold properly. It is important that the electrode holder is the proper size and rating for the carbon electrode to be used.

Semi-automatic units are relatively simple electrode holders mounted on some type of machine carriage. The carriages are often similar to those used with flame cutting track burner systems. The carbon electrode is fed manually by the welder as it is used in the arc.

Full automatic AAC systems are designed to maintain uniform and continuous grooves over long lengths to a tolerance of plus or minus 0.025'' with specified diameters of carbon electrodes. This type of system is voltage controlled and usually has solid state electronic controls.

Carbon Electrodes

Carbon electrodes are suitable for the air carbon-arc process because they ''sublime'' when they reach their melting point of 6,740° F. To define:

Sublime = ''To change from the solid state of matter to the gaseous state without the intermediate liquid state''.

When properly used the carbon electrode should never ''stub'' into the base metal. The resulting groove should be smooth and clean in most metals.

Carbon electrodes used for the AAC process are actually made of a mixture of carbon and graphite materials. There are three basic types of carbon electrodes for use with this process. They are:

DC copper coated carbon electrodes
AC copper coated carbon electrodes
DC plain, or uncoated, carbon electrodes

DC copper coated carbon electrodes are the most popular for use with the air carbon-arc process. The high purity copper coating helps provide longer carbon electrode arc life, better arc stability, and good groove dimensions. These carbon electrodes are available in diameters from 5/32'' to 1''. The copper coated carbon electrodes are designed to operate with direct current, reverse polarity (electrode positive).

AC copper coated carbon electrodes have rare earth materials in the basic carbon matrix. These elements help provide arc stability for the AC welding power. AC carbon electrodes are available in sizes from 3/16'' to 1/2'' diameters. These electrodes are designed to operate with alternating current or with direct current, straight polarity (electrode negative).

DC plain, uncoated carbon electrodes are normally limited to 3/8'' diameters or less. Due to their rapid decomposition under AAC arc conditions, they are not normally used for industrial AAC applications.

General Applications Data

The applications data on the following page is furnished by the Arcair Company, Lancaster, OH, and is used with permission. In all cases where DC carbon electrodes are specified the copper coated electrode is preferred.

Base Metal	Electrode	Current	Polarity
Carbon steel Low alloy steel Stainless steel	DC or AC	DC or AC	DCRP (DCEP) (electrode positive)
Gray cast iron Malleable iron Ductile Iron	DC or AC	DC or AC or DC	DCRP (high amp) DCSP (DCEN)
Copper alloys	AC DC	AC or DC DC	DCSP (DCEN) DCRP (DCEP)
Nickel alloys	AC	AC or DC	DCSP (DCEN)

Figure 206. AAC Electrical Conditions for Some Base Metals.

Air Supply and Hose Requirements

The air pressure required for proper operation of the air carbon-arc process equipment is between 80-100 psi. Smaller diameter carbon electrodes may be used with light duty electrode holders and compressed air as low as 40 psi. The minimum ID for air hose at the 80-100 psi pressures should be 3/8'' for manual and semi-automatic operations. The air hose ID should be 1/2'' minimum for full automatic systems.

AAC and Metallurgy

When properly applied, the air carbon-arc process will have only minor effects on most base metals. Any carbon deposits are usually caused by travel speeds that are too fast or by ''stubbing'' the carbon electrode into the base metal. When working cast or ductile iron it is common to produce a refractory oxide, or slag, at the surface of the groove being made. This oxide layer must be removed before the next AAC pass can be made. This may be done with a slag hammer or multineedle scaling tool.

Air pressures that are below the minimum required by the AAC process will also cause carbon deposits in the base metal. The effect of the carbon will depend on the type of base metal being worked. This will be considered in the section titled ''AAC Process Applications''.

There are no particular problems with machining low carbon steel, or other metals that are not hardenable by heat treating, after they have been worked with the AAC process. Carbon steels with medium or high carbon content, and cast irons, may be non-machinable at their surfaces where the AAC process has been applied.

Since the depth of the hardened layer is only about 0.006''-0.008'', it is relatively easy to set a cutting tool to cut below those depths. For preparation prior to welding, it may be necessary to remove the hardened zone by grinding.

As a general practice, it is correct to recommend grinding a groove joint that has been prepared by the AAC process prior to welding. The depth of the grinding can normally be not more than 0.010''. For some metals, such as aluminum and magnesium, vigorous brushing with a clean stainless steel wire brush will often be sufficient joint surface preparation for welding. It would be preferable, however, to clean the grooved area with a dry router tool before welding.

AAC Process Variables

The variables of the AAC process are relatively few in number but very important to the proper operation of the process. Any considerations must also include factors that are not specifically part of the process. These include the type of base metal, the groove design required, the experience of the welder-operator, and the position of the work to be done.

Actual AAC process variables include the following: the welding power source type (CC or CV/CP), power source output (AC or DC), carbon electrode classification, carbon electrode diameter, carbon electrode shape (round, half-round, flat), speed of travel, finish quality requirements of the groove, and the rating of the electrode holder used. A significant variable is the amperage range within which each diameter of carbon electrode operates. The minimum-maximum amperages for various carbon electrode diameters are shown in Figure 207. It is always best to operate at or near the maximum electrode amperage for superior results.

	Carbon Electrode Diameters In Inches									
	5/32	3/16	1/4	5/16	3/8	1/2	5/8	3/4	1	flat
Min. Amps.	90	150	200	250	350	600	800	1200	1800	300
Max. Amps.	150	200	400	450	600	1000	1200	1600	2200	500

Figure 207. Carbon Electrode Recommended Amperage Ranges.

AAC Process Operating Procedures

It takes skill and common sense to properly operate the air carbon-arc process and equipment. Training is required to permit a welder to learn how to use the AAC process. Proper instruction by competent Instructors will reduce training time to the minimum. Most welders can learn to operate manual and automatic AAC equipment properly in one week of training.

Equipment Preparation

Friends, if the air carbon-arc process equipment isn't set up and connected correctly, the process cannot function the way it should. The basic considerations presented in the following paragraphs will help you get started on the right track! We are assuming that the power source is connected to primary electrical power and that a source of compressed air is available.

1. Connect electrode and work welding cables to the power source output terminals. Tighten connections firmly.
2. Connect the welding electrode power cable to the AAC electrode holder cable assembly. Connect the work cable to the base metal.
3. Connect the air pressure hose to the AAC electrode holder assembly.

4. Insulate the AAC cable and hose connections with the insulating ''boot'' supplied with the AAC unit.
5. Connect the air pressure hose to a source of 80-100 psi air pressure. An air pressure gage and regulator is recommended to regulate the flow pressure. A water vapor trap in the air supply line would be advisable.
6. Select the correct diameter and type carbon electrode.
7. Set the welding power source output to match the maximum amperage requirements of the carbon electrode diameter to be used. REMEMBER: the electrode can draw up to the maximum current it is designed to carry from the power source. Be certain the maximum current required is within the 100% duty cycle capability of the power source.
8. Insert the carbon electrode in the electrode holder. Carbon electrode extension should be not more than 6'', and not less than 3'', beyond the clamping head of the electrode holder. The air jets in the rotatable head should be pointing at the arc end of the electrode. The compressed air jet should hit just behind the arc when it is initiated.
9. Double check all electrical and air connections to be sure they are tightened firmly.
10. The welding helmet lens shade should be a number 12 at the minimum. A shade 14 is preferred for larger diameter carbon electrodes. It is recommended that safety glasses with number 2 shaded lens and side shields be worn under the welding helmet. This will protect the operator from reflected glare and metal sparks inside the helmet. Adequate ear protection should be worn when working with the AAC process.
11. ALWAYS TURN ON THE ELECTRODE HOLDER AIR VALVE BEFORE INITIATING THE CARBON ARC!
12. Be sure that all flammable materials have been removed from the area where the AAC work is to be done. Place protective screens or shields so that any molten metal spatter is restricted to a safe area. Observe all safety rules.

If the air carbon-arc cutting or gouging is to be done in an area where there is still some fire danger, post a fire watch while working.

AAC Basic Use Instructions

1. Turn on the welding power source and set it for the carbon electrode to be used.
2. Turn on the air supply and adjust air pressure to the minimum 80 psi.
3. Insert the carbon electrode in the electrode holder. Electrode extension should be about 6'' beyond the rotatable head.
4. Turn on the air at the electrode holder before striking the arc. Be sure the air stream is hitting just behind the carbon electrode.
5. Initiate the arc by gently touching the carbon electrode tip to the base metal surface. DO NOT DRAW THE ELECTRODE TIP BACK ONCE THE ARC IS STARTED.
6. Maintain a short arc length at all times. Do not touch the carbon electrode to the work after arc initiation. Touching the carbon electrode to the work will cause carbon deposits which are detrimental to the base metal groove quality.

7. The angle of the electrode and speed of travel must be suitable to maintain steady metal removal. Stable travel speeds provide the smooth groove dimensions desired.
8. The sound of a good AAC arc condition is a continuous hissing sound. If you hear a stuttering, interrupted arc sound you can believe the result will be an uneven scalloped groove that will either have to be re-grooved or finish ground out before welding.
9. Maintain the correct carbon electrode angle for the base metal and the working position. A steep electrode angle and slow travel speed will normally produce a deep, narrow groove. A low electrode angle and fast travel speed will produce a relatively wide, shallow groove.
10. The average width of a groove is usually about 1/8''-3/16'' wider than the carbon electrode diameter. A wider groove can be made by using a slight weaving motion with the electrode.
11. The most efficient amperages for AAC carbon electrodes are in the top 20% of their current carrying range. It is always best to use the carbon electrodes at amperages that will produce the highest efficiency of metal removal. For example, a 1/4'' diameter carbon electrode has a current range of 200-400 amperes. It would be most efficient in metal removal in the 320-400 ampere range.
12. The lead angle of the carbon electrode affects the amount of cross sectional area exposed to the arc. The greater the cross sectional area, the higher the amperage demanded from the power source. By controlling the carbon electrode diameter, the amount of cross sectional area can also be controlled. In this way the maximum amperage demand from the power source can be limited.
13. The direction of travel for AAC gouging is a matter of choice for the welder. There must be adequate clearance for the physical movement of the AAC electrode holder along the joint.
14. For vertical operations the carbon electrode should be pointed downward. This will allow the molten metal to follow the natural pull of gravity as it is removed.
15. When cutting or gouging with the AAC process it is necessary to hold the electrode so that excess cross sectional area is not exposed. For example, some cutting operations are done with the carbon electrode held at right angles to the direction of travel and to the base metal surface. It is necessary that the compressed air flow between the carbon electrode and the base metal for satisfactory metal removal. The cutting technique can expose a considerable amount of electrode cross sectional area. It is the responsibility of the AAC operator not to exceed the amperage rating of the welding power source used. Use caution in how the electrode holder, and electrode, is held relative to the base metal.
16. For overhead grooving operations hold the carbon electrode at an angle which will avoid molten metal splattering on the welder.

It will be to the welders advantage to practice with the AAC process before using it on production work. Observe all the general safety rules for electric arc welding. Wear protective clothing, safety glasses with side shields, and ear protection. If the work is in a confined area, use a fresh air respirator.

AAC Process Applications

Carbon Steels

Low carbon and mild steels may be worked with the AAC process following the general rules outlined in the previous paragraphs. There are no special techniques or requirements for these materials.

Cast Iron, Ductile Iron

It is required that DC carbon arc electrodes be used at the highest amperages they can accommodate when working with cast or ductile iron. A very short arc length must be held during the grooving or cutting operation. If the amperage is too low, or the arc length too long, a refractory oxide or slag may be produced on the surface of the groove. This oxide layer must be removed before further AAC work can be done.

AC copper coated carbon electrodes can be used with either AC or DCSP (electrode negative) welding power for grooving ductile and cast irons. This type of electrode and current will create a very clean groove surface.

Stainless Steel (300 series)

All of the austenitic stainless steel alloys can be easily cut or grooved with the AAC process. Distortion may be minimized by fast travel speeds. Always be sure the compressed air jet is directed behind the arc when cutting or gouging. Always turn on the compressed air **before striking the arc.**

Copper and Copper Based Alloys

Copper based alloys are very difficult to groove with the AAC process because of their very high thermal conductivity. DC copper coated carbon electrodes may be used with some success if the base metal is preheated to retard heat losses. The amount of preheat will depend on the type and thickness of copper alloy to be grooved. For some applications DC electrodes have been used with DCSP (electrode negative) with good results.

AC copper coated electrodes are more effective for grooving and cutting as the copper content of the alloy increases beyond about 65%-70%. Again, preheating the base metal to suitable temperatures will assist the grooving operation. In all cases, the preheat must be maintained during the grooving operation or the AAC condition will deteriorate with loss of heat from the base metal.

Aluminum and Aluminum Alloys

Aluminum and aluminum alloys are difficult to groove under the best conditions. The heated aluminum forms dense refractory oxides that make continued operation with AAC almost impossible. The material must be thoroughly cleaned after each pass. DC copper coated carbon electrodes are used with DCRP (electrode positive).

A work angle of approximately 20-25 degrees will provide the smoothest grooves if the speed of travel is proper. In no circumstances should the carbon electrode touch the aluminum during the AAC operation. The carbon would leave a deposit that will seriously interfere with the balance of the grooving operation.

Magnesium and Magnesium Alloys

Magnesium and magnesium alloys may be worked with the AAC process in a manner similar to carbon steel. The major difference is that the speed of travel will be considerably faster with the magnesium and the arc flare will be more pronounced. A fire watch is recommended.

Additional data may be obtained from the manufacturers of the specific metal alloys with which you may be concerned.

Summary

The air carbon-arc method of gouging and cutting is an excellent process for preparing new weld joints. It is also very good for removing old welds or defective and fatigued metal from weld joint areas.

Proper safety precautions must always be taken when using the AAC process. Post fire watch personnel anytime there is possibility of fire due to AAC operations. Shield the area from indiscriminate blowing of molten metal by putting up deflecting shields to stop the molten spatter.

The AAC operator must be supplied with ear and eye protection adequate for the application. Safety is always the first consideration!

CHAPTER 21

Power Sources for the GMAW and FCAW Processes

The welding power used with the gas metal arc welding process is direct current, reverse polarity (electrode positive) according to the process patent issued in 1950. When the process was first developed in 1948 the only DC welding power sources available were rotating armature design units known as DC generators. They were either electric motor driven generators called MG sets or internal combustion engine driven generators. In 1950 the transformer-rectifier welding power source was first presented to the welding public.

Both the electro-mechanical generator type equipment and the static transformer-rectifier type equipment were designed as constant current units. This type of welding power source was specifically designed for the shielded metal arc welding (SMAW) process. It may also be used for other welding processes such as gas tungsten arc welding (GTAW).

Although inherently different in their design concepts both types of welding power sources have the same general output characteristics. The constant current (CC) type power sources have a fixed maximum open circuit of up to a nominal 80 volts. The maximum short circuit current is limited to approximately 150% of the amperage rating of the power source. A typical volt-ampere output curve for a three range power source is illustrated in Figure 208.

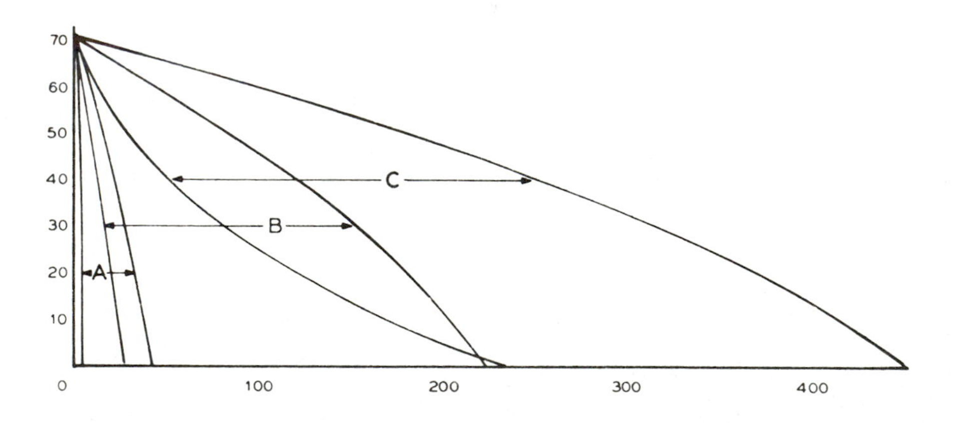

Figure 208. Typical Constant Current Volt-Ampere Curve.

The development of the transformer-rectifier type welding power source was a great improvement for the GMAW process because of the faster rate of response of the unit to changing arc conditions. The motor-generator (MG) welding power sources have an inherently slow rate of response due to the mechanical inertia of the rotating mass of iron and copper conductors. This inertial effect is often termed the "flywheel effect".

For example, a constant current transformer-rectifier type power source has an arc response time two to three times faster than a constant current motor-generator unit. Some benefits of faster response time include better arc starting, decreased stubbing of the electrode at weld initiation, and less cold lapping defects at the beginning of the weld. A comparison of typical motor-generator and transformer-rectifier power source response times is shown in Figure 209.

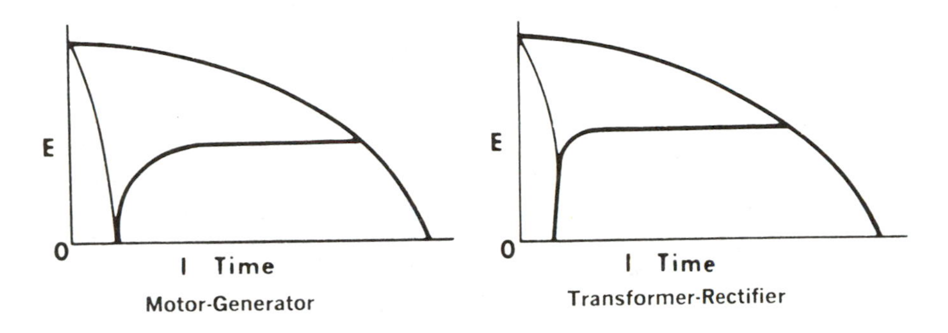

Figure 209. Relative Response Time Curves.

The constant current type DC power sources were all that was available for use with the gas metal arc welding process from 1948 until 1953. The constant voltage/constant potential (CV/CP) type power sources were first displayed to the welding public in 1953.

Constant voltage type welding power sources have a relatively flat volt-ampere output characteristic curve. Their maximum open circuit voltage is normally less than a similarly rated constant current unit. Due to their output characteristics constant potential type power sources have an extremely high maximum short circuit current. This is illustrated in the typical volt-ampere curve shown in Figure 210.

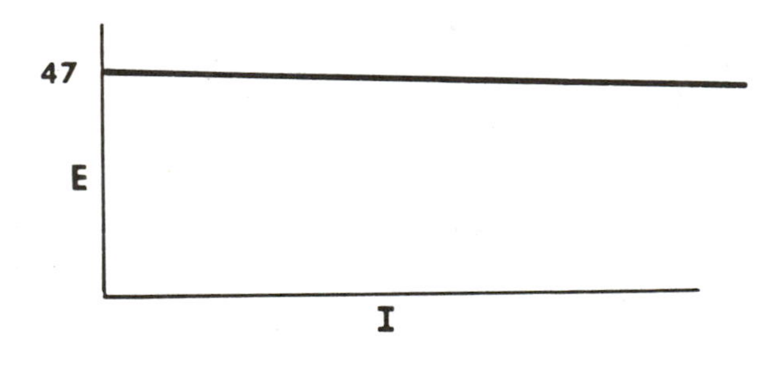

Figure 210. Typical Constant Voltage Volt-Ampere Curve.

If the CV/CP type power source is ever allowed to reach maximum short circuit current, the resistance heating of the conductor coils would be almost instantaneous and catastrophic to the insulation. This would lead to the short circuit destruction of the main transformer within a matter of seconds.

Constant Current (CC) Power Sources

Since constant current type DC power sources were the first units used with the gas metal arc welding process, they will be discussed before proceeding to the constant voltage/constant potential type power sources. This type of welding power source was first developed for use with the SMAW process and flux coated electrodes.

The wire feeder-controls that were developed for use with the constant current type power sources had voltage sensitive drive motors for wire feed speed control. The higher the voltage signal, the faster the motor would run. The lower the voltage signal the slower the drive motor would run.

A voltage signal lead was attached to the workpiece from the drive motor of the wire feeder system. As the arc length became shorter, for example, the arc voltage became lower in value and the wire feed speed would slow down. If the arc voltage became higher because of an increased arc length, the wire feed speed would increase accordingly.

It is obvious that the initial setup of power source and wire feeder-control had to be done with great care to balance out the welding system. It was difficult to obtain a good welding condition even under the best of circumstances. A significant problem with constant current power sources and the GMAW process is the high open circuit voltage and limited maximum short circuit current. This creates the situation of having substantial voltage changes with very little amperage change. The welding amperage range was quite restricted at any setting.

This problem became moot with the development of constant voltage/constant potential type power sources in 1953. At that time the equipment manufacturers changed to a constant speed drive motor for wire feeder-control units.

The constant current type welding power source used with the GMAW process is normally a three phase transformer with some type of rectifier to change the AC to DC. The standard constant current power source design has loose magnetic and inductive coupling in the primary-secondary coil relationship of the main transformer. **Loose magnetic coupling** means that a physical air space exists between the primary coil and the secondary coil. This situation may also be termed "loose inductive coupling" of the coils.

Air is a good resistor to electrical current flow. In addition, **air has poor magnetic permeability.** With the air space between the primary and secondary coils of the main transformer, the magnetic field created in the transformer iron core does not transfer power as efficiently to the secondary coil. This is a designed method of making the constant current power source electrically inefficient. The limited maximum short circuit current necessary for the SMAW and GTAW processes to function properly is assured by this design concept.

It is possible for the welder to make slight changes in both voltage and amperage while welding with constant current power sources. This is done by changing the arc length and, therefore, the arc voltage and amperage relationship.

In the illustration, Figure 211, the only variable shown is the arc length. There is a correlation between arc length and arc voltage. The greater the air space distance between the

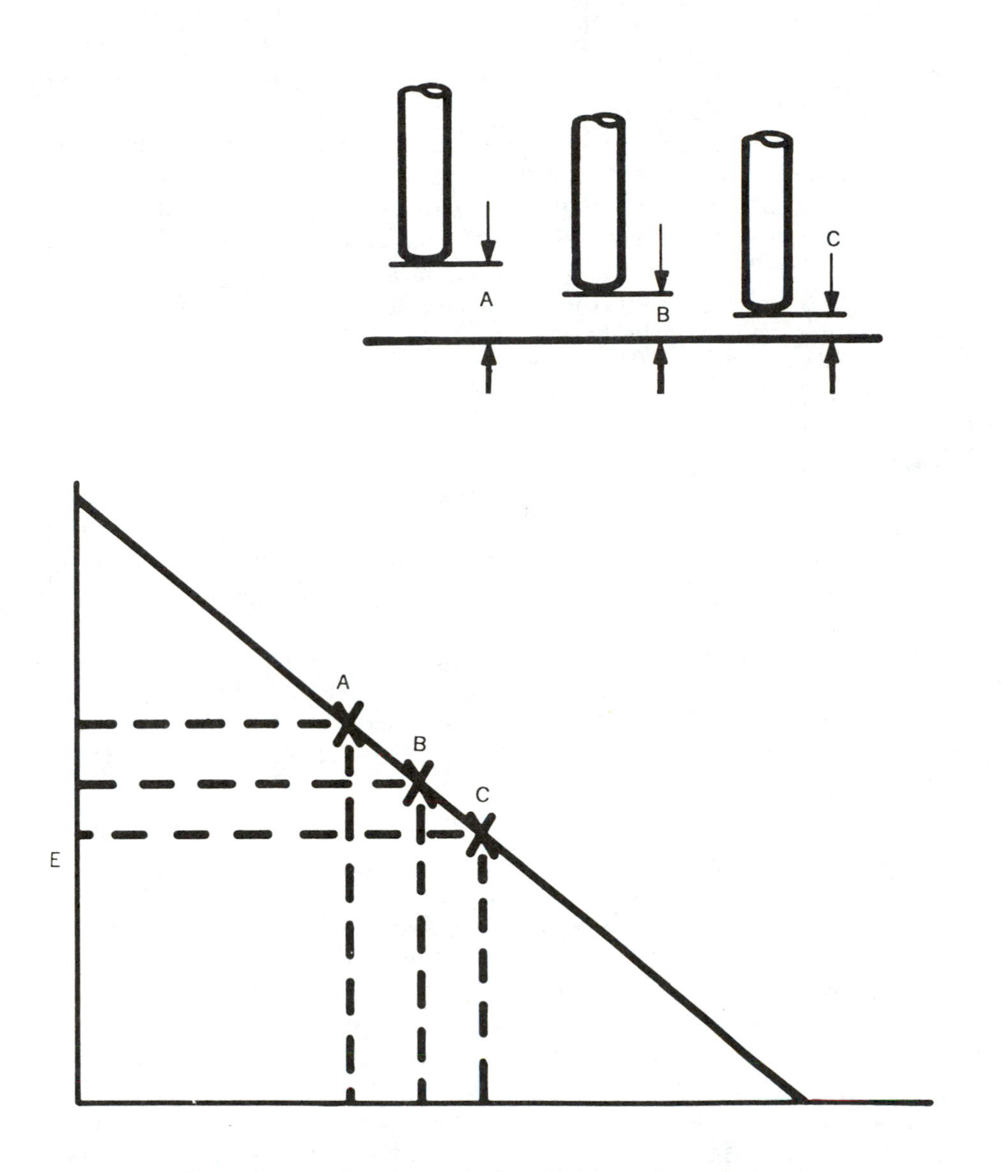

Figure 211. Arc Length-Arc Voltage Correlation

electrode tip and the surface of the base metal, the greater the electrical resistance between the two points. The electrical force that can overcome electrical resistance is voltage.

Voltage is the electrical force that causes amperage to flow in an electrical circuit. It is logical that more voltage would be required to move amperage across the greater arc length (air space). Conversely, as the arc length decreases, arc voltage would also decrease.

In the upper right hand part of the drawing there are three electrodes with different arc lengths. They are labeled "A", "B", and "C". The volt-ampere curve at the lower left side of the drawing also has the three points "A", "B", and "C" shown at various places on the curve. This is how they were determined. Keep in mind that an electric welding arc can be maintained in all three welding conditions.

Begin with the center electrode with "B" arc length. There is some measure of electrical resistance between the electrode tip and the base metal surface. The voltage necessary to overcome the electrical resistance is furnished by the welding power source as "arc voltage".

This is shown as the dotted line **E-B-I** on the volt-ampere curve. For this discussion, we can consider position "B" on the volt-ampere curve as the optimum welding condition. The voltage value represented by E, and the amperage value represented by I, combine to make a balanced welding heat input to the joint.

Now look at the position "A" electrode arc length. There is a longer arc length which means greater electrical resistance across the arc. The increased electrical resistance means that more arc voltage is necessary to sustain the welding arc.

The volt-ampere curve shows a dotted line **E-A-I.** The voltage is higher than at B but the amperage is lower. It is apparent that a longer arc length decreases amperage slightly but increases voltage slightly.

Examining the electrode "C" arc length shows a close, tight arc. There is less arc length and less electrical resistance. The voltage necessary to sustain the welding arc is substantially less than at point "A".

The dotted line **E-C-I** indicates a lower arc voltage but a greater amperage value than at either points "A" or "B". This examination of the various arc lengths and arc voltages does indeed show a specific correlation between the two. It also illustrates that a welder has some latitude in changing the voltage and amperage of the arc while welding with certain processes.

The CC Power Source and Aluminum

It is a historical fact that aluminum was the first metal to be welded with the process we now call Gas Metal Arc Welding. As a matter of fact, the development of the process was required for welding aluminum plate in thicknesses beyond the capacity of the gas tungsten arc welding process. Of course, subsequent developments with DCSP (electrode negative) and helium shielding gas has made heavy aluminum plate welding with the GTAW process commonplace. This was not true in the late 1940's.

Aluminum has a relatively low melting point (1218° F.) and consequently a relatively high melt rate as an electrode material. The metal cannot tolerate high current surges at the electrode tip because of the explosive force that would be generated. The result would be excessive spatter, poor weld quality, and severe surface damage to the base metal. This combination of factors made it almost mandatory that the power sources used with the gas metal arc welding process on aluminum alloys have a limited maximum short circuit current and/or a slower rate of response to welding arc conditions.

It was a stroke of good fortune that the power sources available in 1948 were just what was needed to weld aluminum. They were, and still are, good power sources for spray transfer welding with the GMAW process. Subsequent development of constant voltage power sources with "slope" and "inductance" controls have made aluminum welding with the GMAW process as commonplace as steel welding.

It is interesting to note that the recommendations of various wire feeder-control equipment stated that 3/16" thick aluminum was the minimum thickness of aluminum that could be welded with the GMAW process. All of the present models of wire feeder-controls designed for use with aluminum are capable of welding much thinner gages of aluminum using todays modern techniques.

Conventional constant current power sources may be used with most of the hard surfacing wire feeder-controls presently used in the welding industry. This type of wire feeder-control usually has a voltage sensitive drive motor system. The maximum welding amperage applicable seems to be about 400-450 amperes. The concept of hard surfacing is, of course, to have minimum dilution of filler metal with the base metal. The lower amperages used with the process promote this concept.

In conclusion of this part of the text we can say that, for gas metal arc welding and flux cored arc welding, the constant current type power source is not the best unit available. The preference of the welding industry is for the constant voltage/constant potential type welding power sources.

Constant Voltage/Constant Potential Power Sources

The terms "constant potential" and "constant voltage" are used by various manufacturers to describe a type of welding power source. In truth, the terms "voltage" and "potential" are synonymous in electrical terminology. It must be understood, however, that all power sources have some internal electrical resistance. The resistance will cause some deflection of the output volt-ampere curve so that it is not practical to have a true constant voltage power source. This chapter will refer to them as **constant voltage** or **constant potential type (CV/CP)** power sources.

The first constant voltage type power sources had a relatively flat volt-ampere output characteristic. The voltage drop per hundred amperes of output was only about one and a half volts. The maximum open circuit voltage was 48-50 volts with a mechanically adjustable open circuit voltage capability.

These power sources had relatively unlimited maximum short circuit current which could reach two to three thousand amperes easily. That is why all manufacturers of constant voltage type power sources have recommended that **large stick electrodes** not be used with the CV units. If the stick electrode makes short circuit contact with the base metal, the response of the power source is so fast it will reach maximum short circuit current before the electrode can be broken loose.

Immediately there would be several thousand amperes flowing through coils designed to carry only a few hundred amperes. The resistance heating of the conductor coils would be almost instantaneous, the coil insulation would be destroyed, and the coils would short circuit to the main transformer iron core. This would be apparent at the time by a quantity of smoke and a brilliant blue electrical flash from the interior of the power source. The main transformer has been electrically demolished!

CV/CP Transformer Design

The constant voltage/constant potential type power source has a much different main transformer design than the constant current welding power source. The primary-secondary coil relationship has very tight magnetic and inductive coupling. In fact, the primary and secondary coils are frequently wound together as one coil. (Interleaving and interwinding is also done. The cost of such coil configurations is considerably more and it is usually only used with coils carrying 750 amperes or more).

The main transformer of a typical constant voltage type welding power source is illustrated in Figure 212. Note the close relationship of the primary and secondary coils. It is the absence of air space between the two coils that provides the close magnetic and inductive coupling.

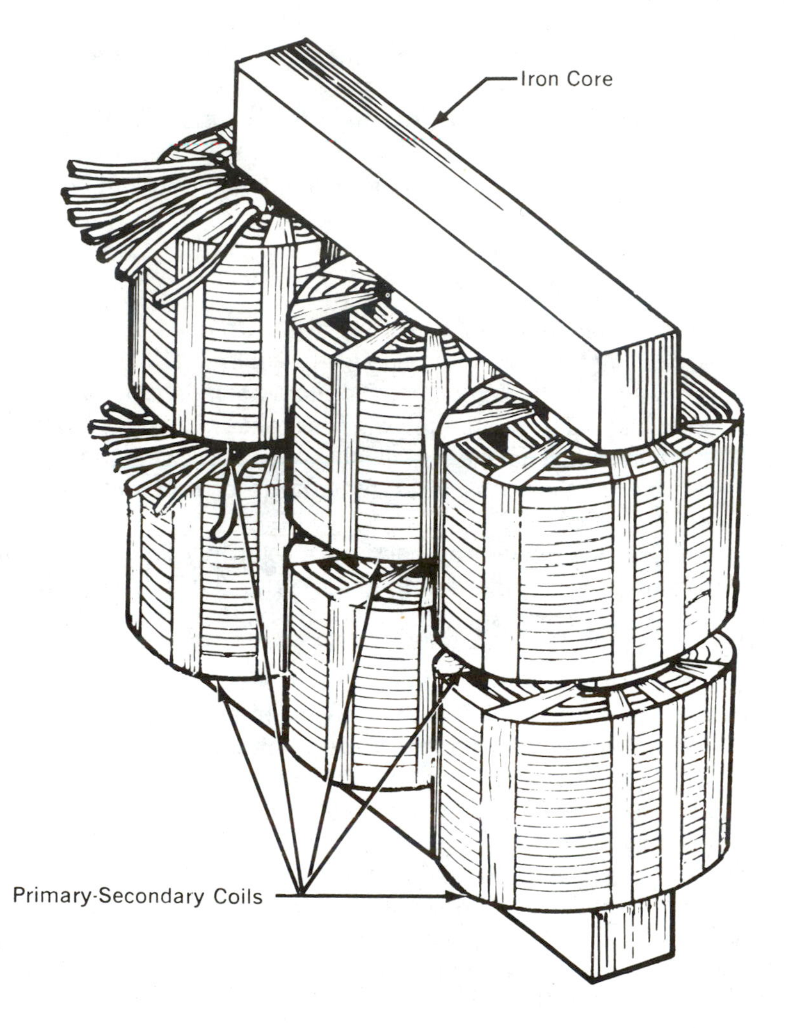

Figure 212. A Typical Constant Voltage Transformer.

Either the primary coil or the secondary coil may be wound to the outside of the total coil. Many of the CV/CP type power sources have mechanical controls for voltage adjustment. In most cases the control is achieved by moving a set of carbon-type brushes over the coil face which has had the insulation removed from the face surface for this purpose. The insulation is removed by machining.

There are a variety of electric controlled and electronic controlled CV/CP power sources available. This type of power source has a fixed open circuit voltage. The output load voltage is set by the control rheostat in most cases. Some of the solid state electronic power sources use the micro-electronics to perform duties previously done with heavy iron cores and power coils. Most solid state electronic controlled power sources use silicon controlled rectifiers (SCR) for both main power rectifiers and for welding contactors.

Amperage Control

Constant potential type power sources have no facilities for amperage control. Their name is indicative of the fact that they will supply amperage as required to maintain a preset load voltage. The actual setting of load voltage depends on the open circuit voltage, the slope setting, and the amount of wire feed speed set on the wire feeder-control.

For example, the welding operation shown in Figure 213 has no value shown for voltage or amperage. The shape of the volt-ampere curve would be determined by the open circuit voltage and the slope setting, if used, on the power source.

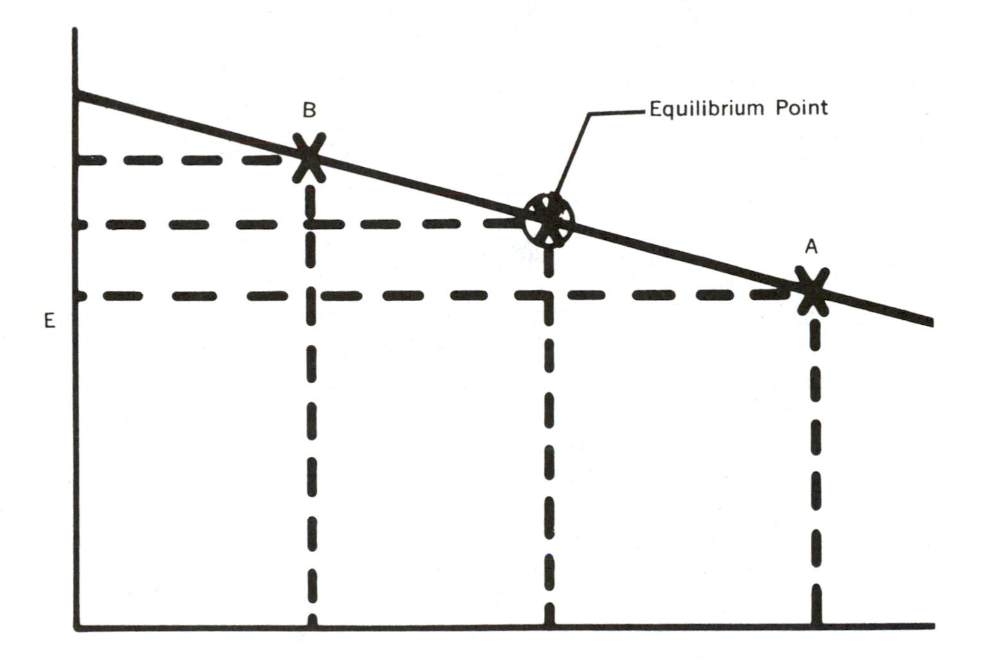

Figure 213. Automatic Current Control, CV/CP Power Sources.

At the point where the wire feed speed and the electrode melt-rate are exactly equal (the equilibrium point), there is a circled "X". The arc voltage is indicated by the dotted line going to the vertical ordinate of the graph (E). The welding amperage is shown by the dotted line going to the horizontal axis of the graph (I).

Remember that arc voltage is determined by arc length. Any variation of arc length will change arc voltage. For a given set of welding conditions this will also adjust the amperage available at the electrode tip. The result is a change in the melt rate of the electrode wire.

At point "A" in Figure 213 there is indicated a lower arc voltage and higher amperage than at the equilibrium point. This could occur only if the welding arc length had been shortened. Momentary shortening of the arc length could be caused by several things including welding over a tack weld.

The result of the lower voltage is higher amperage as shown in the illustration. At this instant of time the welding power source is furnishing considerably more welding current to the arc than it did at the equilibrium point although the wire feed speed remained the same.

Of course, the electrode melt-rate is increased. The higher melt-rate increases the physical arc length and raises the arc voltage. In this manner the welding condition returns to the equilibrium point pre-set on the welding power source and the wire feeder-control system.

Important: As long as none of the other welding condition variables are changed the welding power source will produce output amperage and welding voltage along the plotted volt-ampere curve. This factor causes the volt-ampere relationship to slide up, or down, the volt-ampere curve according to the physical conditions of the welding arc.

If the arc voltage is raised due to the arc being lengthened the state of equilibrium is once again lost as shown at point "B" of Figure 213. The lower welding current is less than that required to maintain a proper melt-rate of the electrode. The electrode therefore drives down to shorten the arc length until once again the point of equilibrium is reached.

GMAW Power Source Summary

From an electrical standpoint it should be noted that all constant current type power sources have inherently **high impedance** transformer design. The impedance is basically provided by the physical air space between the primary and secondary coils of the main transformer. The resulting volt-ampere curve has a steep negative angle between the maximum open circuit voltage and the maximum short circuit current. This type of power source is often called a "drooper".

Constant voltage/constant potential type power sources naturally have a relatively flat volt-ampere output curve. They also have much lower open circuit voltage than CC type power sources. These characteristics are achieved by using a low impedance main transformer design. This simply means there is no air space between the primary and secondary coils of the main transformer. Constant voltage power sources were especially designed for use with the gas metal arc welding process. They are also used for other welding and cutting applications including the FCAW process, the AAC process, the SAW process and others.

Voltages

There are three types of voltages with which we are concerned in welding. They are open circuit voltage, load voltage, and arc voltage.

Open circuit voltage is determined when the welding power source is energized but under no welding load. It is measured at the output terminals of the welding power source with a voltmeter.

Load voltage is also measured at the output terminals of the power source but while the unit is under welding load. The total voltage load affecting the power source is the *load voltage.*

Arc voltage may be measured only at the welding arc. It can actually be set only at the welding arc under welding conditions. Special voltmeters are used to determine arc voltage. For most welding applications it is load voltage that is read on the meters of power sources and shown as "arc voltage" for welding procedure writing.

It is open circuit voltage that is normally set on constant voltage type power sources having mechanical controls. The load, and arc, voltages are determined by other variables in the welding condition. Although the methods of adjusting open circuit voltage may vary for different models of constant potential type power sources the results will always be the same as long as the only variable changed is open circuit voltage.

The electric, and electronic solid state, controlled CV/CP power sources normally have a fixed open circuit voltage. The rheostat actually sets a volt-ampere output characteristic. Adjustment of the wire feed speed control produces load voltage and arc amperage. The basic reason for having electric control is to achieve remote control capability for the welder.

Slope Control

The term "slope" is probably confusing to the person who is trying to understand the control systems on some constant voltage power sources. The first use of the term "slope" was with the gas tungsten arc welding process. The terms "upslope" and "downslope" mean an increase or decrease in the welding current values when welding with the GTAW process. In particular the term "sloping off" means a decrease in the welding current at the end of a GTAW weld for the purpose of crater filling to eliminate crater cracking.

When discussing slope for the gas metal arc welding process (GMAW) it must be understood that it is the shape of the static volt-ampere curve to which we refer. The basic CV/CP volt-ampere curve is flat and has very little slope in it. The volt-ampere curve of the constant current power sources has a lot of slope.

There are two practical methods of putting slope into a constant potential type power source. It may be done by putting some form of resistance in the power source circuit or by putting a reactor system in the secondary AC portion of the power source circuit. Both methods will be examined and their effect on the power source output determined.

Resistance slope control will modify the shape of the static volt-ampere curve but it does nothing else. Usually the resistors are some type of highly resistive metal such as nichrome (a nickel-chromium alloy) or something similar.

This type of slope control is most often used on motor-generator (MG) sets. It is normally a three tap system with minimum, medium, and maximum slope setting indicated on a three position range switch. Resistance slope control is fine for an MG set because the unit has inherently slow response time to changing arc conditions.

Reactor slope control is another thing entirely and it is the system used by manufacturers of most static welding power sources having slope control. In this case, slope is actually caused by impedance and is usually created by the addition of a substantial amount of inductive reactance to the power source circuitry.

The electrical device used for slope control in most constant potential type power sources is a **fourteen turn slope reactor.** (Remember: an electrical turn is one wrap of the conductor wire around the periphery of the coil. A fourteen turn coil would have fourteen physical wraps of the conductor wire around the iron reactor core). More or less slope in the welding power source circuitry could be controlled by varying the amount of reactance in the circuit. Keep in mind that **a reactor inherently opposes change in the welding power circuit.**

The term "impedance" was used in the foregoing paragraphs. Impedance means to slow down but not stop a quantity of something. In this case it is something electrical. There are just two possible values in an electrical circuit which might be impeded and they are voltage and amperage. Since voltage is a force that causes current to flow in an electrical circuit, but which does not flow itself, the amperage flowing in the circuit must be the impeded quantity.

To explain how impedance occurs in this circuit we must investigate the reactor as it is used for slope control. A reactor consists of one or more current carrying coils placed around an iron reactor core. The core material is thin gage electrical steel, insulated on both sides, and is the same quality and thickness used for transformer iron cores.

A variable reactor iron core is illustrated in Figure 214. The slope control shown is typical of some mechanically controlled constant voltage welding power sources. It is usually called "variable slope control" because of the sliding brushes on each reactor coil.

The coils illustrated have 14 electrical turns and are especially designed for slope reactors. One side of the coil insulation has been removed by machining after which the bare copper conductor is polished and mechanically silverplated. Carbon contact brushes move over the face of the coil to provide more or less reactance in the welding circuit. It is the number of **effective electrical turns** in the reactor coil that determines the impedance factor of the reactor.

You may have noticed that there were six individual coils on the main transformer in Figure 212. There are also six coils illustrated in Figure 214 which is the slope reactor. Since both units operate in three phase alternating current circuits, the logical question is, "why six coils on these components?" The answer is quite simple.

The coils that are located one above the other on a single leg of the transformer and reactor cores are in electrical series. This means they function as one coil in each phase. The reason for series stacking of the coils is basically for design convenience. One single coil would make the power source twice as wide as it is thus requiring more floor space. Actually, the power source operates better with the coils in the series configuration.

The Reactor for Slope Control With GMAW

The reactor is normally located in the secondary AC portion of the power source circuitry. It has welding amperage applied through the reactor coils which are wrapped around part of the reactor iron cores. When the alternating current is applied to the reactor coils, a magnetic field is developed. Remember that magnetic field strength depends on three things:

1. The mass and type of iron in the core.
2. The number of effective electrical turns in the coil.
3. The alternating current values flowing in the coil.

It is logical that changing any one of the factors listed would also change the strength of the magnetic field. By making a step-by-step evaluation of the reactor components we can determine their function and the reaction on the other parts of the welding circuit.

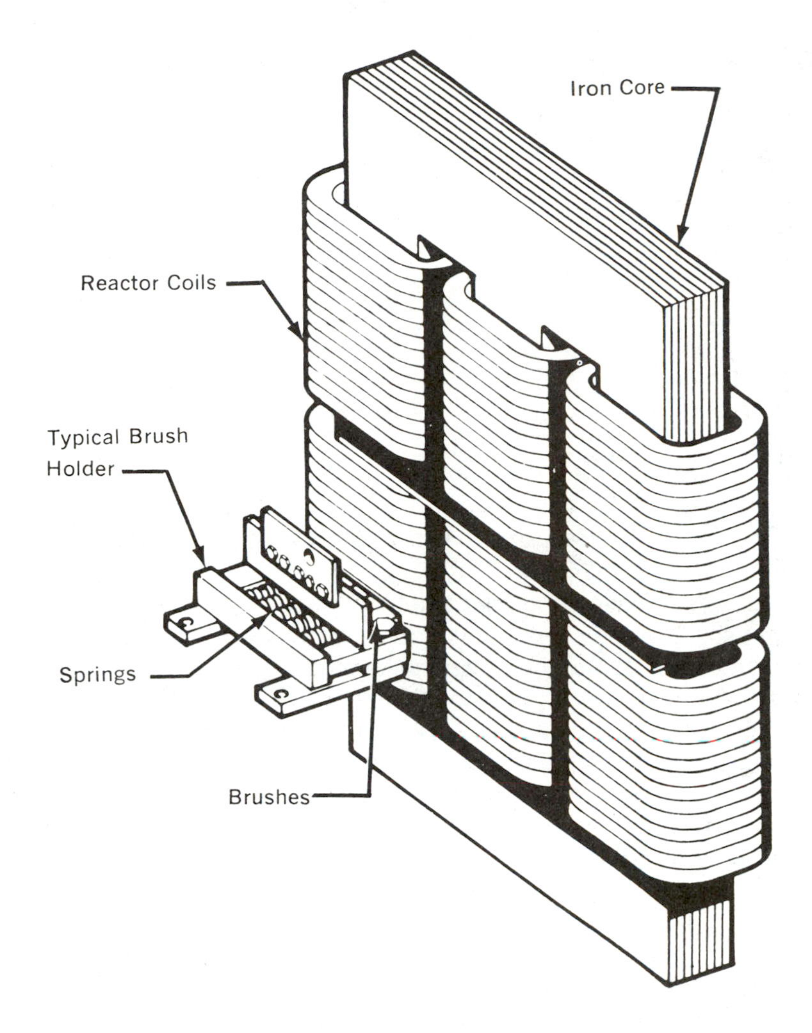

Figure 214. Three Phase AC Slope Control Reactor.

In Figure 215 there is illustrated a profile view of a typical slope reactor coil and iron core leg and a volt-ampere curve. We will use these devices to illustrate how the slope reactor works, what it does to the output amperage, and why it does these things in the circuit.

Consider, please, that the coil profile is just one of the six coils shown in Figure 214. The arrow is pointing to the section of the coil which has had the insulation removed. We will arbitrarily begin with 4 turns of slope in the reactor.

The welding current flows in the reactor coils (only four turns are effective although this is a 14 turn coil). The coil is physically located around one leg of the reactor core. As the current level builds up in the sine wave trace (going from zero to 90 electrical degrees), a magnetic field is created. Energy is used to create the magnetic field but it is not ''used up''. It is, for the moment, stored in the magnetic field.

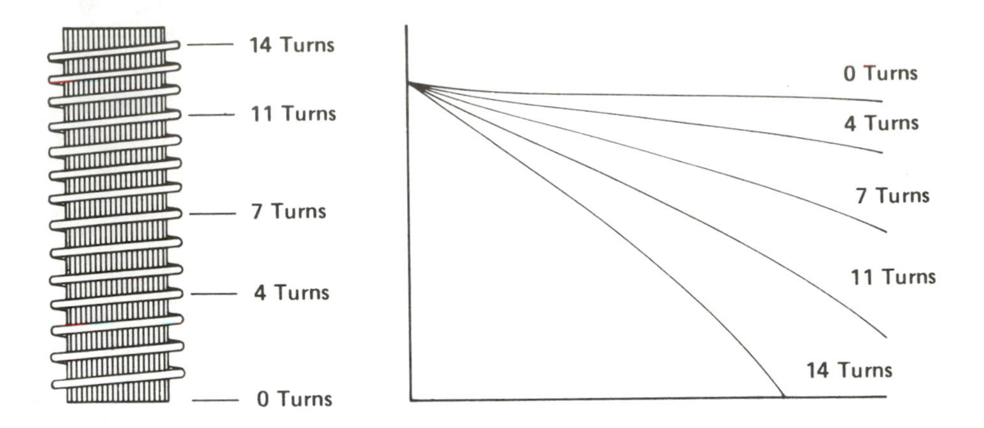

Figure 215. Slope Control Concepts for the GMAW Process.

When the current reaches maximum amplitude, or strength, the magnetic field strength is at maximum also. Now the current value decreases, following the sine wave trace as it goes to zero each half-cycle. As the magnetic field collapses the energy stored in the magnetic field must go someplace. It goes to the only place it can which is the reactor circuit coil where it originated in the first place. This is termed **self-inductance of the reactor coil.**

Recalling the electrical function of a transformer we know when power is induced into the secondary coil it produced a voltage that directed force counter, or opposite, in direction to the impressed AC circuit voltage. A similar counter voltage (counter emf) is created in the slope control reactor. This counter voltage opposes the flow of welding current in the reactor coils which has the effect on the volt-ampere curve shown in Figure 215 at four turns of slope.

If we now set the slope control at seven turns of slope, half of the total reactor coil is in the welding circuit. As current builds up in strength in the AC sine wave, there will again be created a magnetic field. This time, however, the magnetic field is stronger because there are more effective electrical turns in the reactor coil.

The energy used to create the stronger magnetic field is stored momentarily in the magnetic field and, when the current goes to zero as it does each half-cycle, the energy comes back on the reactor coil as a counter voltage. This time it is a stronger counter voltage because it took more energy to build the stronger magnetic field. The volt-ampere curve shows the result of seven turns of slope; a steeper slope to the curve and lower maximum short circuit current.

Progressing to a higher slope value of 11 turns we have approximately three-fourths of the reactor coil now effective in the welding circuit. The magnetic field created will be much stronger than before. Considerably more energy is used to create the magnetic field and that energy is again stored in the magnetic field as it increases to maximum strength. When the current level peaks at either 90 or 270 electrical degrees on the sine wave form, the magnetic field collapses and the energy is returned to the reactor coil as a counter voltage; an impedance holding back the flow of current.

The result is shown in the volt-ampere curve, Figure 215, at 11 turns of slope. It is evident that a very steep angle is created on the volt-ampere curve at 11 turns of slope and the maximum short circuit current is limited still further.

When the entire 14 turns of slope is put into the reactor there is maximum magnetic field strength in the reactor iron core. Maximum energy will be used to create the magnetic field and, when the current goes to zero as it does each half cycle of the AC sine wave form, there will be maximum counter voltage, or **impedance,** to the flow of welding current. The resulting volt-ampere curve is almost like a constant current power source with very limited maximum short circuit current.

It is apparent there is a chain reaction to the function of a reactor for slope control. Current flows in the reactor coil wrapped around the reactor iron core. The energy used to create the magnetic field is momentarily stored in the field. When the magnetic field collapses each half cycle, as shown on the AC sine wave form, the stored energy is returned to the reactor circuit as a counter voltage. The counter voltage is impressed on the reactor coil, where it originated, as an **impedance factor.** The impedance limits the amount of welding current that can flow in the welding circuit.

The results of inducing variable amounts of "slope control" are shown in Figure 216. For the "0" slope setting the volt-ampere output characteristic is relatively flat. By adding electrical turns to the slope reactor, up to 14 turns, the slope curve gradually becomes steeper (more negative) and the maximum short circuit current becomes more limited in value.

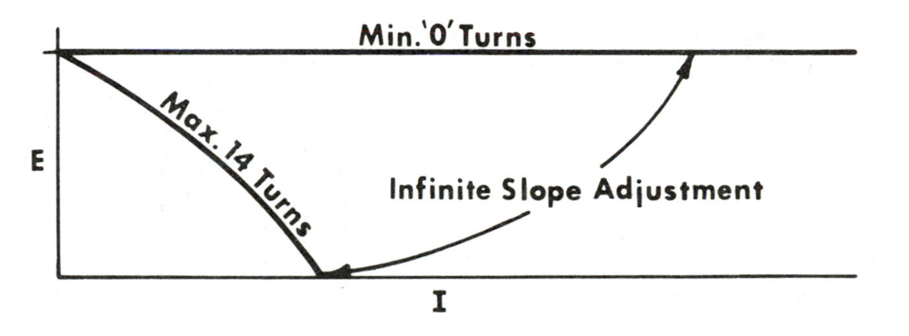

Figure 216. Results of Variable Slope Control.

There are two major results when slope turns of a reactor are added to the welding circuit. They are:

1. The maximum short circuit current is limited in value. (This is apparent in Figure 216 when you look at the volt-ampere curve).
2. The addition of slope turns slows the rate of response of the welding power source to changing arc conditions.

Rate of Response

Earlier in this chapter we asked you to remember that the reactor inherently opposes change in the welding power circuit. This means that, as more of the reactor coils are added to the welding circuit through slope turns, the response of the power source slows down proportionately. In Figure 217 there are illustrated two response conditions. While hypothetical, they do explain response time rather graphically.

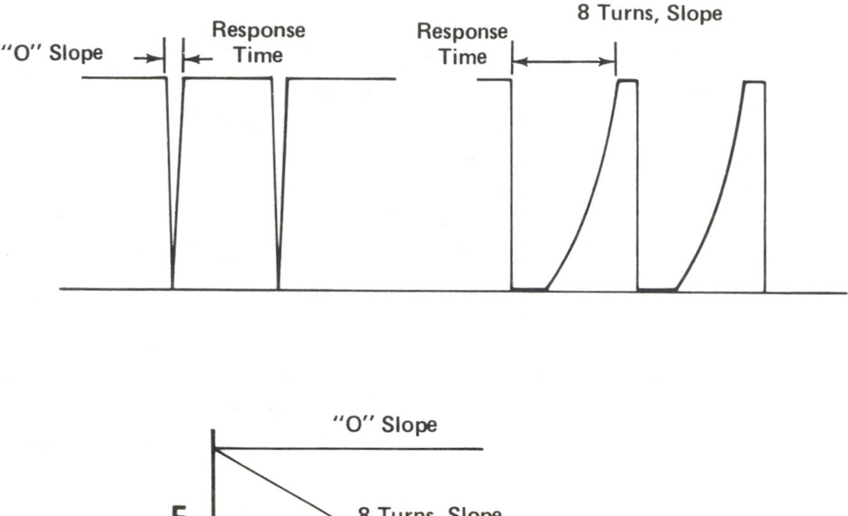

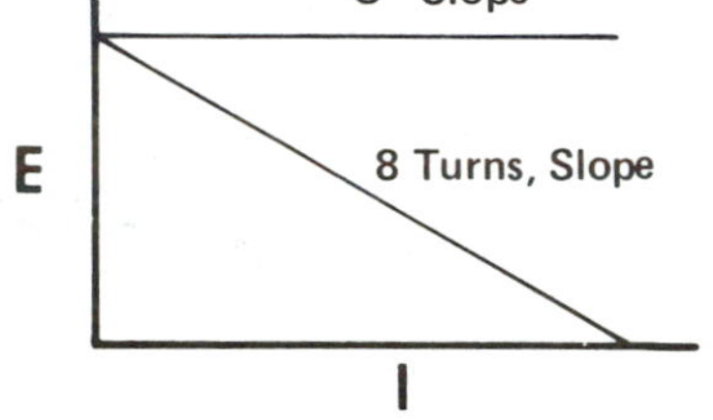

Figure 217. Relative Response Times With Reactor Slope Control.

For example, if there is "0" slope in the welding power source circuitry there would not be much of a short circuit transfer welding condition because the electrode, as it approached the base metal, would cause the arc voltage to drop. With the relatively flat volt-ampere output characteristic, this would literally slide the welding volt-ampere position to a prohibitively high amperage.

The result would be explosively high currents at the electrode tip. This would cause excessive spatter, poor penetration into the base metal, and poor fusion in the weld. The power source response time would be very fast. Most of the arc-on time the power source would be at full welding amperage and voltage. This is apparent in Figure 217.

By adding the correct amount of slope turns to the welding circuit there is a slowing of the response time of the welding power source. The GMAW electrode type and diameter will have a significant bearing on the amount of slope turns added. Figure 217 shows eight turns of slope added which is typical for mild steel and 0.035" diameter electrode material.

The response time is slowed to where an actual short circuit can be made between the electrode wire and the base metal to be welded. Transfer of molten weld metal from the electrode is accomplished during this short circuit arc outage.

"**Response time** is the actual time it takes the welding power source to get from dead short circuit amperage and voltage to welding voltage and amperage".

The rate of rise from dead short circuit to welding voltage and amperage is much slower at 8 slope turns than it is at "0" turns of slope. The time at full welding voltage and amperage is much less than it was before. The result is **limited heat input** to the weld area and the capability to weld in all positions with the short circuit method of metal transfer.

Although the slope settings may change the shape of the volt-ampere curve drastically as more reactance is added to the welding circuit the open circuit voltage remains the same. By varying both the slope settings and the open circuit voltage setting on constant voltage power sources, it is apparent that welding conditions can be changed substantially.

When setting a welding condition on a constant potential type welding power source change only one variable at a time. Adjusting both open circuit voltage and adjustable slope at the same time will make a welding condition very difficult to set.

Some gas metal arc welding techniques require very little slope in the welding power source. Spray type transfer and flux cored arc welding are two good examples where slope is not normally required. The only exception to that rule is for aluminum.

With aluminum it is always best to use as much slope as possible and still obtain the necessary load voltage for spray transfer. The use of slope slows down the rate of response which can totally eliminate spatter in GMAW aluminum welding. Of course, this is predicated on the use of a correct shielding gas, welding techniques and welding procedure.

Slope is absolutely mandatory for the short circuit method of metal transfer. Different slope and response rates are required for different metals and electrode diameters even with this welding technique.

Some Rules

There are a couple of rules of thumb which can be applied here for slope reactor circuits. They are:

1. THE HIGHER THE SLOPE NUMBER, THE LOWER THE MAXIMUM SHORT CIRCUIT CURRENT WILL BE.
2. THE HIGHER THE SLOPE NUMBER, THE SLOWER WILL BE THE RATE OF RESPONSE OF THE POWER SOURCE TO CHANGING ARC CONDITIONS.

In truth, if the average welding person would remember these two "rules of thumb" there would be no problem in understanding reactor slope control for gas metal arc welding!

The Stabilizer or Inductor

There is a control device available on some models of constant voltage type power sources. It is called a **stabilizer** or **inductor.** This control device normally regulates the amount of stabilization in the DC portion of the welding power source circuitry. The control may be either built-in the power source or furnished as an external control system. It is usually an optional control item when purchasing a constant potential type welding power source.

The inductor, or stabilizer, control serves only one function which is to permit a slight changing of the rate of response of the power source without effectively changing the shape of the output volt-ampere curve.

In the opinion of many welding authorities the inductor or stabilizer has limited use with the gas metal arc welding process. It has some value for certain materials with the short circuit transfer method of GMAW. Otherwise, the small amount of inductance achieved is relatively unimportant when you consider the total inductance inherent in the power source circuit.

Points To Remember

1. A **slope reactor** is located in the secondary AC portion of the constant potential power source circuitry. The slope reactor will change both the maximum short circuit current value and the rate of response of the power source.
2. An **inductor** or **stabilizer** is always in the DC portion of the power source circuitry and will only adjust slightly the rate of response of the power source.
3. **Resistance slope control** will only limit the maximum short circuit current output of a welding power source.

Gas Metal Arc Spot Welding

Gas metal arc spot welding has been developed as an industrial tool for fabricating all weldable metals. The NEMA Class 1 power sources used with this production welding technique normally have a minimum 600 ampere, 100% duty cycle output. Some power sources have been developed for sheet metal work that have output ratings as low as 150 amperes, 60% duty cycle. Such units are for light sheet metal work only.

In many cases the top plate thickness of the pieces to be welded requires a power source with a higher open circuit voltage than is usual on constant voltage power sources. The high open circuit voltage is necessary to provide the power to penetrate the top plate in minimum time. Some constant potential type welding power sources are designed with sufficiently high open circuit voltages for this application.

The higher amperage ratings and higher open circuit voltages are also required for most air carbon-arc cutting and gouging applications (AAC).

A typical gas metal arc spot weld is illustrated in Figure 218. The weld interface is almost vertical through the top plate changing into a modified parabola in the lower plate. Depth of penetration is determined as a time function after the welding condition is initially set up.

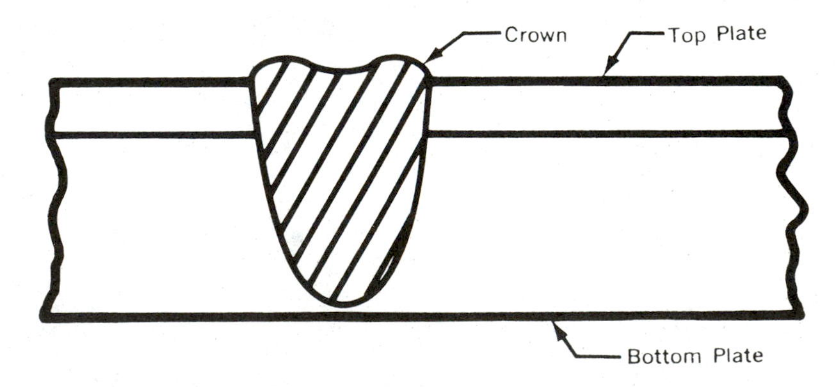

Figure 218. Typical Gas Metal Arc Spot Weld.

Applications of gas metal arc spot welding include tack welding structural members in a weldment, manufacture of articles that must be spot welded from one side only, joining thick to thin plates, etc. The welding technique is fast and clean with no slag residue.

Constant voltage type welding power sources are used in the no-slope, or flat, volt-ampere output characteristic for gas metal arc spot welding. A special GMAW spot control is normally employed with a standard wire feeder-control and gun system. The gun uses specially designed gas nozzles which permit flat, lap seam, or fillet spot welds in joints. The control system includes a timing device which can count welding time in cycles or seconds.

CHAPTER 22

Welding Processes Comparison and Use

It is often necessary to evaluate different welding processes for selection and use to determine which one will do the job at the lowest net cost. Frequently when the gas metal arc welding process is being considered for an application the intent is to replace another welding process already in use. The process in use may be shielded metal arc, gas tungsten arc or submerged arc.

Such evaluations must include the considerations of labor costs, equipment costs, electrode and shielding gas costs, material joint preparation costs, actual arc time costs, weld cleaning time costs and the quality required. The result of any evaluation will show which welding process will perform the operation at the lowest net cost per manufactured item.

Accumulating the facts for a survey of a users shop welding requirements takes time and effort. This is normally done by staff welding engineers although some companies employ independent welding specialist consultants to perform the work. Knowing the proper welding processes and equipment are being applied to the job makes the effort and cost worthwhile.

For this evaluation we will compare the shielded metal arc welding (SMAW) process and the gas metal arc welding (GMAW) process. Other welding processes are not particularly considered although some of the data will be applicable to them as well.

Joint Preparation

For many years there have been standard joint designs used with the arc welding processes. Many of the joint designs were developed for the shielded metal arc welding (SMAW) process. Due to the inherent bulk of the shielded metal arc welding electrodes (because of the flux coatings) the joint groove opening in the base metal had to be very large to accommodate them. The root opening for joints without backing is normally 0''-3/32''.

The root face of the weld joint (the portion of the root having a vertical face) could be no more than about 3/32''. The reason for this is that the penetration of even the deepest digging mild steel electrode (E6010) is only about 1/8'' into the base metal. The penetration depth for all other SMAW electrodes will be less than 1/8''.

It was considered good practice to open up the weld joint angle. The idea was to have as much interface as possible between the base metal and the deposited weld metal. While this concept has been modified in the past few years it is not uncommon to find vee butt joint designs with an included groove angle of 75 degrees or more.

The drawing in Figure 219 shows a typical weld joint for the shielded metal arc welding process. This joint has an included groove angle of 75 degrees. At the right side of the same illustration there is a typical joint design for the gas metal arc welding process. It has a 45 degree included groove angle.

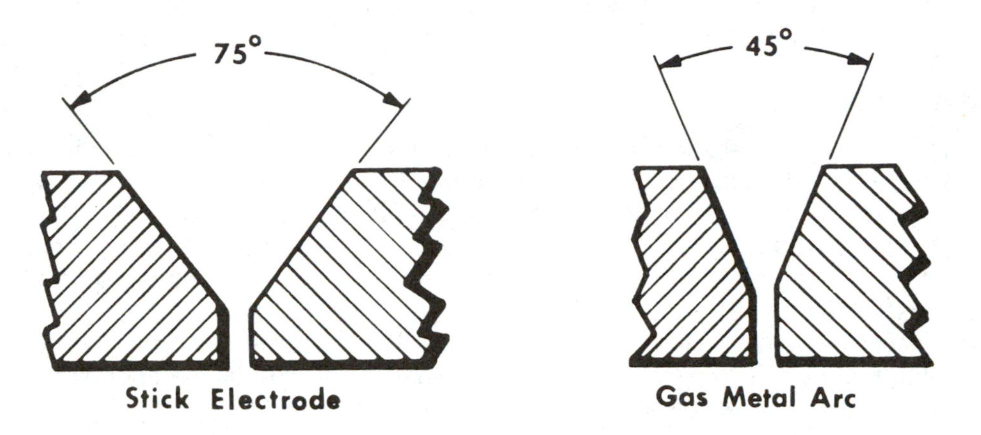

Figure 219. Vee Butt Groove Angle Comparison.

The SMAW vee butt joint design shows a relatively thin root face ("land") area. As was previously stated, the root face for the shielded metal arc welding process is normally not more than 3/32" where full 100% penetration of the weld joint is desired. With a greater root face area the penetration of the weld deposit into the base metal would probably be less than 100%.

In any type of shielded metal arc weldment the joint design should be considered for the most economical use of filler metal. The actual joint design will depend on several factors including base metal type and thickness, position of welding, welding technique used, and type of power source. For example, 3/8" steel plate would normally require some type of joint groove preparation regardless of the welding position used.

The method of edge preparation is also important when considering joint design. Vee butt and bevel groove designs can be easily made with flame cutting techniques and equipment. The most frequently used edge preparation method for carbon steel is the oxygen-fuel gas flame cutting process. The flame cutting equipment is available for manual operation or multi-torch automatic operation. Edge beveling and square edge cutting of shapes in steel are accomplished with relative ease with this type of process equipment.

Although the same factors are applicable to gas metal arc welding the joint designs may be considerably different. For example, it is not uncommon for penetration patterns from one side only in 1/4" mild steel plate to achieve 100% penetration using the buried arc technique and carbon dioxide (CO_2) shielding gas.

The illustration (Figure 219) shows the relative filler metal amounts required for shielded metal arc welding and gas metal arc welding joint designs. It is obvious that more weld filler metal is needed to complete the shielded metal arc weld joint. This means greater heat energy input per linear inch of weld. The result will often be locked-in stress and part distortion.

The gas metal arc welding joint requires much less filler metal for completion. Less heat input from welding results in less possibility of weld distortion from residual stresses.

The cost advantages of a narrower weld joint groove are twofold. There is less base metal removed and less filler metal has to be put in the joint. Cash value of scrap metal is seldom more than about 25% its original cost. By removing less base metal for gas metal arc welding there is a substantial savings in the cost of electrode material to fill the weld joint.

Welding Electrodes

When considering electrode cost per pound it is not uncommon to find that shielded metal arc welding electrodes are about half the cost per pound of gas metal arc welding electrode wire for the same type of application. It is not the purchased price per pound of welding electrodes that should be of concern to the consumer. It is the cost per pound of the deposited weld metal that really counts! An examination of comparative deposition efficiencies of the two process electrode types will help clarify the point.

The losses by weight of electrode materials are substantial when using the shielded metal arc welding process. Basing the facts on 5/32'' diameter electrodes the approximate electrode losses by weight will be as follows:

Stub end loss (based on 2'' electrode stub ends)	17%
Spatter and flux covering losses	27%

The total is 44% loss by weight of the purchased pounds of SMAW electrodes. As noted the figures are based on 2'' electrode stub ends which are seldom achieved in welding shops. Most fabricators and contractors, when estimating SMAW electrode efficiencies, calculate no more than 30-35% of the purchased pounds of electrodes will be deposited as weld metal.

There are also electrode losses with the gas metal arc welding process but they are minimal. For example, we will consider a welding application using 0.035'' diameter mild steel electrode wire. The base metal is mild steel, the shielding gas is carbon dioxide (CO_2), and the short circuit transfer technique of GMAW is being used.

Let us assume that the welder cuts off the equivelent of 1/2'' from the end of the electrode wire each time he completes a weld. The assumed quantity of 1/2'' is probably twice as much as the welder would normally cut off in actual practice.

The number of inches of electrode wire per pound is shown in Data Chart 2 of the appendix. Finding mild steel at the left side of the chart, we move to the right until we come to the column headed 0.035'' diameter. There we find that there are 3,650 inches per pound of 0.035'' diameter mild steel electrode wire.

Based on the loss of 1/2'' of electrode wire each time a weld is completed, it is apparent that the welder can make 7,300 individual welds before there is a pound of waste electrode. The cost per deposited pound of weld metal is normally less expensive with the GMAW process than with the SMAW process.

Weld Cleaning

Many manufacturing people consider that **actual welding time** used in making an SMAW weldment is the major factor in welding costs. That is not normally true. The greatest cost of shielded metal arc welding is actually the **cleaning and finishing of parts** after the welding operation is completed. This fact has been proven in both U. S. Government and private industry surveys of costs in the welding industry.

There is very little requirement for weld cleaning and finishing with the gas metal arc welding process. With the exception of flux cored electrodes used in the FCAW process, there is no flux and therefore no slag removal required. If the welding process and welding procedure are correctly set up, there should be little or no weld spatter from the welding operation. Chipping, grinding and finishing will be at a minimum.

Operator Training

The need to be trained in any welding process is mandatory if the welder is to be considered a professional craftsman. The need to be trained in the gas metal arc welding process after having achieved journeyman status in one or more arc welding processes can be a concern to welders. Some are reluctant to try learning to use the GMAW process because it is new to them and they are afraid they might fail to master it correctly the first time. Their fears are unfounded.

It is true that the shielded metal arc welding process and gas tungten arc welding process require much time and practice to learn. For example, it is common for SMAW courses to require about 600 working hours before the welder is considered trained to even minimum standards. Often GTAW courses require 80-120 additional working hours for training. Most gas metal arc welding courses require only about 40-60 working hours to learn the process. In many cases where a person has welded before, the training time is only a matter of two or three days. Even totally unskilled operators have been taught to use the GMAW process in less than 60 working hours.

The relative ease of training welding operators to use the gas metal arc welding process is simple to explain. Much of the control required for welding with the process is incorporated into the constant voltage type welding power source and the wire feeder-control equipment. The only basic variables the welder must watch are the gun angle relative to the workpiece, electrode extension from the contact tube, the speed of travel, and the gas shielding pattern. As experience is gained in the use of the GMAW process and equipment, the welder will quickly master the more difficult techniques used with the process. These include vertical-up welding, horizontal groove welds, and overhead welding.

Continuous Electrode Controlled Welding

One of the outstanding features of the gas metal arc welding process is the elimination of weld starting and stopping due to changing electrodes. Many shielded metal arc weld defects are caused by non-metallic slag inclusions, crater cracking, cold lapping, overlap and the like which occur when the welder stops to change electrodes with the SMAW process. Decreasing the number of weld starts and stops will certainly decrease the possibility of such defects occurring.

The gas metal arc welding power source and equipment controls several variables which permit the welder to weld as fast as he is capable of moving and seeing. (At about 40 inches per minute the human eye can no longer focus on the arc area and the weld seam). The welder has the opportunity to set the welding condition on the gas metal arc feeder-control system and the constant voltage type power source before beginning the actual weld.

The welding condition will remain stable until open circuit voltage, wire feed speed, slope, gas flow rate or some other variable is physically changed. This pre-set condition permits the welder to concentrate on the weld path to travel, the welding puddle, and the welding technique he is using. There is high current density at the welding arc and the electrode tip. This allows the welder to have controlled penetration, controlled bead dimensions, and a stable arc plasma and arc column. This all results in high quality, high speed gas metal arc welds.

Current Density

Current density may be defined as the amperage per square inch of electrode cross sectional area. At a specific amperage, for example, there would be much higher current density with 0.030'' diameter electrode wire than there would be with 0.045'' diameter electrode wire. The explanation is in the cross sectional area of the two electrodes as shown:

0.030'' dia. = 0.00071 square inches of area.
0.045'' dia. = 0.00160 square inches of area.

The current density may be easily calculated for any diameter of welding electrode. Dividing the welding amperage used by the electrode cross sectional area in square inches results in amperage per square inch of electrode. As an example, Figure 220 shows the calculation for a specific electrode diameter at a specific amperage.

welding current = 100 amperes
GMAW electrode dia. = 0.030'' dia.

$$\frac{100 \text{ amperes}}{0.00071 \text{ inches}^2)} = 140{,}845 \text{ amps/inches}^2$$

Figure 220. Typical Current Density Calculation.

The current density is 140,845 amperes per square inch of electrode.

Data Chart 3 in the appendix of this book shows information for calculation of current density for electrodes 0.020'' dia. to 3/8'' dia. Since every electrode type and diameter has a minimum-maximum current range within which it may operate correctly, it is well to know the current density at which each electrode operates best.

If current density is too high for the electrode type and diameter there is the possibility of electrode burnback into the contact tube. Low current density, on the other hand, could cause ''stubbing'' of the electrode into the base metal. To define:

Stubbing = ''the electrode has made short circuit contact with the base metal and there is no welding arc present''.

A condition called ''roping'' may occur in conjunction with the stubbing condition. In a roping condition the electrode wire becomes red hot due to the short circuit condition of the electrode with the base metal. The heating of the electrode is caused by the electrical resistance of the electrode wire to welding current flow. The electrode will continue to feed from the gun as long as the trigger is pulled. The electrode is in a semi-plastic state so it will literally coil, or "rope", on the surface of the base metal. Stubbing and/or roping may be caused by excessive wire feed speed.

Arc Energy Differences, GMAW and SMAW

A high current density arc has more arc energy concentrated at one point, the electrode tip. For example, gas metal arc welding has high current density and shielded metal arc has low current density. The gas metal arc column is sharp and incisive while the shielded metal arc column is relatively soft and widespread. The relative arc columns and typical deposit configurations are illustrated in Figure 221.

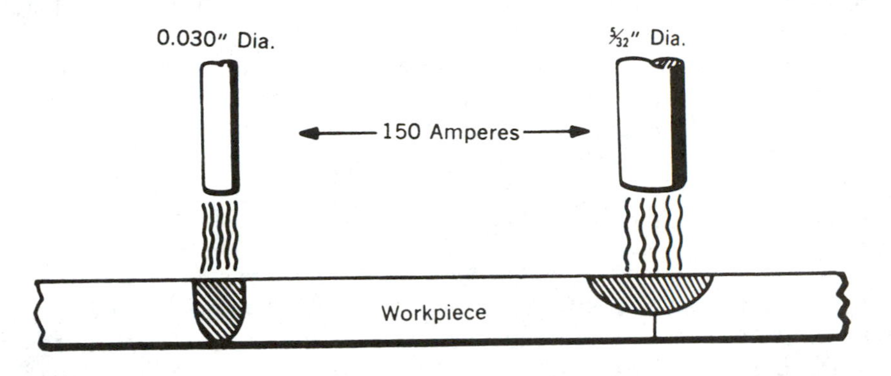

Figure 221. Weld Deposit Comparisons, Same Amperage.

It is logical that the weld deposit made with shielded metal arc will have a wider, shallower, and more parabolic shape than a deposit made with the gas metal arc welding process. The welding speed of travel with the SMAW process would be slower, with more heat applied per linear inch of weld, than with the gas metal arc welding process.

Since there is a greater volume of molten metal deposited, per linear inch of weld, with the shielded metal arc welding process the heat input differential of the two welding processes is considerable. There is greater penetration with the GMAW process because of the higher current density at the electrode tip. It is logical that the bead width-to-depth ratio will be less with the GMAW process than with the SMAW process. Typical comparative weld deposits are illustrated in Figure 222.

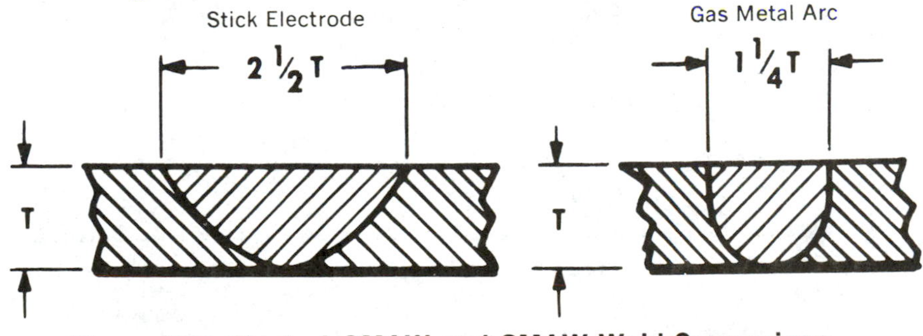

Figure 222. Typical GMAW and SMAW Weld Comparison.

A normal weld made with the shielded metal arc welding process on 1/4'' carbon steel would have a top bead width of approximately 2 1/2 - 3 ''T'' where ''T'' equals the thickness of the plate being welded. Using the gas metal arc welding process on the same plate thickness, with either CO_2 or an argon-CO_2 shielding gas mixture, the top bead width would be about 1 - 1 1/4 ''T''. Joint preparation is considered to be correct for both joints.

Other Welding Benefits From GMAW

Other benefits of the gas metal arc welding process include less heat affected zone (HAZ) width, less possibility of grain growth in the base metal, and less distortion of the weldment due to thermal expansion and contraction. With GMAW, more of the heat is applied and used in the weld deposit instead of dissipating in the base metal.

Due to the speed of travel, and the fact of welding heat being concentrated in a small area of the workpiece, there is less heat energy input per linear inch of weld with the GMAW process. This assumes a comparison with the shielded metal arc welding process when welding over a specified bead length on specific base metal.

Gas Metal Arc Welding Applications

It has been estimated that approximately 75-80% of all welding performed today is on some type of carbon steel material. Probably 75% of the total steel welded is classed as low carbon mild steel. The gas metal arc welding process has successfully competed with other welding processes in the welding industry for a good share of this work. The reason for its wide acceptance is because it welds faster, cleaner, and more efficiently where applicable. The process has found applications from sheet metal to heavy structural welding applications.

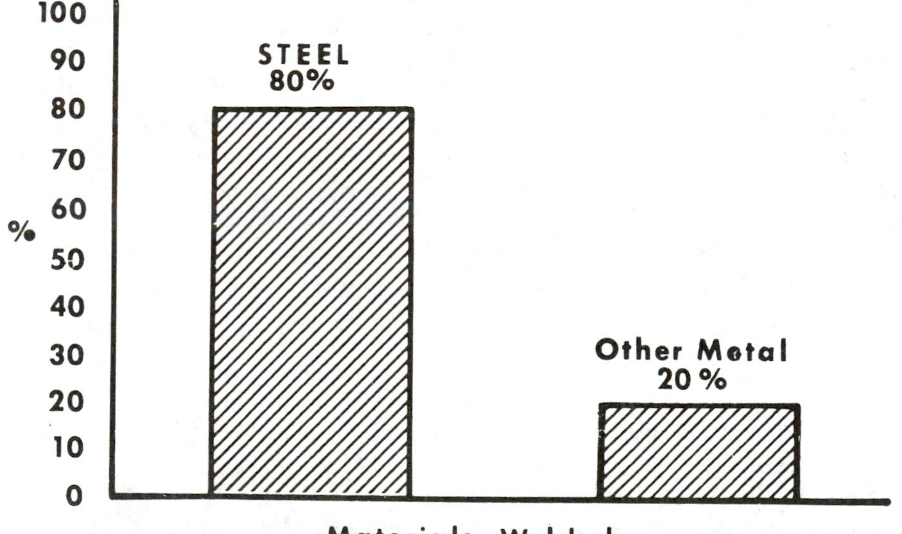

Figure 223. Metals Welded by Approximate Percentages.

Steel Welding With GMAW

All of the known methods of metal transfer may be used for welding carbon steel and low alloy steel with the gas metal arc welding process. Spray type transfer, using an argon-rich shielding gas mixture, is fast, clean and essentially defect-free. Penetration into the base metal is excellent for both groove and fillet joints.

The only limitations on true spray transfer is that it is normally used only for the flat groove or flat and horizontal fillet weld joints. The weld puddle is very fluid so it is not normally recommended for out of position welding.

The buried arc method of metal transfer has been finding much use in industry lately to replace certain flux cored electrode applications. The reason is the fumes from the FCAW process and therefore costly fume control systems necessary to remove them. It costs less to do the job with the buried arc technique of gas metal arc welding.

The short circuit method of metal transfer is designed primarily for welding steel materials 1/4'' thick or less. The shielding gas may be either welding grade carbon dioxide (CO_2) or an argon-CO_2 gas mixture.

Solid electrode wires for short circuit transfer are classified in the AWS Specification A5.18 for carbon steel and A5.28 for low alloy steels. Selection of solid filler metal should be based on the requirements of the individual application and the shielding gas.

The short circuit method of metal transfer is used in a variety of welding applications. These include all types of sheet metal fabrication, pipe welding (particularly root passes), angle iron frames, and certainly any applicable out-of-position welding work. Some nominal welding condition voltages and amperages are listed in the appendix of this book.

Porosity

In any type of welding there is always a percentage of failures for one reason or another. A weld defect can cause re-work of the part and, in some cases, can actually cause the part to be scrapped. Since re-work is always expensive (almost invariably it will cost a minimum of 10 times the original weld cost) it is to be avoided whenever possible.

Some of the weld defects that can cause re-work include non-metallic slag inclusions, lack of interpass fusion, lack of joint sidewall fusion, inadequate penetration, cracks, and porosity. Probably the single most common defect is porosity.

Porosity is one of the recurring problems faced when welding any type of base metal with any welding process. There are some general rules which may help to minimize porosity problems when welding with the gas metal arc welding process. The suggestions certainly do not include all the possible solutions for eliminating porosity in welds. However, due consideration to the points made will certainly help decrease the possibilities of porosity in GMAW welds.

1. Welding speeds that are too fast may cause either partial, or complete, loss of the shielding gas pattern in the arc area and cause porosity.
2. Electrode current densities that are too high will often cause porosity because of the excessive heat of the molten metal from the electrode. In some cases there are severe oxidation losses of both the deoxidizers and alloying elements as they move across the arc column.

In this case it is possible that the electrode is too small in diameter for the amperage used. If this is true the next larger electrode diameter should be tried at a slightly adjusted amperage. If the smaller diameter electrode wire must be used it will be necessary to reset the welding condition variables and decrease the welding amperage.

3. The shielding gas, or gas mixture, must be the correct type for the metal being welded and the type of metal transfer used. The correct flow rate must also be used or unsatisfactory results will occur. It is mandatory that shielding gases for welding be uncontaminated and at the proper dew point.
4. It is important that the welding electrode be maintained in the center of the shielding gas flow pattern. If the electrode is off center in the gas nozzle it can cause erratic arc behavior and weld porosity. A buildup of weld spatter in the gas nozzle can also cause the gas flow pattern to be disturbed.

In all carbon steel welding there is the possibility of a silicate residue developing on the weld surface. This is particularly noticeable when welding steels with a high silicon content or when using a highly deoxidized electrode material. The residue appears as a glassy brown substance that is somewhat difficult to remove. The silicate residue **should** be removed between welding passes in a multipass weld. It **must** be cleaned off prior to any painting or plating of the welded part. Removal may be accomplished with chipping tools or by grit blasting.

Aluminum

The use of aluminum in structural work and shipbuilding has increased greatly in recent years. The metal is light in weight yet has the capability of being alloyed to achieve strengths comparable to low carbon mild steel.

Aluminum has excellent thermal and electrical conductivity. It is factors such as these that make aluminum a highly desirable metal for many applications. Its high strength-to-weight ratio has made it an invaluable material for the aircraft and missile industries. Not the least of its attributes are its low weight, neat appearance and ability to be formed into almost any shape desired.

Aluminum and most of its alloys may be welded by most of the gas metal arc welding techniques. It does take some skill to weld aluminum out of position with the short circuit method of metal transfer. I normally do not recommend using this technique on aluminum because of the possibility of trapping oxide in the weld. Argon or argon-helium mixtures are normlly used as shielding gases for welding aluminum with the GMAW process.

A consideration of prime importance when welding aluminum is the cleanliness of the base metal. Aluminum forms a heavy refractory oxide at the metal surface in all circumstances. The oxide may cause irregularities in the weld deposit if not removed prior to welding. Aluminum oxide has a melting point in excess of 4,000° F. while the base metal melts at approximately 1218° F.

Pre-weld cleaning may be accomplished with either mechanical or chemical methods. A power stainless steel wire brush may be used for removing the oxide although a metal scraper will often do the job faster. Be sure the wire brush is really removing the oxide and not just polishing it.

Chemical cleaning is another method used for removing surface oxides from aluminum. A system of cleaning solutions are often used in the following sequence:

1. Sodium hydroxide bath (approximately 150°-175° F.).
2. Hot water rinse.
3. Chromic acid bright dip.
4. Hot water rinse.
5. Hot air dry (optional).

The important thing to remember is that aluminum oxides must be removed from the welding zone before the welding process begins.

Aluminum plate that has been sheared presents a unique problem to welding the material. The action of shearing aluminum is approximately 35% shear and 65% controlled tear of the metal. The sliding action of the oiled shear blade past the aluminum metal edge will usually result in some fold-over of the material shear surface. There is a definite possibility that oil or other foreign particles will be trapped in the sheared edge of the metal.

A sheared edge should always be degreased and at least a 0.025'' thickness removed from the sheared edge with a dry router or some other effective tool. This should certainly be done before any welding is attempted with the GMAW process on the aluminum base metal.

Spray transfer of aluminum is a fast welding technique. Most welding is done in the flat or horizontal positions although it is not unusual to weld out-of-position with spray transfer and relatively small diameter aluminum electrode wire. By small diameter we mean electrode diameters of 0.035'' or less.

The method used in spray transfer welding of aluminum is normally the forehand technique. The gun is pointed in the direction of travel at about a 10-15 degree angle. This assures correct gas shielding of the weld area. The correct gun angle will help eliminate spatter while welding.

CHAPTER 23

Plasma Arc Cutting

Plasma arc cutting is a method of severing metals using an electric arc and industrial gases at high velocity through a constricted nozzle orifice. The plasma arc cutting (PAC) torch and power source are specially designed for the process. All of the commercially used metals may be cut with the PAC process.

The plasma torch features a tungsten electrode which is recessed inside the constricting nozzle of the PAC torch. The orifice gas, usually low cost nitrogen, is dispensed through the constricting orifice of the water-cooled nozzle as an ionized gas. The ionized orifice gas is emitted as a high temperature, high speed plasma stream through the nozzle constricting orifice.

When the constricted arc and plasma column strike the base metal they impinge on a very small area. The **high heat** and force of the arc plasma melts the base metal. The **high velocity** of the plasma gas literally blows the molten metal away thereby creating a kerf. The kerf is usually square on one side but the second side may have an angle of up to seven degrees. It is possible to minimize the kerf angle by using a water shield when plasma cutting. To define:

Kerf = "The width and shape of a space left after a cutting operation or process has been performed".

Some plasma arc cutting equipment uses a secondary "shielding gas" in addition to the orifice "plasma" gas. This may be carbon dioxide (CO_2) for most metals. The intent of the secondary shielding gas is to assist the high speed plasma gas in removing the molten metal from the kerf. It also helps reduce the spatter on the nozzle front end of the plasma torch. The secondary shielding gas helps cool the torch which improves efficiency and adds extra life to consumable components in the system. The ability to pierce plate material efficiently is another benefit of a secondary shielding gas.

Special PAC Terminology

There are a variety of special terms used with the plasma arc cutting (PAC) process and equipment. Some of them are defined here.

Electrode = Tungsten electrode, usually 2% thoria-bearing.

Orifice gas = The gas that flows in the plenum chamber around the tungsten electrode. It is thermally ionized in the arc. The ionized orifice gas expands and flows through the constricted orifice as a high velocity plasma gas.

Constricting nozzle = A water-cooled copper nozzle surrounding the tungsten electrode and the plenum chamber of the PAC torch. It contains the **constricting orifice.**

Constricting orifice = The small hole, of restricted diameter, through which the plasma gas and plasma arc pass at high velocity.

Electrode setback = The distance the tungsten electrode is recessed behind the constricting orifice of the constricting nozzle. The setback distance is measured from the outer face of the nozzle.

Shielding gas nozzle = The outer nozzle on dual flow PAC torches. The outer shielding gas is contained between the constricting nozzle and the shielding gas nozzle.

Shielding gas = The secondary gas is used to help remove molten metal from the kerf and protect the orifice end of the PAC torch.

Orifice throat length = The physical measurement of the constricted orifice from the inner edge to the outer edge.

Torch stand-off distance = The distance from the outer face of the constricting orifice nozzle to the base metal surface.

Basic Plasma Arc Cutting Equipment

The plasma arc cutting (PAC) process equipment consists of the following basic components:

Power source, DC output, straight polarity (electrode negative).
Control console with high frequency system. (May be built-in as part of the power source).
PAC torch with hoses and cable assembly; constricted torch nozzle with water cooling.

A typical manual plasma arc cutting system is shown in Figure 224. Note the 25' hose and cable assembly attached to the manual torch. The small tool kit box is for spare parts necessary for operation of the torch.

Plasma arc cutting systems are available with amperage ranges from 100 amperes to 1,000 amperes. High amperage operation may be achieved with two or more power sources in parallel. The plasma arc cutting (PAC) torch amperage capacity is the limiting factor.

Both manual and automatic PAC torch systems are presently available from various manufacturers. The equipment is quite easy to set up for either type of arc cutting operation. Automatic torches may be used on the same shape cutting equipment as flame cutting torches. Cutting speed of travel must be sufficiently fast to make clean cut edges. Photo-electric tracers, numerical control systems, or computer controlled systems are excellent for single or multi-torch automatic PAC cutting applications.

Figure 224. Typical Plasma Arc Cutting System.

The PAC Power Sources

Plasma arc cutting power sources are specially designed with high open circuit voltages. Specific power sources are actually designed for certain types and models of PAC torches. The output characteristics are essentially a constant current type power source. Open circuit voltage may range from 100-400 volts. This type of power source is classed as a "special" power source in the NEMA EW-1 Standard. The limitations on open circuit voltage which applies to welding power sources does not apply to PAC power sources.

The power sources designed for higher amperage torches and heavier metals also have the highest open circuit voltage. The 400 open circuit volts provide rapid arc initiation and power for piercing materials up to 2" thick. Some power sources are designed to operate from either their standard voltages and amperages or through an optional low ampere grid. On some models, open circuit may be varied for different metals and thicknesses.

Many power sources designed for the PAC process have built-in high frequency systems, gas valves and solenoids, and contactor control. Care should be used in selecting the PAC system for your application. The metal types and thickness to be cut are very important factors as well as the speed of travel in cutting. Be sure you have all the facts before purchasing a PAC unit.

Process Conditions and Gas Selection

The selection of the orifice plasma gas is normally dependent on the type and thickness of base metal to be cut. The required quality of the kerf surface is also a factor to consider. There are a number of industrial gases and gas mixtures to select from for each of the metals. The plasma arc cutting process was developed using inert gas as the plasma medium for non-ferrous metals. Subsequently other gases were tried and found acceptable for certain applications.

According to manufacturers recommendations and AWS literature the following conditions are used with specific plasma gases and/or shielding gases for the PAC process. In all cases the manufacturer of the PAC equipment should be consulted for their recommendations concerning the procedure and equipment setup for specific metal and metal thicknesses. The secondary voltage for all materials and thicknesses is calculated at 200 load volts.

CARBON STEEL				
Thickness, inches	**Speed in./min.**	**Orifice diameter***	**Amperes DCEN**	**Secondary power, KW**
1/4"	200	1/8"	275	55
1/2"	100	1/8"	275	55
1"	50	5/32"	425	85
2"	25	3/16"	550	110

*Orifice plasma gas flow rates will vary with orifice diameter. The 1/8" orifice dia. uses approx. 200 CFH. The 3/16" dia. orifice uses about 300 CFH. The gases used may be compressed air, nitrogen, nitrogen with oxygen added below the electrode area, or nitrogen with up to 10% hydrogen added. Consult the equipment manufacturer recommendations for each application.

Figure 225. Typical PAC Conditions for Carbon Steel.

For the various metal thicknesses the torch standoff distance will range from 1/4" to 1/2". The thicker the metal to be plasma cut, the greater the standoff distance. Check the manufacturers recommendations and follow their instructions for setup and operation of the PAC equipment.

ALUMINUM ALLOYS				
Thickness inches	**Speed in./min.**	**Orifice diameter****	**Amperes DCEN**	**Secondary power, KW**
1/4"	300	1/8"	300	60
1/2"	200	1/8"	250	50
1"	90	5/32"	400	80
2"	20	5/32"	400	80
3"	15	3/16"	450	90
4"	12	3/16"	450	90
6"	8	1/4"	750	170

**Plasma gas flow rates will vary from 100 CFH for 1/8" orifice to about 250 CFH for 1/4" orifice. The gases used are nitrogen, argon, and argon with hydrogen additions to about 35%. See the manufacturers recommendations for each application.

Figure 226. Typical PAC Conditions for Aluminum Alloys.

It is well to keep in mind that actual operating conditions are usually established by the PAC equipment manufacturers to suit their particular torch models and equipment. Regardless of what is stated in these charts and tables, check the recommendations and instructions provided by the manufacturer of your PAC equipment for the metals to be cut. You will find that experience with the equipment will permit some variations in the conditions and plasma gas flow rates.

STAINLESS STEELS				
Thickness inches	**Speed in./min.**	**Orifice diameter*** **	**Amperes DCEN**	**Secondary power, KW**
1/4"	200	1/8"	300	60
1/2"	100	1/8"	300	60
1"	50	5/32"	400	80
2"	20	3/16"	500	100
3"	16	3/16"	500	100
4"	8	3/16"	500	100

***Orifice plasma gas flow rates range from 100 CFH for 1/8" orifice to about 200 CFH for 3/16" dia. orifice. The orifice plasma gases used include nitrogen and argon plus up to 35% hydrogen additions. See the manufacturers recommendations for each application.

Figure 227. Typical PAC Conditions for Stainless Steels.

Plasma Arc Cutting Procedures

There are several variables that should be considered when setting up a procedure for the plasma arc cutting process. The PAC equipment manufacturer will have established conditions for certain metals and thicknesses when using specific models of torches. These conditions, or process variables, will interrelate as seen in the following list:

Power source model and rating.
PAC torch model and rating.
Type of outer shielding (gas, water, none).
Constricting nozzle design and orifice diameter.
Orifice gas, regulators and flow rates.
PAC torch standoff distance.
Amperage and voltage used.

Some of these variables have been shown in the Figures 225, 226, and 227 for carbon steel, aluminum, and stainless steels. Keep in mind that the charts should be used only as initial guides to set PAC conditions. Modifications can then be made to suit your specific applications. For example, it is possible that slower travel speeds will provide better quality kerf edges.

PAC Power Source

The power producing portion of the PAC power source is basically a constant current transformer-rectifier unit. It is unique because it has higher open circuit voltage (up to 400 volts) than any welding power source is permitted to have. The rating of the power source may range from 100 to 1,000 amperes. If greater amperages are required two or more power sources may be placed in parallel operation.

Various controls including high frequency for the pilot arc may be built-in the PAC power source. In many models of power sources the coolant system is also built-in so that the plasma arc torch cannot be activated without cooling water flowing.

PAC Torch Model

The PAC torch model design is a critical part of the total PAC system. For example, the PAC torch will limit the amount of amperage that may be used with the system. The PAC torch is usually part of the overall PAC system design including the power source and control system.

The PAC torch body is designed to center the electrode holder and tungsten electrode concentric to the centerline for the constricted orifice in the water cooled nozzle. The plasma gas flows around the electrode and is heated by the plasma arc. The gas expands and exits the torch plenum chamber at high velocity through the constricted orifice of the gas nozzle. The plasma jet, often called the "stinger", may be 8"-10" in length beyond the end of the gas nozzle.

PAC torches are available in both manual and automatic modes. The manual torch designs may include a 70° head and a 90° head. The automatic torch is designed for machine mounting and has no head angle. It is shaped similar to an automatic flame cutting torch.

Torch heads are available for using a secondary "shielding gas" to prevent aspiration of outside atmosphere into the plasma gas and to protect the hot base metal as it is being plasma cut. Another type of torch uses water as the secondary shielding medium. The work life of the water cooled torch gas nozzle is extended due to the cooler working conditions attained with this equipment.

Shielding Gas, Regulators and Flow Rates

Please do not confuse the term "shielding gas" with the term "orifice plasma gas". They are normally two separate gases which perform two entirely different functions.

The shielding gas type depends on the procedure that has been developed for the type of metal to be plasma cut. In many cases the shielding gas is carbon dioxide (CO_2). For certain materials the shielding gas may be argon or argon plus hydrogen. The regulator must be a gas regulator and NOT a gas flowmeter. A flowmeter will not permit correct gas flow regulation. Flow rates must be as developed for each PAC procedure.

Constricting Nozzle Design and Orifice Diameter

This is actually a related part of the PAC torch design. The *diameter of the orifice* is most important to the orifice gas flow rate as well as the base metal thickness to be plasma arc cut. The design of the constricting nozzle, and nozzle orifice, is the basis for how the torch will operate and what amperages it is able to carry.

Orifice Gas, Regulators, and Flow Rates

The orifice gas type is determined by the base metal to be plasma arc cut. For example, titanium and zirconium may be properly cut with the PAC process using argon gas. These two materials would be embrittled at the cut edge if reactive gases were used. Carbon steel and stainless steel normally use nitrogen as the orifice gas. See the PAC equipment manufacturers recommendations for the specific metal you have to cut.

The orifice gases, as well as the shielding gases, must be controlled through regulators rather than flowmeters. Only regulators designed to accommodate the specific orifice gases will control the gas flow correctly.

Typical flow rates for the orifice gases are normally shown in the PAC equipment manufacturers literature. If in doubt, don't hesitate to contact the manufacturers representative for additional information.

PAC Torch Standoff Distance

The PAC torch standoff distance is the distance between the working end of the constricting nozzle and the surface of the base metal to be plasma arc cut. This will normally be in the range of 1/4'' to 1/2'', depending on the thickness and type of metal being cut. Torch standoff distance can best be set when developing the plasma arc cutting procedure. Too great a torch standoff will cause the top edges of plasma cuts to melt and become rounded. Procedures should be in writing with all aspects of the operation detailed carefully.

Amperage and Load Voltage

The amperage and load voltage recommendations of the equipment manufacturer should be followed when developing a plasma arc cutting procedure. Modification of the recommended settings may be made as the specific application requires. Write the procedure and verify it with both PAC shop and supervisory personnel. Copies of plasma arc cutting procedures should be recorded and maintained with other welding procedures by the quality control department.

PAC Applications Data

Manual plasma arc cutting is normally done at low amperages and travel speeds. Low-amp units will operate very satisfactorily between 50-100 amperes. Stainless steel and aluminum thicknesses up to one inch may be cut with manual torches. Carbon steel and low alloy steel up to about 1/2'' thick may be cut with the PAC process but flame cutting is more cost effective for these thicknesses.

Manual plasma arc cutting operations are excellent for cutting materials that are in-place on weldments. For example, cutting holes in stainless steel or aluminum cylindrical or spherical tanks is very easy with the PAC process. The extended hose and cable assembly furnished with most PAC torches permits great mobility with the unit.

Automatic plasma arc cutting applications are reasonably simple. Once the **plasma arc cutting procedure** is set on the power source and other equipment, all the operator has to do is push the starting button. The equipment takes over and follows the programmed procedure. It is always wise to run the torch over the intended cut at least once *without power* to be sure that the intended cutting path is followed correctly.

Travel Direction and Kerf

The **direction of travel** when making a plasma arc cut is very important to the quality of the kerf edges. The reason is that most PAC torches create a rotation, or **swirl,** in the orifice plasma gas. This provides a more efficient transfer of arc heat energy to one side of the kerf.

For example, if the plasma rotation is **clockwise** as is common in most standard PAC torches, the heat input to the right side of the kerf will be more efficient than on the left side. The "right" and "left" sides of the kerf are always considered to be based on the direction of travel of the cut. The result will be a reasonably vertical edge on the right side of the kerf and a slight (5°-7°) bevel on the left side.

If there is to be a scrap side of the cut, it is always important that it be on the beveled edge of the kerf. For the clockwise plasma gas rotation, the scrap side would be the left side of the kerf.

For those applications where the left side is the desired square side, PAC torch manufacturers provide reverse rotation, or reverse "swirl", components for their torches. In this case, the rotation would be counter-clockwise with the more efficient transfer of arc energy to the left side of the kerf.

Caution: All PAC torch manufacturers caution that the torch should never touch the work piece while the plasma arc is working. This may cause severe damage to the torch nozzle and constricting orifice. Again, read the manufacturers instructions before operating the PAC equipment. Step-by-step instructions for installation and operation of the PAC equipment is provided with each PAC system. In addition, most manufacturers provide a troubeshooting guide to follow if the PAC equipment malfunctions.

Safety With Plasma Arc Cutting

The AWS/ANSI Standard Z49.1 "Safety In Welding And Cutting" is applicable to the plasma arc cutting process. Special precautions must be taken to prevent eye damage from the cutting arc. Hearing protection is recommended for those in the immediate area of PAC operations.

Water tables designed for use with the plasma arc cutting process will diminish the noise level to some degree. Special torch equipment is designed for use with water tables. This includes the water muffler which is designed to minimize the noise of plasma arc cutting operations.

It is necessary to check where the arc stream of molten metal will go while the plasma arc cut is being made. Careful planning will eliminate possible fires due to careless operation of the PAC process. For portable cutting operations it is always wise to post a fire watch near the plasma cutting operation.

PAC Cut Quality

The kerf surface cleanliness and smoothness are of concern when considering quality of the cut. In addition, the volume of dross at the bottom edges of the cut, kerf edge squareness, and sharpness of the top kerf edges all are important to quality in cutting. The material being cut and the operating conditions will normally dictate the quality of a plasma arc cut.

Speed of travel is always desireable from the production standpoint. The faster the cut, the higher the production. With plasma arc cutting, the faster the cutting speed of travel, the less likely to meet the higher quality standards that might be desired.

Good plasma arc cutting quality can be obtained fairly easily on stainless steel and aluminum up to about 3'' thickness. For carbon steel, good quality cuts with very little dross may be obtained on materials up to 1 1/2'' thick.

Summary

Plasma arc cutting is a good process for quality cutting of stainless steel, aluminum, and other metals. Cutting procedures should be developed and tested before being used on production parts.

Epilogue

This book has been written for you. It contains considerable data about the fundamentals of welding and cutting. However, it does not contain all the information there is, or ever will be, about ANYTHING!

I wanted to share some of the information I have learned over a period of years. If you will use it as a springboard of knowledge, you can progress far beyond where I will ever be! This is important to you and to me.

For you it is the opportunity to gain job satisfaction, understanding of your fellow workers, and pride in accomplishment. Along with this goes a good reliable income to take care of you and your family.

For me it is the knowledge that what has been given to me in trust I have nurtured and passed along to others. There are words and thoughts of many people in this book. I just put them together in some sense of order. I am privileged to have been of service.

God bless you all and may He walk with you all the days of your lives.

—Ed Pierre

APPENDIX

DATA CHART INDEX

DATA CHART 1

Applicable Electrical Laws

Coulombs Law.

"Charged bodies attract, or repel, each other with a force that is directly proportional to the product of the charges on the bodies and inversely proportional to the square of the distance between them".

Newtons Law (Gravity).

"Every object attracts every other object with a force that is directly proportional to the product of the masses of the objects and inversely proportional to the square of the distance between them".

Ohms Law.

"The value of the current, in amperes, in any electrical circuit is equal to the difference in potential, in volts, across the circuit divided by the resistance in ohms of the circuit".

Lenz's Law.

"The induced emf of any circuit is always in such a direction as to oppose the effect that produces it".

DATA CHART 2

Wire Calculation Data

In calculating welding costs it is necessary to know the cost of the welding wire. A helpful figure is the number of inches in a pound of a specific type and diameter electrode. This can be calculated by the formula:

$$\text{Inches per pound} = \frac{\text{Linear inches per cubic inch}}{\text{Pounds per cubic inch}}$$

Linear inches per cubic inch is a constant value regardless of the material. Pounds per cubic inch is the unit of density of the material. These constants are shown in table form:

Table I

Diameter of Wire (solid). Decimal	Fraction	Linear inches per cubic inch.
0.020	....	3,180
0.025	....	2,190
0.030	....	1,415
0.035	....	1,040
0.040	....	796
0.045	....	629
0.062	1/16	332
0.078	5/64	208
0.093	3/32	148
0.125	1/8	81

Table II

Wire Type	Pounds per cubic inch.
Magnesium	0.063
Aluminum	0.098
Aluminum Bronze (10%)	0.275
Stainless Steel (4xxx)	0.280
Mild Steel	0.285
Stainless Steel (3xxx)	0.290
Silicon Bronze	0.308
Copper-Nickel (60-40)	0.320
Nickel	0.321
Deoxidized Copper	0.325

DATA CHART 2 (cont.)

With these constants several useful calculations can be made. As an example, suppose 1/16" diameter aluminum electrode wire is being fed at 300 inches per minute and the welding time of the joint is ten minutes:

FROM TABLES I AND II—

$$\text{Inches per pound} = \frac{332 \text{ (table I)}}{.098 \text{ (table II)}} = 3{,}390 \text{ inches per pound}$$

Length of wire used = 300 in. per minute X 10 minutes = 3,000

$$\text{Weight of wire} = \frac{3{,}000}{3{,}390 \text{ in./lb.}} = 0.885 \text{ lb.}$$

If the aluminum wire cost $1.30 per pound the cost of the deposited metal is $1.15.

The chart below lists the number of inches per pound of the various wire types and diameters.

INCHES PER POUND—MATERIAL

Fraction						1/16	5/64	3/32	1/8
Decimal	0.020	0.025	0.030	0.035	0.045	0.062	0.078	0.093	0.125
Mg.	50500	34700	22400	16500	9990	5270	3300	2350	1280
Al.	32400	22300	14420	10600	6410	3382	2120	1510	825
Al. Br.	11600	7960	5150	3780	2290	1220	756	538	295
SS-400	11350	7820	5050	3720	2240	1180	742	528	289
Mild St.	11100	7680	4960	3650	2210	1160	730	519	284
SS-300	10950	7550	4880	3590	2170	1140	718	510	279
Si. Br.	10300	7100	4600	3380	2040	1070	675	480	263
Cu-Ni	9950	6850	4430	3260	1970	1040	650	462	253
Ni.	9900	6820	4400	3240	1960	1030	647	460	252
DO-Cu.	9800	6750	4360	3200	1940	1020	640	455	249

Mg. = Magnesium
Al. = Aluminum
Al. Br. = Aluminum Bronze
SS-400 = 400 series Stainless Steel
Mild St. = Mild Steel
SS-300 = 300 series Stainless Steel
Si. Br. = Silicon Bronze
Cu-Ni = Copper-Nickel
Ni. = Nickel
DO-Cu. = Deoxidized Copper

DATA CHART 3

Current Density Calculation Chart

Wire Diameter	Decimal Equivalent	Area Inches²	Wire Diameter	Decimal Equivalent	Area Inches²
0.020	0.020	0.00031	3/16	0.1875	0.0276
0.025	0.025	0.00051	13/64	0.2031	0.0324
0.030	0.030	0.00071	7/32	0.2187	0.0376
0 035	0.035	0.00096	15/64	0.2344	0.0431
0.045	0.045	0.00160	1/4	0.2500	0.0491
3/64	0.047	0.00173	17/64	0.2656	0.0553
1/16	0.0625	0.00307	9/32	0.2812	0.0621
5/64	0.0781	0.0048	19/64	0.2969	0.0692
3/32	0.094	0.0069	5/16	0.3125	0.0767
7/64	0.109	0.0093	21/64	0.3281	0.0845
1/8	0.125	0.01227	11/32	0.3437	0.0928
9/64	0.1406	0.0154	23/64	0.3594	0.1014
5/32	0.1562	0.0192	3/8	0.3750	0.1105
11/64	0.1719	0.0232			

Current Density.

Current density is calculated by dividing the electrode area, in square inches, into the welding amperage value used. The result is the amperage per square inch of electrode. For example:

7/64" dia. wire = $0.0093\,\overline{)\,500\text{ amps}}$ = 53,762 amps/inches².

DATA CHART 4

Properties of Metals

Element	Symbol	Melting Point °F.	Coefficient of Exp. °F.	Elect. Cond. % Pure Cu	Lb/Cu. In.
Aluminum	Al	1218	0.0000133	64.9	0.098
Antimony	Sb	1167	0.00000627	4.42	0.239
Beryllium	Be	2345	0.0000068	9.32	0.066
Bismuth	Bi	520	0.00000747	1.50	0.354
Cadmium	Cd	610	0.00000166	22.7	0.313
Chromium	Cr	2822	0.0000045	13.2	0.258
Cobalt	Co	2714	0.00000671	17.8	0.322
Copper	Cu	1981	0.0000091	100.0	0.323
Gold	Au	1945	0.0000080	71.2	0.697
Iron	Fe	2795	0.0000066	17.6	0.284
Lead	Pb	621	0.0000164	8.35	0.409
Magnesium	Mg	1204	0.0000143	38.7	0.063
Mercury	Hg	—38	0.	1.80	0.489
Molybdenum	Mo	4748	0.00000305	36.1	0.368
Nickel	Ni	2646	0.0000076	25.0	0.322
Platinum	Pt	3224	0.0000043	17.5	0.774
Selenium	Se	428	0.0000206	14.4	0.174
Silver	Ag	1761	0.0000105	106.00	0.380
Tellurium	Te	846	0.0000093	. . . .	0.224
Tin	Sn	450	0.0000124	15.0	0.264
Tungsten	W	6170	0.0000022	31.5	0.698
Vanadium	V	3110	0.	6.63	0.205
Zinc	Zn	787	0.0000219	29.1	0.258

DATA CHART 5

AMERICAN WELDING SOCIETY

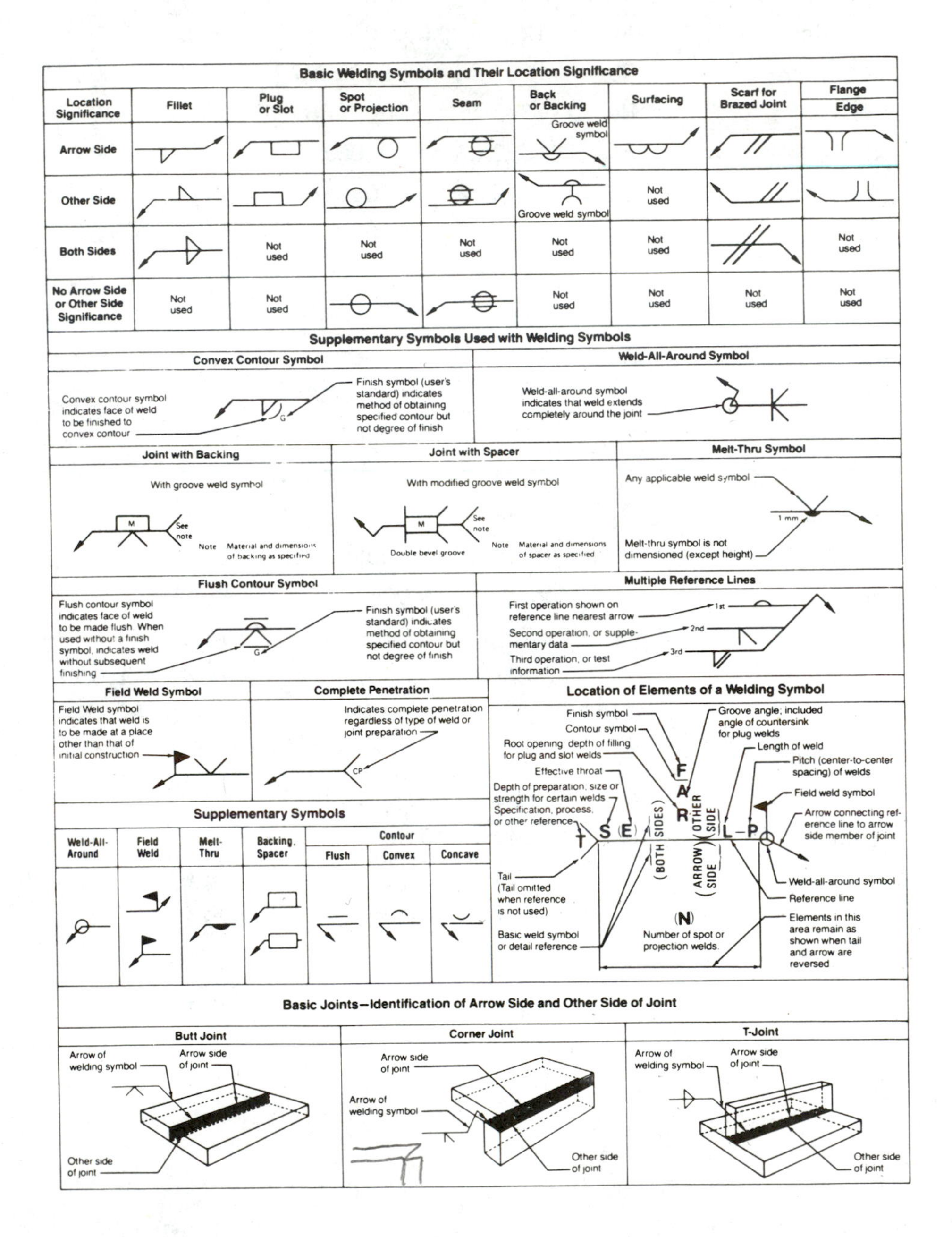

DATA CHART 5 (cont.)

STANDARD WELDING SYMBOLS

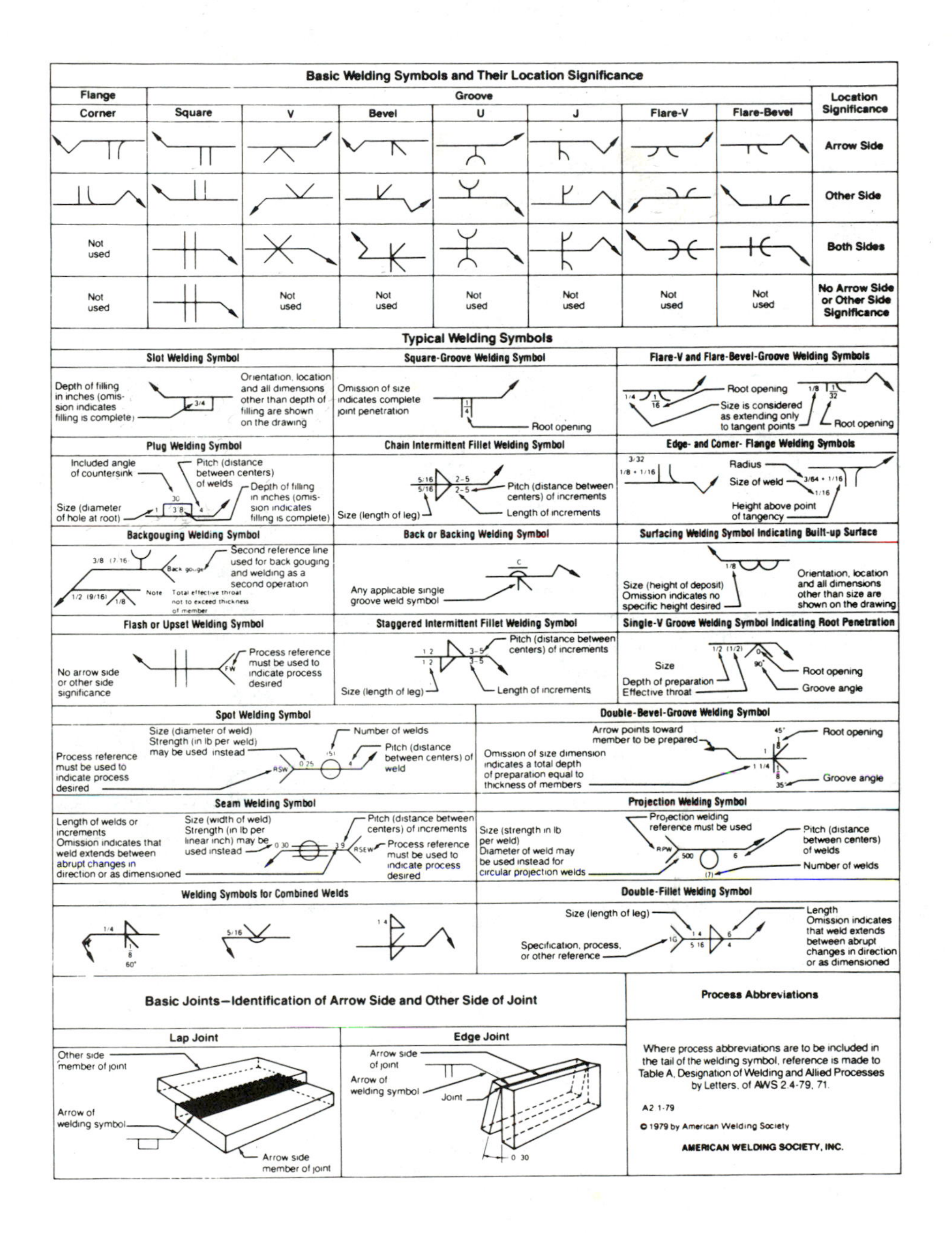

DATA CHART 6

Metric Conversion Tables

The unit of length of the metric system, the meter, is intended to be one ten-millionth part of the distance from the equator to the pole. The meter is divided into ten parts called decimeters. Each decimeter is divided into ten equal parts known as centimeters and each centimeter is divided into ten millimeters. A millimeter is one thousandth part of a meter.

The metric units of length, mass and capacity are sub-divided decimally using the Latin prefixes of deci-, centi-, and milli-. The Greek prefixes deka, hecto, kilo and myria are used to indicate the multiplication of the units by ten.

U. S. to Metric	Metric to U. S.
Linear	
1 inch = .0254000 meters	1 meter = 39.3700 inches
1 foot = .304800 meters	1 meter = 3.28083 feet
1 yard = .914400 meters	1 meter = 1.09361 yards
1 mile = 1609.35 meters	1 kilometer = .62137 miles
= 1.60935 kilometers	
Square	
1 sq. inch = 6.452 sq. centimeters	1 sq. cm. = .1550 sq. inches
1 sq. foot = 9.290 sq. decimeters	1 sq. meter = 10.7640 sq. feet
1 sq. yard = .836 sq. meters	1 sq. meter = 1.196 sq. yards
Cubic	
1 cu. inch = 16.387 cu. cm.	1 cu. cm. = .0610 cu. inches
1 cu. foot = .02832 cu. meters	1 cu. meter = 35.314 cu. feet
1 cu. yard = .765 cu. meters	1 cu. meter = 1.308 cu. yards
Capacity	
1 fluid ounce = 29.57 milliliter	1 centiliter = .338 fluid ounces
1 quart = .94636 liters	1 liter = 1.0567 quarts
1 gallon = 3.78544 liters	1 hectoliter = 26.417 gallons

DATA CHART 7

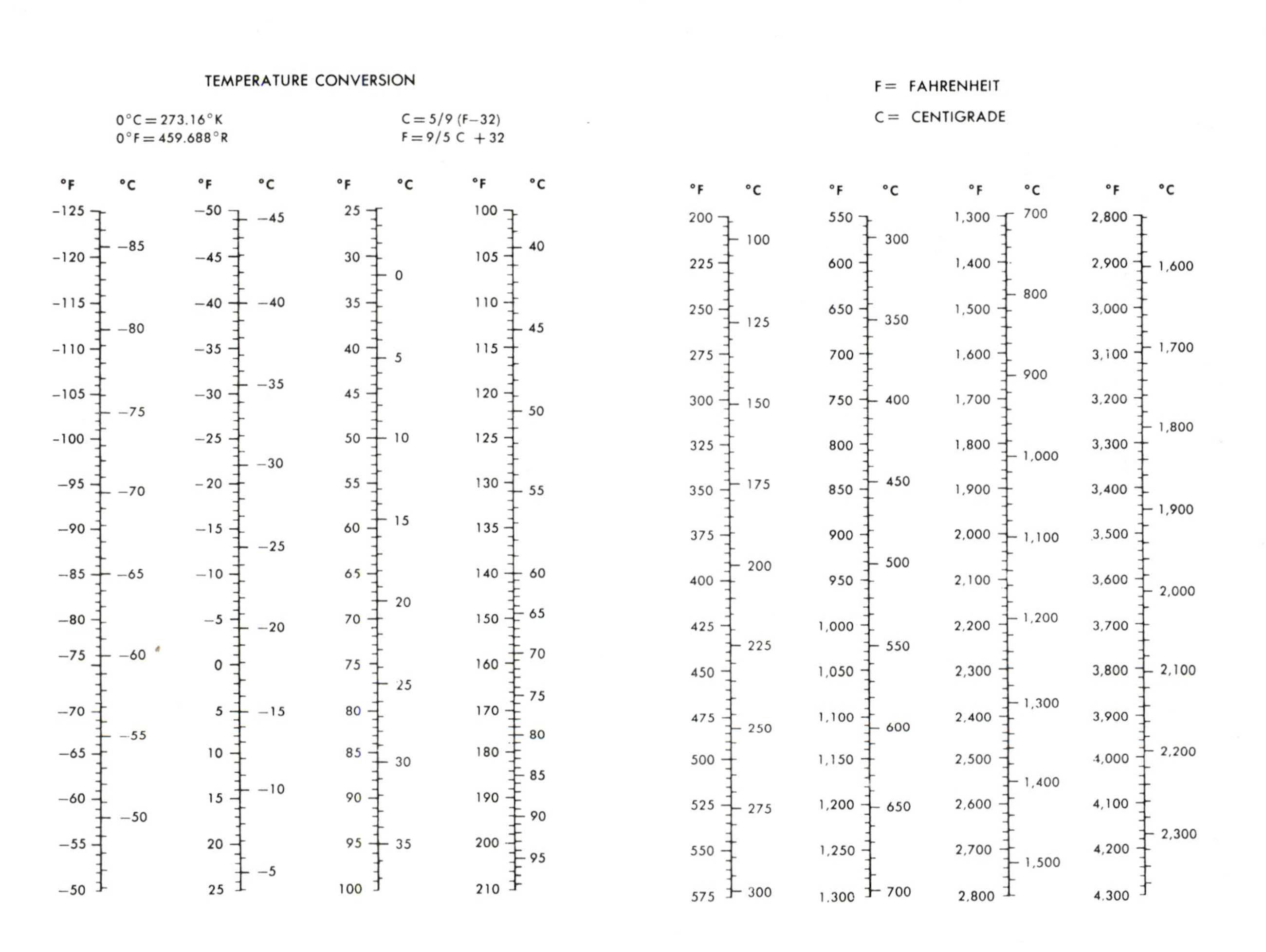

TEMPERATURE CONVERSION
0°C = 273.16°K
0°F = 459.688°R
C = 5/9 (F−32)
F = 9/5 C +32
F = FAHRENHEIT
C = CENTIGRADE
°F
°C

DATA CHART 8

Welding Cable Size, Amperage, and Length Correlation

(Nominal 4 volts drop, 60% duty cycle)

	**DC Amperes										
Total Length	**100**	**150**	**200**	**250**	**300**	**350**	**400**	**450**	**500**	**550**	**600**
50'	6	6	4	4	3	2	2	1	1	1/0	1/0
75'	6	4	3	2	1	1	1/0	2/0	2/0	2/0	3/0
100'	4	3	2	1	1/0	2/0	2/0	3/0	3/0	4/0	4/0
125'	4	2	1	1/0	2/0	3/0	3/0	4/0	4/0	*250	250
150'	3	1	1/0	2/0	3/0	3/0	4/0	250	250	*300	300
175'	2	1	2/0	3/0	3/0	4/0	250	250	300	300	---
200'	2	1/0	2/0	3/0	4/0	250	300	300	---	---	---
225'	1	1/0	3/0	4/0	250	250	300	---	---	---	---
250'	1	2/0	3/0	4/0	250	300	---	---	---	---	---
300'	1/0	3/0	4/0	250	300	---	---	---	---	---	---
350'	2/0	3/0	250	300	---	---	---	---	---	---	---
400'	2/0	4/0	300	---	---	---	---	---	---	---	---

*MCM cable

**For AC welding power, use next larger size.

DATA CHART 9

Mechanical Properties of Mild Steel at Elevated Temperatures*

Temp. °F.	Ultimate Strength psi	Yield Point psi	Elastic Limit psi	Modulus of Elasticity psi	Elongation in 2 inches %	Reduction of area %
100	54,000	39,000	36,000	30,700,000	48	68
200	56,000	43,000	33,000	29,000,000	43	66
300	58,000	46,000	28,000	27,300,000	39	65
400	60,000	47,000	22,000	25,600,000	37	64
500	63,000	46,000	16,000	23,900,000	37	64
600	66,000	44,000	14,000	22,200,000	40	66
700	60,000	40,000	12,000	20,500,000	44	68
800	52,000	37,000	10,000	18,000,000	49	76
900	43,000	32,000	8,000	17,100,000	55	84
1000	34,000	27,000	7,500	15,400,000	60	90
1100	25,000	21,000	5,000	13,700,000	64	94
1200	19,000	17,000	3,500	12,000,000	68	96
1300	14,000	13,000	2,000	10,300,000	72	98

Welding induces residual stresses in the weldment. These stresses may be reduced by post-weld thermal stress relief. The residual stress remaining in the weldment after thermal stress relief will depend on the rate of cooling. Uneven cooling of the part from stress relief temperatures may set up additional stresses and cancel the effects of the thermal stress relief cycle. After a part is stress relieved, the rate of cooling should be constant and consistent with the type of material being worked. Stress relieving temperatures, and cooling rates, should be based on manufacturers recommendations and appropriate regulatory bodies standards.

* American Welding Society Welding Handbook, 6th Edition.

DATA CHART 10

Typical Stress Relieving Temperatures*

Material	Soaking Temperature, °F.
Carbon Steel	1100-1250
Carbon—½% Moly steel	1100-1325
½% Chrome—½% Moly steel	1100-1325
1% Chrome—½% Moly steel	1150-1350
1¼% Chrome—½% Moly steel	1150-1375
2% Chrome—½% Moly steel	1150-1375
2¼% Chrome—1% Moly steel	1200-1375
5% Chrome—½% Moly (Type 502) steel	1200-1375
7% Chrome—½% Moly steel	1300-1400
9% Chrome—1% Moly steel	1300-1400
12% Chrome (Type 410) steel	1350-1400
16% Chrome (Type 430) steel	1400-1500
1¼% Mn—½% Moly steel	1125-1250
Low-alloy Chrome-Ni-Moly steel	1100-1250
2%-5% Nickel steels	1050-1150
9% Nickel steel	1025-1085

* American Welding Society Welding Handbook, 6th Edition. The charted temperatures are to be used as a guide only. Actual stress relieving temperatures used will depend on local codes and established procedures.

DATA CHART 11

American National Standards Institute

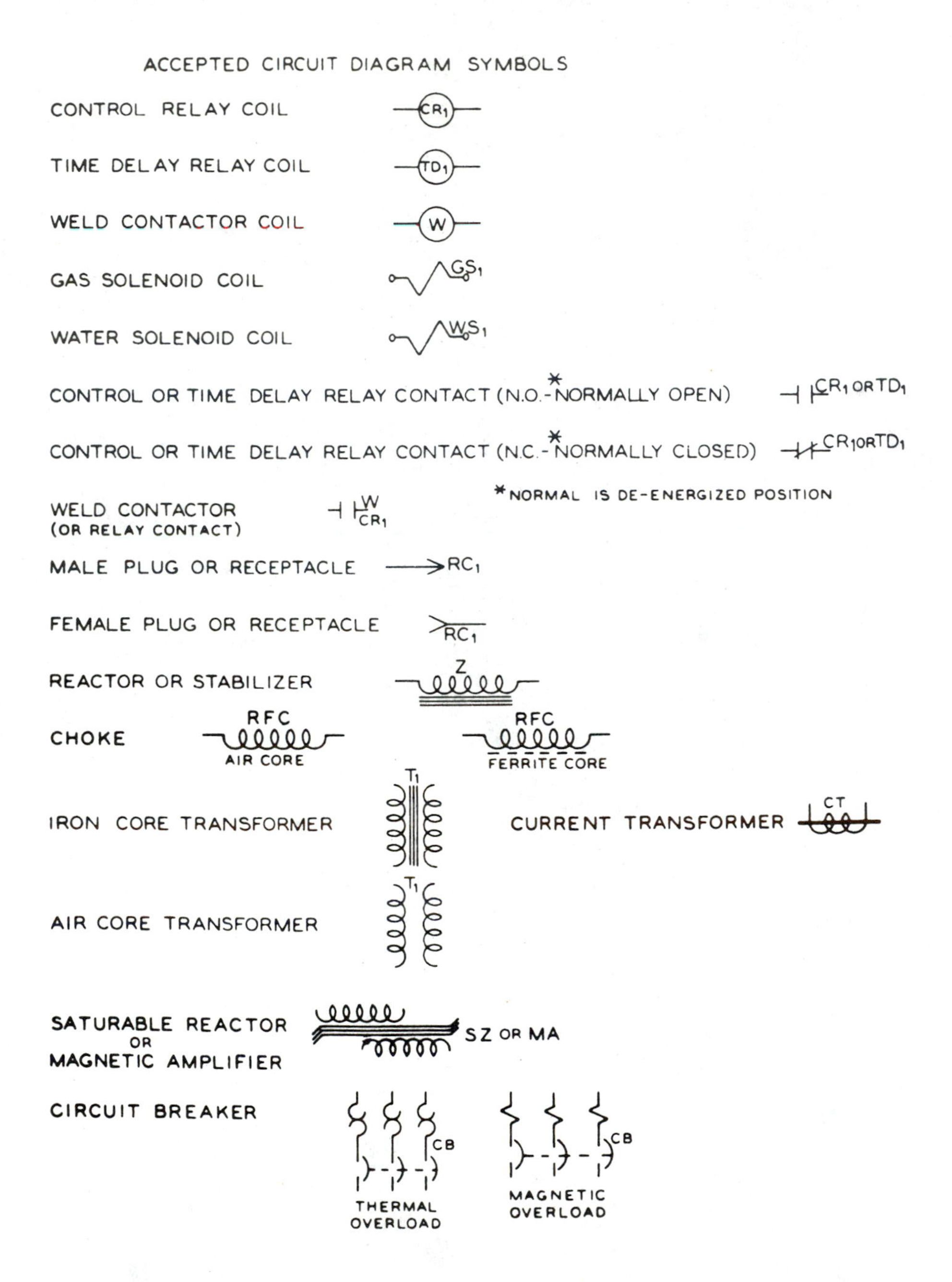

DATA CHART 11 (cont.)

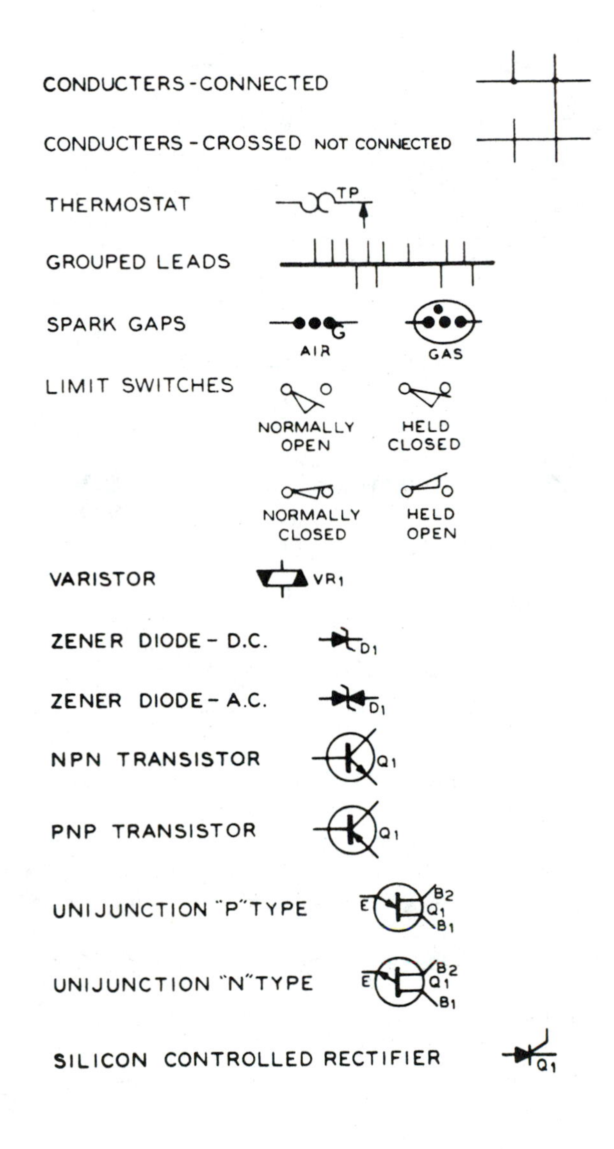

DATA CHART 11 (cont.)

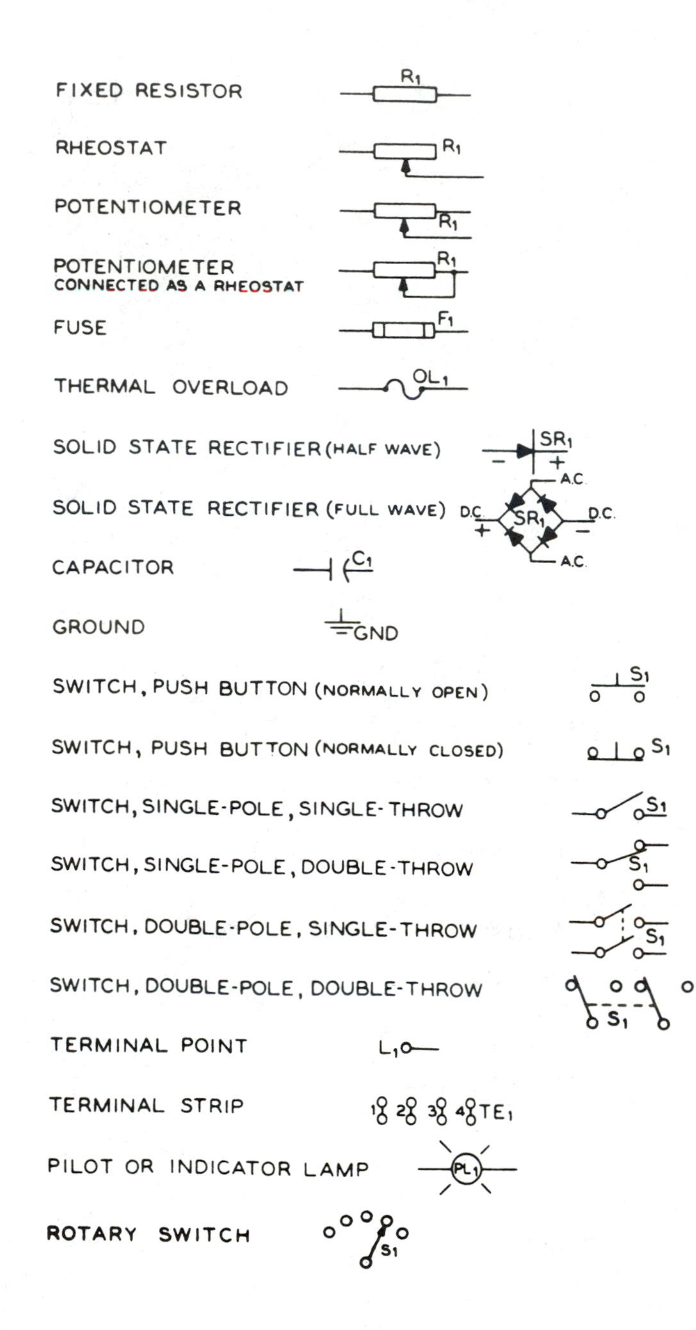

DATA CHART 12

Fractions, Decimals, Millimeters

Fraction	Decimals	Millimeters	Millimeters	Decimals	Fraction
1/64	.0156	0.3969	13.0969	.5156	33/64
1/32	.0313	0.7938	13.4938	.5313	17/32
3/64	.0469	1.1906	13.8906	.5469	35/64
1/16	.0625	1.5875	14.2875	.5625	9/16
5/64	.0781	1.9844	14.6844	.5781	37/64
3/32	.0938	2.3813	15.0813	.5938	19/32
7/64	.1094	2.7781	15.4781	.6094	39/64
1/8	.125	3.1750	15.8750	.625	5/8
9/64	.1406	3.5719	16.2719	.6406	41/64
5/32	.1563	3.9688	16.6688	.6563	21/32
11/64	.1719	4.3656	17.0656	.6719	43/64
3/16	.1875	4.7625	17.4625	.6875	11/16
13/64	.2031	5.1594	17.8594	.7031	45/64
7/32	.2188	5.5563	18.2563	.7188	23/32
15/64	.2344	5.9531	18.6531	.7344	47/64
1/4	.250	6.3500	19.0500	.750	3/4
17/64	.2656	6.7469	19.4469	.7656	49/64
9/32	.2813	7.1438	19.8438	.7813	25/32
19/64	.2969	7.5406	20.2406	.7969	51/64
5/16	.3125	7.9375	20.6375	.8125	13/16
21/64	.3281	8.3344	21.0344	.8281	53/64
11/32	.3438	8.7313	21.4313	.8438	27/32
23/64	.3594	9.1281	21.8281	.8594	55/64
3/8	.375	9.5250	22.2250	.875	7/8
25/64	.3906	9.9219	22.6219	.8906	57/64
13/32	.4063	10.3188	23.0188	.9063	29/32
27/64	.4219	10.7156	23.4156	.9219	59/64
7/16	.4375	11.1125	23.8125	.9375	15/16
29/64	.4531	11.5094	24.2094	.9531	61/64
15/32	.4688	11.9063	24.6063	.9688	31/32
31/64	.4844	12.3031	25.0031	.9844	63/64
1/2	.500	12.7000	25.4000	1.000	1

DATA CHART 13

Hardness Conversion Table for Carbon and Alloy Steels

(All values approximate)

Brinell (carbide ball)	Rockwell "C" scale	Rockwell "B" scale	Carbon Steel Tensile ksi
264	27	---	128
258	26	---	125
253	25	---	122
247	24	---	120
243	23	---	117
240	--	100	116
234	--	99	112
222	--	97	106
210	--	95	101
200	--	93	96
195	--	92	93
185	--	90	89
176	--	88	85
169	--	86	81
162	--	84	78
156	--	82	75
150	--	80	72
133	--	75	66

DATA CHART 14

Inches to Millimeters

Inches to Millimeters

1 Inch = 25.4 Millimeters   1 Millimeter = .0394 Inches

Inches	0″	1″	2″	3″	4″	5″	6″	7″	8″	9″	10″	11″	12″	Inches
	.000	25.400	50.800	76.200	101.600	127.000	152.400	177.800	203.200	228.600	254.000	279.400	304.800	
1/32	.794	26.194	51.594	76.994	102.394	127.794	153.194	178.594	203.994	229.394	254.794	280.194	305.594	1/32
1/16	1.588	26.988	52.388	77.788	103.188	128.588	153.988	179.388	204.788	230.188	255.588	280.988	306.388	1/16
3/32	2.381	27.781	53.181	78.581	103.981	129.381	154.781	180.181	205.581	230.981	256.381	281.781	307.181	3/32
1/8	3.175	28.575	53.975	79.375	104.775	130.175	155.575	180.975	206.375	231.775	257.175	282.575	307.975	1/8
5/32	3.969	29.369	54.769	80.169	105.569	130.969	156.369	181.769	207.169	232.569	257.969	283.369	308.769	5/32
3/16	4.763	30.163	55.563	80.963	106.363	131.763	157.163	182.563	207.963	233.363	258.763	284.163	309.563	3/16
7/32	5.556	30.956	56.356	81.756	107.156	132.556	157.956	183.356	208.756	234.156	259.556	284.956	310.356	7/32
1/4	6.350	31.750	57.150	82.550	107.950	133.350	158.750	184.150	209.550	234.950	260.350	285.750	311.150	1/4
9/32	7.144	32.544	57.944	83.344	108.744	134.144	159.544	184.944	210.344	235.744	261.144	286.544	311.944	9/32
5/16	7.938	33.338	58.738	84.138	109.538	134.938	160.338	185.738	211.138	236.538	261.938	287.338	312.738	5/16
11/32	8.731	34.131	59.531	84.931	110.331	135.731	161.131	186.531	211.931	237.331	262.731	288.131	313.531	11/32
3/8	9.525	34.925	60.325	85.725	111.125	136.525	161.925	187.325	212.725	238.125	263.525	288.925	314.325	3/8
13/32	10.319	35.719	61.119	86.519	111.919	137.319	162.719	188.119	213.519	238.919	264.319	289.719	315.119	13/32
7/16	11.113	36.513	61.913	87.313	112.713	138.113	163.513	188.913	214.313	239.713	265.113	290.513	315.913	7/16
15/32	11.906	37.306	62.706	88.106	113.506	138.906	164.306	189.706	215.106	240.506	265.906	291.306	316.706	15/32
1/2	12.700	38.100	63.500	88.900	114.300	139.700	165.100	190.500	215.900	241.300	266.700	292.100	317.500	1/2
17/32	13.494	38.894	64.294	89.694	115.094	140.494	165.894	191.294	216.694	242.094	267.494	292.894	318.294	17/32
9/16	14.288	39.688	65.088	90.488	115.888	141.288	166.688	192.088	217.488	242.888	268.288	293.688	319.088	9/16
19/32	15.081	40.481	65.881	91.281	116.681	142.081	167.481	192.881	218.281	243.681	269.081	294.481	319.881	19/32
5/8	15.875	41.275	66.675	92.075	117.475	142.875	168.275	193.675	219.075	244.475	269.875	295.275	320.675	5/8
21/32	16.669	42.069	67.469	92.869	118.269	143.669	169.069	194.469	219.869	245.269	270.669	296.069	321.469	21/32
11/16	17.463	42.863	68.263	93.663	119.063	144.463	169.863	195.263	220.663	246.063	271.463	296.863	322.263	11/16
23/32	18.256	43.656	69.056	94.456	119.856	145.256	170.656	196.056	221.456	246.856	272.256	297.656	323.056	23/32
3/4	19.050	44.450	69.850	95.250	120.650	146.050	171.450	196.850	222.250	247.650	273.050	298.450	323.850	3/4
25/32	19.844	45.244	70.644	96.044	121.444	146.844	172.244	197.644	223.044	248.444	273.844	299.244	324.644	25/32
13/16	20.638	46.038	71.438	96.838	122.238	147.638	173.038	198.438	223.838	249.238	274.638	300.038	325.438	13/16
27/32	21.431	46.831	72.231	97.631	123.031	148.431	173.831	199.231	224.631	250.031	275.431	300.831	326.231	27/32
7/8	22.225	47.625	73.025	98.425	123.825	149.225	174.625	200.025	225.425	250.825	276.225	301.625	327.025	7/8
29/32	23.019	48.419	73.819	99.219	124.619	150.019	175.419	200.819	226.219	251.619	277.019	302.419	327.819	29/32
15/16	23.813	49.213	74.613	100.013	125.413	150.813	176.213	201.613	227.013	252.413	277.813	303.213	328.613	15/16
31/32	24.606	50.006	75.406	100.806	126.206	151.606	177.006	202.406	227.806	253.206	278.606	304.006	329.406	31/32

DATA CHART 15

Pounds to Kilograms—Kilograms to Pounds

Pounds to Kilograms

1 Pound = 0.453592 Kilogram

Pounds	Kilograms	Pounds	Kilograms	Pounds	Kilograms	Pounds	Kilograms	Pounds	Kilograms
0.1	0.05	1	0.45	10	4.54	100	45.36	1000	453.59
0.2	0.09	2	0.91	20	9.07	200	90.72	2000	907.18
0.3	0.14	3	1.36	30	13.61	300	136.08	3000	1360.78
0.4	0.18	4	1.81	40	18.14	400	181.44	4000	1814.37
0.5	0.23	5	2.27	50	22.68	500	226.80	5000	2267.96
0.6	0.27	6	2.72	60	27.22	600	272.16	6000	2721.55
0.7	0.32	7	3.18	70	31.75	700	317.51	7000	3175.14
0.8	0.36	8	3.63	80	36.29	800	362.87	8000	3628.74
0.9	0.41	9	4.08	90	40.82	900	408.23	9000	4082.33
1.0	0.45	10	4.54	100	45.36	1000	453.59	10000	4535.92

Kilograms to Pounds

1 Kilogram = 2.204622 Pounds

Kilograms	Pounds	Kilograms	Pounds	Kilograms	Pounds	Kilograms	Pounds	Kilograms	Pounds
0.1	0.22	1	2.20	10	22.05	100	220.46	1000	2204.62
0.2	0.44	2	4.41	20	44.09	200	440.92	2000	4409.24
0.3	0.66	3	6.61	30	66.14	300	661.39	3000	6613.87
0.4	0.88	4	8.82	40	88.18	400	881.85	4000	8818.49
0.5	1.10	5	11.02	50	110.23	500	1102.31	5000	11023.11
0.6	1.32	6	13.23	60	132.28	600	1322.77	6000	13227.73
0.7	1.54	7	15.43	70	154.32	700	1543.24	7000	15432.35
0.8	1.76	8	17.64	80	176.37	800	1763.70	8000	17636.98
0.9	1.98	9	19.84	90	198.42	900	1984.16	9000	19841.60
1.0	2.20	10	22.05	100	220.46	1000	2204.62	10000	22046.22

DATA CHART 16

AWS Filler Metal Specifications

(Some titles abridged)

A5.1 Carbon Steel Covered Arc Welding Electrodes.
A5.2 Iron and Steel Oxyfuel Gas Welding Rods.
A5.3 Aluminum Covered Arc Welding Electrodes.
A5.4 C-R Chromium and Chromium-Nickel Covered Arc Welding Electrodes.
A5.5 Low-Alloy Steel Covered Arc Welding Electrodes.
A5.6 Copper and Copper Alloy Covered Electrodes.
A5.7 Copper and Copper Alloy Bare Welding Rods and Electrodes.
A5.8 Brazing Filler Metal.
A5.9 C-R Cr and Cr-Ni Steel Bare and Composite Electrodes and Rods.
A5.10 Aluminum Bare Welding Rods and Electrodes.
A5.11 Nickel and Nickel Alloy Covered Electrodes.
A5.12 Tungsten Arc Welding Electrodes.
A5.13 Solid Surfacing Welding Rods and Electrodes.
A5.14 Nickel and Nickel Alloys Bare Welding Rods and Electrodes.
A5.15 Welding Rods and Covered Electrodes for Welding Cast Iron.
A5.16 Titanium and Alloy Bare Welding Rods and Electrodes.
A5.17 Carbon Steel Electrodes and Fluxes for Submerged Arc Welding.
A5.18 Carbon Steel Filler Metals for Gas Shielded Arc Welding.
A5.19 Magnesium Alloy Welding Rods and Bare Electrodes.
A5.20 Carbon Steel Electrodes for Flux Cored Arc Welding.
A5.21 Composite Surfacing Welding Rods and Electrodes.
A5.22 Flux Cored C-R Chromium and Chromium-Nickel Electrodes.
A5.23 Bare Low-Alloy Steel Electrodes and Fluxes Submerged Arc Welding.
A5.24 Zirconium and Zirconium Alloy Bare Welding Rods and Electrodes.
A5.25 Consumables for Electroslag Welding of Carbon and HSLA Steels.
A5.26 Consumables for Electrogas Welding of Carbon and HSLA Steels.
A5.27 Copper and Copper-Alloy Gas Welding Rods.
A5.28 Low Alloy Steel Filler Metals for Gas Shielded Arc Welding.
A5.29 Low Alloy Steel Electrodes for Flux Cored Arc Welding.
A5.30 Consumable Inserts.

INDEX

—A—

—B—

—C—

—D—

—E—

—F—

—G—

—H—

—I—

—J—

—K—

—L—

—M—

—N—

—O—

—P—

—Q—

—R—

—S—

—T—

—U—

—V—

—W—

—Y—

—Z—

Student Survey

Welding Processes and Power Sources, 3rd edition

Edward R. Pierre

Students, send us your ideas!

The author and the publisher want to know how well this book served you and what can be done to improve it for those who will use it in the future. By completing and returning this questionnaire, you can help us develop better textbooks. We value your opinion and want to hear your comments. Thank you.

Your name (optional) ______________________ School ______________

Your mailing address ____

City ______________________ State ___________ Zip ___________

Instructor's name (optional) ________________ Course title ______________

1. How does this book compare with other texts you have used? (Check one)

   ☐ Superior ☐ Better than most ☐ Comparable ☐ Not as good as most

2. Circle those chapters you especially liked:

   Chapters: 1 2 3 4 5 6 7 8 9 10 11 12 13 14 15 16 17 18 19 20 21 22 23

   Comments:

3. Circle those chapters you think could be improved:

   Chapters: 1 2 3 4 5 6 7 8 9 10 11 12 13 14 15 16 17 18 19 20 21 22 23

   Comments:

	Excellent	Good	Average	Poor
Readability of text material	( )	( )	( )	( )
Logical organization	( )	( )	( )	( )
General layout and design	( )	( )	( )	( )
Up-to-date treatment of subject	( )	( )	( )	( )
Match with instructor's course organization	( )	( )	( )	( )
Illustrations that clarify the text	( )	( )	( )	( )
Selection of topics	( )	( )	( )	( )
Explanation of difficult concepts	( )	( )	( )	( )

5. List any chapters that your instructor did not assign. ______________________

____

6. What additional topics did your instructor discuss that were not covered in the text?

____

____

7. Did you buy this book new or used? ☐ New ☐ Used
   Do you plan to keep the book or sell it? ☐ Keep it ☐ Sell it
   Should your instructor continue to assign this book? ☐ Yes ☐ No
   Did you purchase the Study Guide and Exercises for the text? ☐ Yes ☐ No

8. After taking the course, are you interested in taking more courses in this field?

   ☐ Yes ☐ No

   Did you take this course to fulfill a requirement, or as an elective?

   ☐ Required ☐ Elective

9. What is your major? ______________________________

   Your class rank? ☐ Freshman ☐ Sophomore ☐ Junior ☐ Senior ☐ Other, Specify:

10. Is shop required in conjunction with the course? ☐ Yes ☐ No

    If so, is the shop separate, or combined with the lecture?

    ☐ Separate ☐ Combined

11. Kindly rate the coordination of the Study Guide & Exercises with the text. (Check one)

    ☐ Excellent ☐ Good ☐ Average ☐ Poor ☐ Very poor

12. GENERAL COMMENTS:

May we quote you in our advertising? ☐ Yes ☐ No

Please remove this page and mail to:

Mary L. Paulson
Burgess Publishing Company
7108 Ohms Lane
Minneapolis, MN 55435

THANK YOU!